Quantitative EEG Practical Guide

Jesús Pastor

Quantitative EEG Practical Guide

Springer

Jesús Pastor
Clinical Neurophysiology
Hospital Universitario de la Princesa.
Instituto de Investigación Biomédica.
Madrid, Spain

ISBN 978-3-032-05456-2 ISBN 978-3-032-05457-9 (eBook)
https://doi.org/10.1007/978-3-032-05457-9

This Springer imprint is published by the registered company Springer Nature Switzerland AG
The registered company address is: Gewerbestrasse 11, 6330 Cham, Switzerland

To my beloved wife, Mónica, my daughter, Eugenia, and my son, Gonzalo.

Preface

This book is written from a fundamentally personal perspective, which gives it specific characteristics. The starting point is the need to face daily diagnostic challenges that, remarkably, change from day to day, with patients who present variable or, at times, clearly abnormal clinical or bioelectrical features.

Throughout my more than 30 years as a clinical neurophysiologist, my goal has been to make the most accurate and objective possible diagnosis. Of course, accuracy is an unattainable limit, and objectivity is impossible for an agent (a person) endowed with subjectivity. However, like any field of Science, both are desirable goals. Clinical Neurophysiology is based on established Sciences, such as Neuroscience, and fields of Physics, such as Electromagnetism and Biophysics. Therefore, it is a realm that lends itself well to the rigor and mathematical approach that these subjects permit.

This textbook is an introduction to quantitative EEG (qEEG) in the routine daily practice and reflects the experience of a clinical neurophysiologist dedicated to performing resting state EEG, continous EEG in ICU, video electroencephalography for the diagnosis of epilepsy and preoperative assessment and electrocorticography (ECoG).

The text has been conceived to show the application of some numerical analysis to most of the clinical problems faced in a highly complex university hospital in Madrid (Spain). It's not, therefore, a comprehensive description of analytical methods, because the reader can find more detailed and founded books whose authors are more expert and authorized. It's a textbook wrote by a physician and mainly directed to physicians engaged in the diary problems of accurate and efficient diagnosis of diverse pathologies of the brain. However, it is problematic, or even impossible, to understand profound and extensively the power of qEEG without a deep understanding of some mathematical methods. In fact, the main difference between a true expert in a field and a technician is that the former can offer an adequate justification for its judgments, meanwhile the technician is able to apply an algorithm (even extremely complex) without a deeper knowledge. This is the reason I decided to include an extensive part with the mathematical foundations of the numerical methods used. I'm conscious that the number of formulas in a book is inversely proportional to the readers, but I accept the challenge. Probably, after an initial arid period of intense effort, some physicians (I would like to think that the majority!) feel themselves confident enough to read dense papers about numerical analysis and, what is more important, try to use a proper analytical script. Nowadays there are a lot of toolbox and languages, some of them commercial and other free, that can be overpowered with the help of the innumerable resources obtained from the literature and Internet. By resume, my goal is to provide physicians with the background required to understand the principles of numerical analysis and show what can be expected from its use in their quotidian work.

The book is divided in three parts dotted of inner coherence and obviously related with the other. The first part, including Chaps. 1, 2, 3, 4, 5 and 6, describes the physiology of the brain, the origin of the EEG, the biophysical properties of the system, and the description of the main numerical methods used in our work. Chapter 4, Analysis in the frequency domain, is a detailed description of the most used method (i.e., the Fourier transform). It is slightly technical and arduous, but it is a cornerstone for a profound understanding of the technique. Nevertheless, all clinical chapter can be understood without an intense fight with the formulas. There are many other numerical methods used in the literature, but I have described only those used in our daily practice.

The second part, Chaps. 7, 8, 9, 10, 11, 12, and 13, describes several examples of qEEG used in true patients. I have tried to show cases where the information obtained by visual EEG is clearly overcome by the information obtained through qEEG. This is the main message of the book: qEEG improves, in an extraordinary way, the diagnostic capacity of conventional EEG. The last Chap. 13 is speculative and only reflects a personal bet.

Finally, the last part includes nine mathematical topics which thoroughly substantiates and justifies the ideas presented in the first two parts.

I would like to insist that the aim of this book is to convey to physicians the extraordinary diagnostic power and objectivity provided by qEEG, even in the most seemingly trivial recordings. If I have not achieved this goal, it is only due to my limitations in expressing a deep conviction about the irreplaceable usefulness of qEEG.

For MATLAB® and Simulink® product information, please contact: The MathWorks, Inc. 3 Apple Hill Drive, Natick, MA, 01760-2098 USA

Tel: 508-647-7000

Fax: 508-647-7001

E-mail: info@mathworks.com

Web: www.mathworks.com

Madrid, Spain Jesús Pastor

Competing Interests The author has no competing interests to declare that are relevant to the content of this manuscript.

Acknowledgments

I would like to acknowledge the collaboration of numerous colleagues in the development and discussion of the method. We have presented and analyzed cases in clinical sessions at the Clinical Neurophysiology department of the University Hospital La Princesa, in Madrid (Spain), where the comments of staff members and residents have been very important. However, it would be unfair if I didn't highlight, above all, the collaboration and support of Dr. Lorena Vega-Zelaya, who has spent countless hours discussing and testing the methods used and the results obtained (as well as its help in choosing the acronym) and collaborating in the elaboration of the papers from our group used in this work.

Discussions with colleagues in seminars, courses, and conferences have also been of great help in refining the method. Finally, it would not have been possible finish this work without the patience and supporting of my family, who have had to endure so many moments of absence.

This work was in part financed by a grant from the Instituto de Salud Carlos III (ISCIII) PI23/00393 and was partially supported by FEDER (Fonds Europeen de Development Economique et Regional).

Contents

Part II Practical Application of qEEG

Abbreviations

AD	Alzheimer's disease
AP	Action potential
ARAS	Ascending reticular activating system
BiPD	Bilateral independent periodic discharge
BPA	Burst of periodic activity
cEEG	Continuous electroencephalography
CJD	Creutzfeld-Jakob disease
coh	Coherence
CSF	Cerebrospinal fluid
CT	Computed tomography
DLB	Dementia with Lewy bodies
EEG	Electroencephalography/electroencephalogram
EM	Electromagnetic theory
EMG	Electromyogram
EOG	Electrooculogram
EPSP	Excitatory postsynaptic potentials
ES	Epileptic seizure
ET	Electrostatic theory
F	Frontal lobe
FFT	Fast Fourier transform
FIRDA	Frontal intermittent rhythmic delta activity
FMT	Frontal midline theta rhythm
FS	Focal seizure
FTD	Fronto-temporal dementia
GABA	Gamma-aminobutyric acid
GPFA	Generalized paroxysmal fast activity
GRDA	Generalized rhythmic delta activity
GS	Generalized seizure
HE	Hepatic encephalopathy
ICU	Intensive care unit
INPH	Idiopathic normal pressure hydrocephalus
IPSP	Inhibitory post synaptic potentials
IRDA	Intermittent rhythmic delta activity
KC	K-complex
LGS	Lennox-Gastaut syndrome
logPS	Logarithmic transform of the power spectrum
MEG	Magnetoencephalography

mP	Maximum power for sigma band
MR	Magnetic resonance
NCSE	Nonconvulsive status epilepticus
p	Probability
PDA	Polymorphic delta activity
PDD	Parkinson's disease dementia
PDR	Posterior dominant rhythm
pF	Maximum peak of frequency for sigma band
PNDA	Persistent nonrhythmic delta activity
PO	Parieto-occipital lobe
PS	Power spectrum
PSG	Polysomnography
qEEG	quantified electroencephalography
rsEEG	Resting state EEG
SCN	Suprachiasmatic nucleus
SE	Status epilepticus
SEM	Standard error of mean
SS	Sleep spindles
SSE	Shannon's spectral entropy
T	Period (s)/temporal lobe. Context dependent
Th	Thalamus
ThC	Thalamo-cortical neurons or fibers
TIRDA	Temporal intermittent rhythmic delta activity
UE	Uremic encephalopathy
VD	Vascular dementia
v-EEG	Video-electroencephalography
VSW	Vertex sharp-waves
WE	Wernicke's encephalopathy

Part I

Fundamentals and Method

1 Introduction to Electroencephalography

1.1 Why Should We Study How the Brain Works?

The physiological function of the brain underlies reasoning, emotions, perception, and behavior. Furthermore, abnormal functioning of this system is responsible for neurological and psychiatric diseases. Therefore, the objective assessment of brain function and malfunction is critical for numerous medical and scientific fields [1, 2]. The dynamics of brain activity reflects interaction between different, and usually anatomically segregated, neural networks organized to process information [3]. The brain is an organ that has evolved to manage information through electrical activity. Therefore, one of the best ways to access this information processing is by recording and analyzing the bioelectrical activity generated by the brain. To do this, we have two fundamental methods, the electroencephalography (EEG[1]) and the magnetoencephalography (MEG).

[1]The acronyms EEG and MEG can be used both for the techniques described and for the recordings made using them, which are the electroencephalogram and the magnetoencephalogram.

The EEG is the fastest, cheapest, and most portable method to measure brain activity in a non-invasive way. The temporal resolution is fitted to the nervous system dynamics and can be easily implemented in the hospital environment [4, 5].

1.2 Quantitative EEG in Clinical Practice Nowadays

Electroencephalography is one of the oldest ancillary diagnostic methods currently used in medicine. It was described one century ago by the German psychiatrist Hans Berger [6]. Since then, its use has rapidly spread, and practically every hospital in the world has an EEG device. However, although EEG is probably the fastest, cheapest, and most straightforward method to obtain neurophysiological information from the human brain in a non-invasive way [7], its use, at least in clinical practice, is excessively restricted to the diagnosis of epilepsy. Nevertheless, it should always be remembered that the primary function of the cerebral cortex is to exchange information by generating bioelectrical signals, not only in epilepsy.

J. Pastor, *Quantitative EEG Practical Guide*, https://doi.org/10.1007/978-3-032-05457-9_1

The brain is the organ with a network organization par excellence [4]. The work of the brain and the result of this process cannot be understood without the concept of network. Therefore, the partial or collective malfunction of the system will imply a bad processing of information and the appearance of symptomatology. Often, a network malfunction doesn't necessarily depend on the actual anatomical loss of a part; it's enough that communication between the different parts would be inadequate. In other words, a lack of synchronization between the parts of a network is enough for a network malfunction. Nevertheless, the process of synchronization in the EEG is very difficult (if not impossible!) to evaluate by the naked eye, even for the most trained neurophysiologist. In fact, subtle variables like this are difficult to assess in practice. Yet, other variables are difficult to be evaluated. The definition of epileptic spikes, probably the most used concept in EEG, is problematic, because even for qualified electroencephalographers, agreement in the identification of spikes could be between 50% and 75% [8]. Nonetheless, despite this limitation, EEG has been seen as a method to ascertain epilepsy.

In addition to this excessive restriction to epilepsy, EEG is sometimes reported as a subjective method, depending strongly on the interpreter [9–13]. There has been an attempt to reduce inter-rate variability among interpreters by the introduction of a consensus for EEG interpretation [14–17]. Although these consensus and classifications introduce some objectivity in EEG analysis, the variables are mostly qualitative or binary (present/absent), and specificity is therefore not very high.

In the past few decades, clinicians, neuroscientists, physicists, mathematicians, and engineers (among other experts) have sought a way to overcome these flaws to increase the true field of EEG application and reduce subjectivity in diagnosis. Some research has been devoted to combining EEGs and imaging techniques, mainly magnetic resonance imaging [18–20]. Another direction has been to develop mathematical tools to increase reliability and deepen the information extracted from neurophysiological recordings. Collectively, this approach is called quantitative EEG (qEEG [5, 21]). A canonical formulation toolbox is not needed to define qEEG. Instead, we can say that every EEG recording that uses any type of post hoc numerical method to obtain a result about frequency spectrum, synchronization, network dynamics, or anything else from a bioelectrical magnitude can be considered qEEG. This approach has been increasing in popularity since the introduction of digitalization in electroencephalography. However, the use of qEEG in clinical practice is far from being generalized, except for ICUs, where long-term monitoring (continuous EEG, cEEG) and qEEG are slowly increasing [22–28].

It's difficult to identify all the causes explaining why qEEG is not so widespread in clinics as in biomedical research. Probably the complexity of mathematical methods and the lack of friendly analytical software would be relevant. In fact, many of the papers describing very interesting and promising results in the identification of objective biomarkers in today's difficult-diagnosing pathologies (at least at the very beginning of the symptomatology) use mathematical concepts and developments of moderate to high difficulty for clinicians. In a complementary way to this situation, the use of deep learning or classifying methods without a straightforward root in the pathophysiology, although undoubtedly successful and efficient, does not inspire enough confidence in the medical staff.

However, little by little, most clinicians are conscious that there is not only important, but indispensable, a knowledge deep enough in numerical methods, because the help that they can offer is out of discussion. In the last decades, the term endophenotype has been coined to denote a measurable component of the pathology along the pathway between genotype and phenotype [2] and is equivalent to the concept of biological marker. This concept is pivotal in the development of personalized medicine, because it links the genetic and environmental conditions for the illness with its manifestation. The quest for these morphological, bioelectrical, or biochemical endophenotypes is growing every day in clinical and scientific literature and also in every proposal funds. EEG is especially well placed to identify some of these endophenotypes for different pathologies, further from epilepsy, including psychiatric illnesses.

Some authors claim that the short time required for a cEEG review is one of the most given reasons for the use of qEEG in the ICU and other diagnostic fields. However, we have taken a different approach to qEEG during cEEG or standard EEG for ambulatory/hospitalized patients: instead of just simplifying seizure detection (another manifestation of the excessive focus on epilepsy); our aim is to obtain a comprehensive and efficient view of bioelectrical brain physiology/physiopathology in the most objective way.

In this way, it is important to keep in mind that qEEG is a tool designed to help the clinician in a deeper understanding of brain dynamics and consequently achieve a better diagnosis. However, as a tool or numerical device, qEEG probably wouldn't be enough by self for diagnosis human pathologies without the interpretation of results by a neurophysiologist (at least we hope so) and this is the main reason because clinicians need to be confident in the interpretation of results obtained by numerical methods and they need to be skillful to relate these results with brain pathology. With both aims in mind, we will be able to increase the field of utility for EEG, clearly out of epileptology, and gain an objective measure of very relevant brain variables of great significance in human pathology.

Nowadays there are several packages (MATLAB®, PHYTON®, R®), some of them free and others commercial, that can be relatively easy to learn and use by clinicians. A big number of scripts are available on the Internet or from commercial suppliers (e.g., EEGlab). Any physician (after all a postgraduate) sufficiently motivated should be capable of gaining enough confidence and competency in the management of numerical toolbox to ameliorate its diagnosis and can orientate these devices to its main interests (epilepsy, dementias, critically ill patients, psychiatry, etc.).

It is difficult to offer a concise summary and deep enough of all the numerical methods published in human pathology. On the other side, that is not our motivation. In this book, we try to describe one of the possible approaches to qEEG but one that has proved to be very useful in clinical practice because we developed the program with the specific goal of manageability and straightforward result interpretation, because it's based on the homeostatic system of the brain responsible for the EEG. Since the first version (2016), we have modified it according to our interest and we have used it systematically in all kinds of patients, including pre-operative assessment of epilepsy, conventional rsEEG, emergency, or critically ill patients.

We do not pretend that this would be the best method to implement in clinical practice. Probably there would be a different approach, as much as clinicians interested in, but it's an example of the utility and (at least we hope so) can motivate other neurophysiologists and physicians to go further in the knowledge and encourage them to implement or use numerical systems.

However, the first step to introduce qEEG is to briefly introduce the physiological foundations of EEG. It's out of the scope of this book to describe in detail the origin of bioelectrical activity by the neurons. Instead, we refer the reader to excellent monographies and general books [29–34].

1.3 A Brief Description of the Cortex

We can enunciate a series of general principles in the processing of information by the cerebral cortex [35, 36]:

- All cortical regions have input and output information. This means that, strictly speaking, there are no terminal or initial centers of operations or behaviors.
- Each cortical area carries out multiple operations, providing multiple outputs to other

areas, each of which will be interested in the result of a given processing.
- There is no cortical area that is not related to a subcortical center. It is not clear whether the information received by other cortical areas is the same as that sent to subcortical regions.

For all these reasons, and although it is evident that there are topographical connections (point to point) between different cortical regions, it is very likely that a large part of the connections between areas are made through operational connections, that is, the final area will receive from the initial area only the result of the information processing carried out by the initial part.

From an evolutive, functional, and architectural point of view, two types of cortices can be distinguished: the archipallium or allocortex and the neopallium or isocortex.

1.3.1 Archipallium or Allocortex

This is the oldest evolutionary component of the cortex. It can be further subdivided into two parts, paleocortex (formed by three layers), receiving afferences from the cranial nerve I (olfactory), and reaching the olfactory trigone, a region clearly linked to olfactory stimuli, and archicortex (formed by four layers), which forms the hippocampus and is separate from the neopallium by the hippocampal fissure.

1.3.2 Neopallium or Neocortex

This is the newest evolutionary part of the brain and the larger surface. Information in the neocortex is processed through layers in an interconnected network of neurons called macrocolumns (∅ = 3 mm) or modules, which are a set of neurons and interneurons highly interconnected [37]. These units contain about 1000 minicolumns. The proposed diameter for these columns would be 200–300 μm, although for other authors it would be 400 μm [7]. An increase in cortical surface area, therefore, provides a greater capacity to process information [35]. The overall density of neurons is relatively constant, regardless of the area (with the exception of V1): about 100,000 cells/mm^2 of cortical surface.

The typical width of the cerebral cortex is 2–4 mm and is formed by six layers which are [38], from the most superficial to the deepest:

1. Molecular (acellular): occupied by dendrites from lower layers and axons.
2. Outer granules: small spherical cells (granules). Efferents in feed-forward loops to corticocortical projections. Afferents: receive corticocortical feed-back projections from hierarchically superior areas.
3. External pyramids: contain several cell types, but mainly pyramids, which are larger in the deeper regions. Their efferents are corticocortical (feed-forward). They do not receive specific afferences.
4. Internal granules: This is the layer through which information enters both from other cortical regions (in corticocortical feed-forward connections) and from subcortical regions (cortical-thalamus). It is, therefore, particularly large in primary sensory regions, especially in the primary visual cortex, where three sublayers can be distinguished, which, by forming striae, give it the name of striate cortex.
5. Internal pyramids: typically contain pyramids larger than those of layer III. Its efferences are toward subcortical regions (pyramidal pathway, in primary motor cortex) or corticothalamic regions and toward cortical regions (feed-back connections).
6. Multiform: contains cells with multiple morphologies. Its axons project to subcortical and corticocortical regions (feed-back). In fact, the main afference of the relay cells of the thalamic nuclei comes from this cortical layer, being up to an order of magnitude higher than the peripheral afferences.

In addition to these hierarchical ascending and descending links, there are also lateral connections between adjacent and more distant regions, associating cortical areas of the same hierarchical level. Lateral connections can originate in all layers leaving a cortical area (all layers except 1 and 4) and terminate in the entire width of the target cortical column.

The excitatory neurons in the cortex are divided into two main types: stellate and pyramidal cells. Both neurons within the local module of the cortex are connected to each other forming a reverberation circuit, responsible for the maintenance of local activation. The feedback control of this activity is regulated by the presence of inhibitory neurons receiving excitatory inputs directly from the thalamus and indirectly from the surrounding excitatory circuits.

The ability of the cortex to process information, associate it with emotional states, store it as memory, and initiate actions is modulated by five structures located deep in the encephalon. These structures, therefore, will be relevant for the explanation of EEG dynamics [39]:

1. The basal ganglia (caudate, putamen, pallidum, subthalamic, and substantia nigra). They regulate movements and contribute to certain forms of cognitive processes such as skill learning. They receive information from all parts of the cortex but send it, through the thalamus, only to the frontal lobe.
2. Hippocampus and associated cortices (entorhinal). Responsible for the formation of long-term memory and spatial location, among other functions.
3. Amygdala. This complex set of nuclei is involved in the emotional analysis or motivational significance of sensory stimuli and in the coordination of actions of a variety of brain systems. Receives input from cortical sensory regions and, in turn, projects to the neocortex, basal ganglia, hippocampus, hypothalamus, and brainstem.
4. Thalamus (Th). It is the main relay nucleus for the incoming information from the outer body surface and environment. Thalamus participates from the Ascending Reticular Activating System (ARAS) which modulates the general cortical excitability. This complex structure is composed of more than 50 nuclei [40] and is surrounded by the reticularis nucleus. This nucleus forms a kind of shield covering the dorsal thalamus and is formed by inhibitory GABAergic neurons that project back to the thalamus. In turn, it receives excitatory projections from the thalamus, the cortex, and inhibitory projections from the brainstem. Sensory-related nuclei can be divided into first-order, receiving sensory information directly from peripheral senses (except olfactory), and high-order relay nuclei, which receive highly processed sensory information.
5. Ascending Reticular Activating System. It was originally viewed as a heterogeneous collection of neurons diffusely distributed throughout the brain stem reticular formation. However, this system is formed by different nuclei of specific neurotransmitters, widely distributed throughout the diencephalon and cortex (Fig. 1.1). The major part of the ARAS consists of monoaminergic and cholinergic neurons, found mainly in four regions [40]:
   - 5.1. The locus coeruleus, which contains noradrenergic cells.
   - 5.2. The dorsal and medial raphe nuclei, which contain serotonergic and dopaminergic neurons in a lesser degree.
   - 5.3. The pedunculopontine and laterodorsal tegmental nuclei, which contain cholinergic nuclei.
   - 5.4. The tuberomammilary nucleus, containing histaminergic neurons.

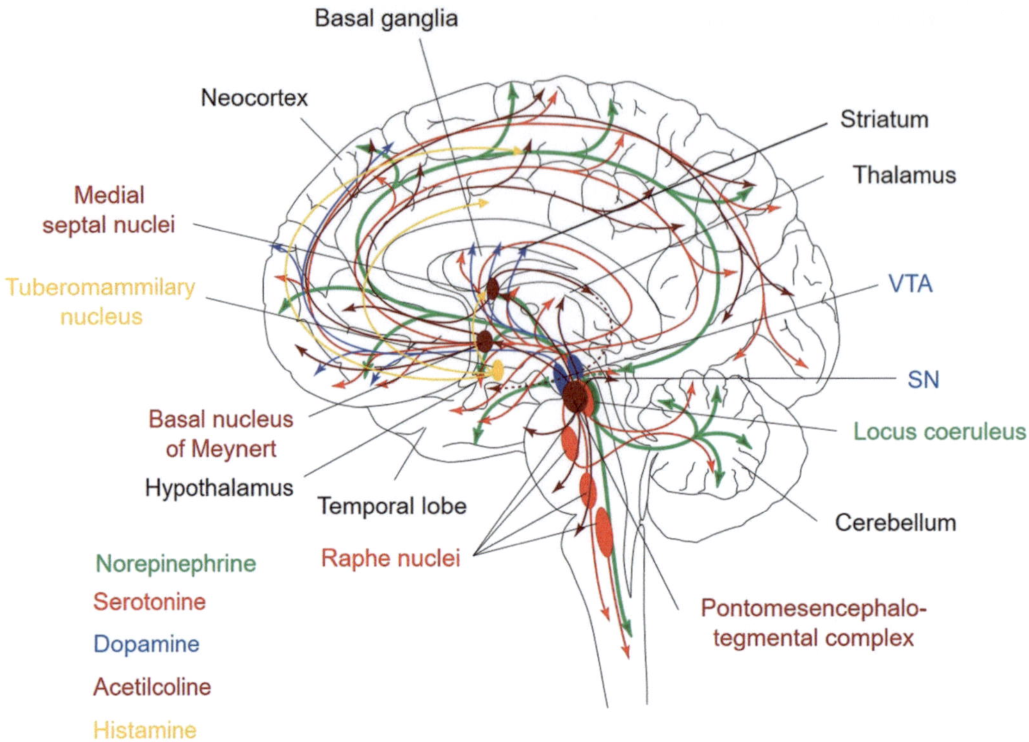

Fig. 1.1 Major cell groups in the ARAS human brain. SN subthalamic nucleus, VTA ventral tegmental area

1.4 The Bioelectrical Brain, an Overview

A detailed description of the brain is out of the scope of this work because all the readers would be familiar with the anatomy and basic physiology. Excellent descriptions can be found in the references [38, 41]. Our concern is related to the bioelectrical activity generated by this organ. And, although we shall discuss in detail some biophysical properties of the brain-skull system in Chap. 2, in this paragraph we will offer a slight landscape of the origin of bioelectrical activity.

The cells responsible for the generation of bioelectrical potentials are nearly exclusively the neurons.[2] The main objective of these cells is to codify and process information by means of changes in their membrane potential. See references [30, 32, 33] for a detailed description of the process. As we will see in the next chapter, an electrical current is generated by the presence of an electric field ($\vec{E}$) which, at the same time, is generated by a transmembrane potential (V) originated by the selective distribution of ions. This V behaves like the electromotive force of a battery. When conditions are adequate (e.g., ionic channels open), different ion species will move according to their electric charge to their reversal potential, resulting in a current. The collective effect of thousands of these small currents generated by individual neurons creates a macroscopic current (although very small in magnitude) travelling through different regions of brain and adjacent structures, as meninges, cerebrospinal fluid (CSF), bone, or skin, until it reaches the electrodes placed at the surface of the scalp.

The cerebral cortex is the main structure generating EEG, although many other structures are

[2]Of course, there are numerous evidences that astrocytes can be responsible of bioelectrical properties of the brain. However, its participation in EEG is not very important, so we can ignore.

involved in its regulation. The cortex is a folded structure fluctuating in thickness from about 2 to 5 mm, with an estimated surface area of roughly 1600 to 4000 cm^2, containing the amazing number of about 10^{10} neurons [7]. Considering that every neuron receives around 10^4 to 10^5 synapses, the number of total synapses can be 10^{14}–10^{15} for a human brain. This is an impressive number, bearing in mind that the number of stars in the Milky Way (our galaxy) is estimated to be $2–4 \times 10^{11}$.

The signal picked up by EEG electrodes is the overall amount of extracellular current elicited by the neural tissue, approximately near the electrode. This current has contribution mainly from the cortex, also this is modulated by different neural structures [42]. The bioelectrical sources of the EEG are:

1. Synaptic Activity

The dendrites and soma of a neuron form a tree-like structure, with a conducting interior, surrounded by a relatively insulating membrane [33]. Along the entire structure, there are tens of thousands of synapses. Neurotransmitters act on synaptic receptors (basically AMPA and NMDA) to induce excitatory currents (excitatory post-synaptic potentials or EPSP, Fig. 1.2b) or inhibitory currents, mainly acting through GABA receptors (inhibitory post-synaptic potentials or IPSP). The flow of cations into the intracellular space produces a "sink" current. According to the conservation of energy law, the sink current of the active region must be balanced. Inactive regions close to the "sink" act as a "source" of current for the active region, inducing a flow of ions in the opposite direction from the intracellular to the extracellular space. Although the currents are closed circuits, from the extracellular point of view there is a net influx/efflux of net current. The duration of postsynaptic potentials can last around 20–30 ms, which corresponds to a frequency of 30–50 Hz.

Depending on the location of the sink current and its distance from the source current, a dipole will appear, whose electric potential decays by $1/r^2$ as we shall discuss in detail in the next chapter.

2. Action Potentials (AP) or Spikes

These generate the most intense currents across the neuronal membrane. They are mediated mainly by Na^+, K^+, Cl^-, and Ca^{2+} currents, are of short duration and large amplitude near the soma. Its dynamic is usually described by the Hodgkin-Huxley model [43] which explains the main features of this waveform, e.g., existence of a threshold, the presence of absolute and relative refractory periods, and overshoot above 0 mV [44]. It's important to keep in mind that AP is an all-or-nothing process, that is, once started the process, there is no way to modify it (Fig. 1.2c).

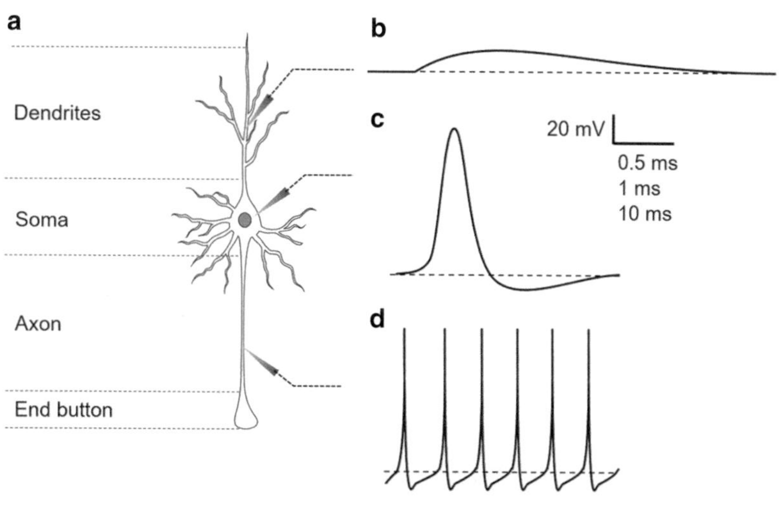

Fig. 1.2 Bioelectrical sources of EEG. (**a**) Scheme of a cortical neuron and its main structures. (**b**) Excitatory post-synaptic potential (EPSP) generated at dendrites. (**c**) Action potential (AP) generated at soma and axon hillock and codified. The output of the neuron (**d**) is transmitted through the axon as a burst of AP

Sometimes, synchronous APs from several neurons can contribute significantly to the high-frequency components of the local field potential in the cortex. Nevertheless, its short duration (around 1–2 ms) usually difficult an efficient synchronization and their contribution to EEG typically is low. In fact, although AP currents contribute in a high degree to local extracellular potentials recorded through microelectrodes (e.g., during deep brain stimulation surgery), its contribution to distant scalp potentials is much smaller than synaptic currents and, therefore its contribution to EEG is low. However, it is the dynamic or AP (e.g., frequency, timing) that originates the activity at pre-synaptic buttons and, consequently is the basis of the presence of synaptic activity in the post-synaptic neuron.

The integration of the input in the form of EPSP and IPSP from the whole dendritic tree is carried out in the soma and axon hillock and codified as a burst of AP of defined frequency and timing, being transferred via the axon (Fig. 1.2d) until the end button.

In practice, the main source of EEG signal is the synaptic activity, and, in a lesser degree, action potentials originated from the cerebral cortex, at least from the cortex located at the convexity of the brain, because structures in the mesial regions and basal surfaces cannot be easily recorded. Not only because the postsynaptic activity is spreading located along the neuron and spikes are more concentrated around the soma and axon hillock, but also because the longer duration of postsynaptic activity facilitates synchronization of neural elements at frequencies below 70 Hz, the goal of EEG in clinics.

In a first approximation, it is useful to consider that every cortical macrocolumn can be considered as a functional unit in the processing of information and, in the same way, as the bioelectrical unit in the EEG origin. Therefore, the overall activity of the synapsis of a given volume of tissue (formed by a set of columns) will produce a net change in the extracellular current in that volume whose propagation by means of volume conduction will be recorded at the scalp. These synaptic action fields [7] can be considered as the true sources or generators of EEG.

We can suppose that the net result from all the bioelectrical activity from a column, in a define time interval, can be approximated by a dipole whose direction and magnitude (not dependent from the anatomical dimensions, but the separation of sources and sinks by the addition of synaptic activity) can be considered as the basic bioelectrical source for the EEG. The summation of big amounts of these dipoles from spread surfaces of cortex will produce the potential recorded by the surface electrodes. Therefore, we can think the cortex as a bidimensional surface (although folded by the presence of sulci) of dipoles. See The dipolar mode in the EEG in Chap. 2. The dynamics and interactions between these cortical columns during the processing of neural information are the substrates of the dynamical changes in the EEG.

Although the synaptic fields can be considered as the EEG generators, there is no univocal relationship between the EEG and this activity. This is because there is a significant amount of synaptic activity that cannot be recorded through EEG, as occurs with columns located at opposite sides of the same sulcus, that can cancel each other, or anatomical structures named closed fields (e.g., the amygdala in the temporal lobe). Besides, the physical properties of the brain-skull system preclude the recording of ranges of frequencies above beta/gamma activity (see next chapter).

1.5 The Homeostatic System of the EEG

The EEG is the multivariate determination (in time and space) of the electrical potentials generated by the brain and recorded on the surface of the scalp. In clinical practice, EEG oscillatory activity is divided into four bands according to their oscillation frequency: delta (δ), theta (θ), alpha (α), and beta (β). There are other faster and slower bands that are difficult to obtain with enough quality in daily clinical practice and therefore we will not discuss these brain activities. See Fig. 1.3 for a scheme of these structures that will be discussed in the next paragraphs.

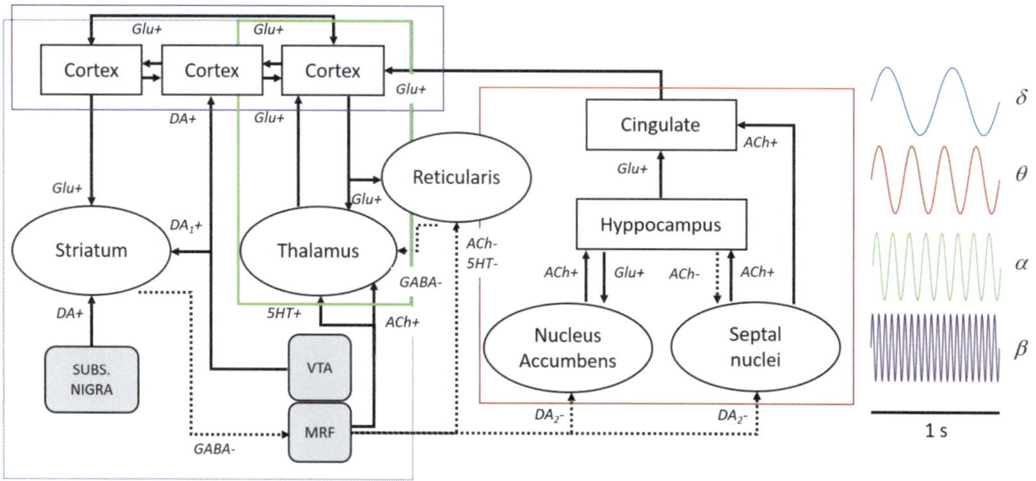

Fig. 1.3 Homeostatic system of the EEG. Although alpha activity is colored in green, the presence of theta as posterior dominant rhythm is due to the same structures working defectively. Refer to this figure for all the rhythms described next

This complex neuroanatomical homeostatic system is probably genetically determined and regulates basal levels of local synchronization, global interactions between different regions, spectral composition, and periodic sampling of signal space [2, 5, 45, 46].

The activity of the brain encompasses a continuum in frequency between near 0 and 1000 Hz, although not all of these are picked up in clinics. Although the cerebral cortex is the main anatomical structure generating these potentials [7], the regulation of the different EEG bands is performed by different sets of encephalic structures. This aspect is of extraordinary relevance because the specific involvement of different structures in different pathologies will lead to specific changes in the diverse numerical variables that can be obtained from the EEG.

For any rhythm to be recordable from the scalp, it must fulfill the following requirements [2]:

1. The sources have to be near the surface but not too deep in the brain.
2. The sources must be synchronized over relatively large spatial scales.
3. The currents must be strong enough on the skin of the skull.

Obviously, not all the cortical bioelectrical activities satisfy the three requirements which means that only a part of the whole brain activity is recorded at the EEG.

Briefly, we will describe the neural origin of the different frequency bands in the brain.

An extensive review of all EEG rhythms and their variants is beyond the scope of this brief, schematic summary. I refer the reader to the literature for detailed information [47–49].

1.5.1 Delta Band (δ)

It is defined as those waveforms whose frequency is between 0 and 4 Hz and comprises less than 10% of the normal waking EEG by age 10 years. In the wakeful states, delta can be considered a normal finding in the very young and in the elderly. The normal elderly may have few irregular delta complexes in the temporal regions. It is similar to temporal theta in the distribution, often more in the left than right temporal regions, but normally is present for less than 1% of the time. Some delta is normal in people older than 60 years, mainly at the onset of drowsiness and in response to hyperventilation. This rhythm defines the slow-wave sleep (N3) and therefore is the main component during that period. Excessive generalized delta is abnormal and indicates an encephalopathy that is etiology unspecific. Focal arrhythmic delta usually indicates a structural lesion involving the white matter of the ipsilateral hemisphere, especially when it is continuous and unreactive [50].

There are two types of physiological δ activity, but here we only describe the observed during wakefulness.

Slow waves lower than 4 Hz are generated within the cortex, characterized by a sequence of up and down states of a cortical neural network. Up state is characterized by activation, increasing the discharge rate of cortical neurons. On the contrary, the down state is characterized by a suppression of neural activity mediated by a strong hyperpolarization. The collective activity of neurons in the same state elicits an alternate opposite local field potential that can be recorded at the scalp. This oscillation, through the putamen/globus pallidus, in turn inhibits the brainstem nuclei responsible for the ARAS and modulates the activity generated in the Th. Anatomical or functional disconnection of the cortex from the Th (e.g., impairment of white fibers typically) results in significant enhancement of this band. Although both states are generated by intracortical mechanisms, the influence of TC activity is highly suspected. The appearance of these oscillations can be summarized into several categories [48], such as:

1. Intermittent rhythmic delta activity (IRDA), which can be classified according to its scalp location in frontal (FIRDA), temporal (TIRDA), or occipital rhythms (OIRDA). These are regular, intermittent, generally brief, and monomorphic bursts, clearly different from the surrounding background. Frequently, Intermittent generalized rhythmic delta activity (GRDA) is observed.
2. Polymorphic delta activity (PDA) can be considered as abnormal when a consistent asymmetry is observed and it is usually associated with asymmetric superimposed faster frequencies and lack of change with alertness. The finding of maximum slowing with minimal faster frequencies indicates a focal injury of white matter.
3. Persistent nonrhythmic delta activity (PNDA), usually found in severe focal injuries of cortex as stroke or brain trauma. This pattern can be considered as a kind of PDA more stable along the time and its significance is similar.

Reversible generalized PDA can be observed in encephalopathy, syncope, and ischemia or post-ictal state. However, irreversible PDA or PNDA can be observed in post-anoxic state, brain infection or inflammation, and in some phases of dementias. It's important to know that there is a relationship between the serum glycaemia and the onset of PDA. In fact, serum glucose lower than 35 mg/100 ml gives rise to generalization spreading of PDA.

Regarding FIRDA or OIRDA, its presence is usually associated with developing or resolving the pathology more than chronic illness, although it can be observed during metabolic dysfunction or an increase in intracranial pressure. However, TIRDA is a more specific pattern because it is highly associated with mesial temporal lobe epilepsy which can be regarded as epileptic activity.

1.5.2 Theta Band (θ)

The theta band is the frequency band oscillating between 4 and 8 Hz. There are different types of theta activity in the daily EEG. Let's see the most relevant.

1.5.2.1 Frontal Midline Theta Rhythm

After solving problem tasks, a train of rhythmic waves was described at 6–7 Hz with a maximum amplitude at frontal midline, anterior to the vertex [2, 51]. The presence of this frontal midline theta rhythm (FMT) can also be induced by emotions, concentration, and during mental tasks, especially mathematical operations. During the resting state with closed or open eyes, the FMT rhythm is only observed in approximately one-third of normal awake, young adults, usually with amplitude lower than 15 μV at the frontocentral region.

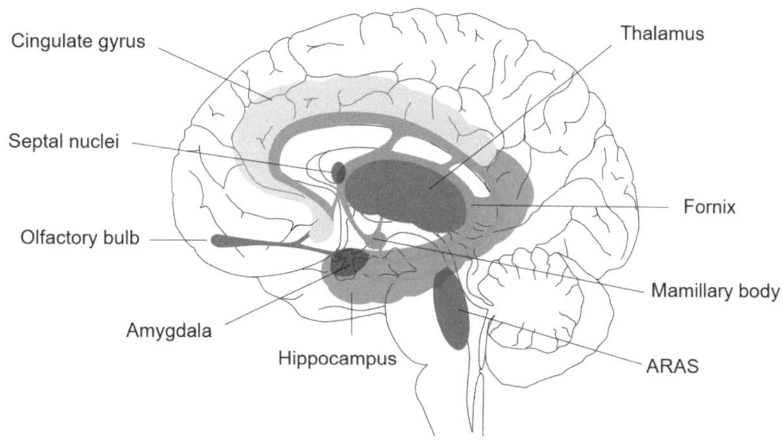

Fig. 1.4 Scheme of limbic system and structures involved in the FMT rhythm

It is debated the sources of FMT, and it has been postulated that alternatively activated at 40–120 grades phase shift of prefrontal medial superficial cortex and anterior cingulate cortex can be responsible for this rhythm [52]. Nevertheless, it has also been postulated that a scatter activity in wide cortical areas of frontal lobes of both hemispheres can be the source of FMT [53] (Fig. 1.4).

1.5.2.2 Other Theta Rhythms

Theta activity during wakefulness is normally enhanced by hyperventilation and drowsiness. Intermittent 4–5 Hz activity in both temporal lobes, or even with a lateralized predominance, may occur in about one-third of the asymptomatic elderly and cannot be considered as abnormal [50].

The presence of theta activity in posterior regions is an abnormal function of the thalamocortical neurons responsible for the alpha rhythm, working at a relatively hyperpolarized resting state [54, 55]. Therefore, we cannot consider that theta activity in older or encephalopathic patients would be produced by specific networks, other than those responsible for alpha activity, but working at a lower frequency.

1.5.3 Alpha Rhythms (α)

The first recording of alpha rhythm was performed by Hans Berger nearly a century ago [6] and coined the term. He also showed that alpha waves were blocked upon eye opening and during some kind of mental activities. Its frequency ranges from 8 to 13 Hz and there are several variants across the scalp, depending on their location and reactivity, although the name of alpha rhythm is reserved for the occipital one.

1.5.3.1 Occipital Alpha Rhythm

During normal development, an 8 Hz alpha frequency appears by 3 years of age and remains stable between 8 and 12 Hz even during normal aging into the later years of life. This activity is maximum at the occipital pole and is not necessarily synchronized between both hemispheres. Voltage asymmetries of higher than 50% should be regarded as being abnormal, especially when the left side is greater than the right.

It is enhanced with eyes closed and is blocked (desynchronized) when the subject opens his eyes. At 8 years old the alpha rhythm is present and there is a slight increase in frequency from 8 to 20 years old and a slight decrease with aging. Usually, the posterior dominant rhythm (PDR) in healthy adults is around 10 Hz, although individual variations in frequency or width of EEG spectra can be very important [56]. It is best observed during relaxed wakefulness and has a side-to-side difference lower than 1 Hz. Unilateral failure of the alpha rhythm to attenuate reflects an ipsilateral abnormality (Bancaud's phenomenon). Normally, alpha frequencies may transiently increase immediately after eye closure (alpha squeak).

Alpha variants include harmonic forms that are one-half (slow alpha) or two times (fast alpha) the main frequency with similar distribution and reactivity. Alpha variants may have a notched appearance. Paradoxical alpha occurs when alertness results in the presence of alpha, and drowsiness does not [49]. Sometimes the alpha rhythm is absent, giving the EEG a low voltage appearance. However, these low-voltage records are found in 4–10% of normal adults. However, the low-amplitude records can be found in alcoholics and drug addicts.

There is increasing evidence showing that alpha rhythms are generated at the thalamus [54, 55, 57], although the participation of occipital cortex is debated [58]. In fact, alpha, theta, and also delta frequencies are elicited by acetylcholine and glutamate at thalamic lateral geniculate thalamo-cortical (ThC) neurons [54]. In mammal models, a convincing model of interplay between ThC relay cells and interneurons, coupled by gap junctions and chemical synapses, has been described [57]. Therefore, alpha rhythms can be considered as epiphenomena that reflect thalamic regular bursting transferred to the cortex. ThC relay neurons have a relatively broad range of resting potentials, mainly regulated by glutamate. The most depolarized resting potentials give rise to faster alpha activity (upper to 10 Hz), meanwhile the hyperpolarization decreases the mean frequency of bursts, given place at theta activity in the cortex (<7.5 Hz). Therefore, the presence of slow posterior activity observed in several pathologies (e.g., encephalopathy) is due to a hyperpolarization of ThC relay neurons.

Alpha activity is considered as an idling rhythm of the corresponding cortical areas. In fact, during occipital alpha rhythm, there is a decrease in metabolism of occipital cortex. The appearance of alpha rhythms in specific sensory areas would inhibit the sensory information processing in those areas [59].

This rhythm, in fact, is not unitary and is generated at three different regions, as we shall discuss next. See Fig. 1.5.

1.5.3.2 Parietal Alpha Rhythm

In some cases, a high amplitude alpha component can be observed at the sagittal parietal area (maximum at Pz). Usually, it can be enhanced by closed eyes, although a decrease can be observed in some subjects. This rhythm is independent of

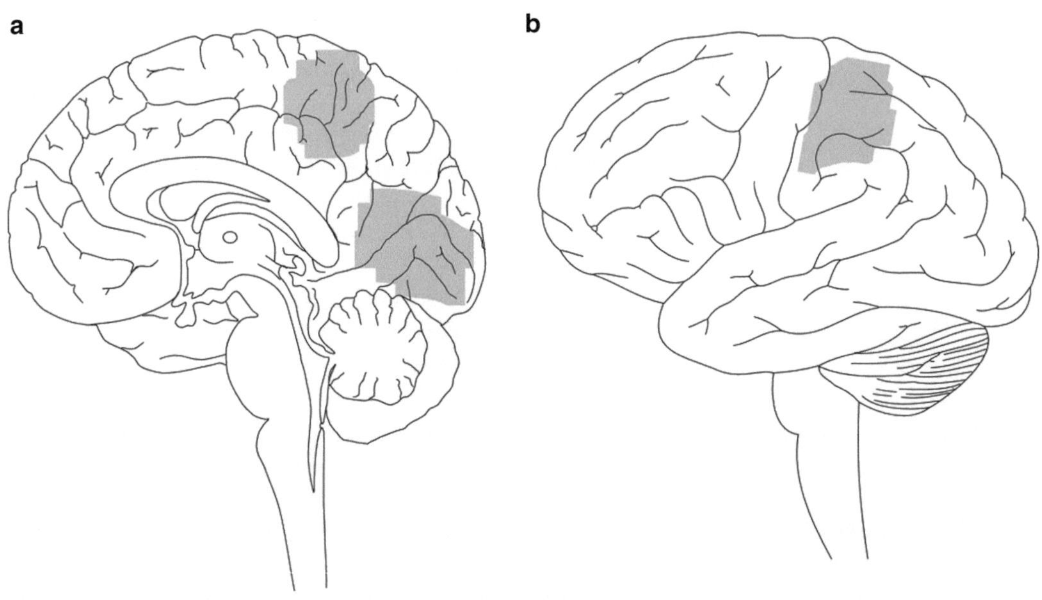

Fig. 1.5 Origin of the three kinds of alpha rhythms. (**a**) Mesial view of the brain showing the origin of the parietal (anterior) and occipital (posterior) alpha rhythms. (**b**) Lateral view of the brain showing the cortical origin of the post-central mu-rhythm

the occipital rhythm abovementioned. Its frequency is lower than the occipital one, and the overall power increases with mental tasks when compared with the eyes closed condition.

Their cortical sources are located at the mesial surface of the parietal lobe, cranial and frontal, placed respect to the sources of the occipital alpha rhythm. See Fig. 1.5a.

1.5.3.3 Mu-Rhythms (μ)

This rhythm is named by its negative shape that reminds the Greek letter μ (mu). It is recorded above the Rolandic fissure with a maximum at C3 and C4 regions. Their origin at the sensory-motor cortical strip explains why they are blocked by the corresponding contralateral hand movement or touching. Although the imagination of movements can suppress the mu-rhythm, these rhythms are independent for both hemispheres. See Fig. 1.5b.

1.5.4 Beta Rhythms (β)

Beta rhythms are frequencies higher than 13 Hz and lower than 30 Hz. It was firstly described by Berger and associated with focused attention. It is a common rhythm, and normally observed with a voltage lower of 20 μV. In fact, voltages beyond 25 μV in amplitude can be considered as abnormal. Beta activity during wakefulness normally increases during drowsiness and with mental activation. Persistently reduced voltages of more than 50% suggest a cortical gray matter abnormality within the hemisphere having the lower voltage. Nevertheless, minor asymmetries may simply reflect skull irregularities [50].

The beta band is conventionally divided into different sub-bands: low-beta (13–20 Hz) and high-beta (21–30 Hz).

At least two distinct beta rhythms can be identified:

1.5.4.1 Rolandic Beta Rhythm

It can be observed as spontaneous activity during eyes closed and open conditions in healthy subjects, and the maximum is located over the motor-sensory strip. Its frequency is about 20 Hz but cannot be considered as a harmonic of the alpha mu-rhythm because it has been shown to have different sources for both kinds of waveforms [60]. Suppression of this rhythm is associated with finger movement (motor-related task) and is usually followed by a rebound of beta synchronization. Although the cortical interactions are important in the building of this rhythm, the thalamic regulation is more important than for the frontal rhythm.

It is interesting to note that, in contrast to the power increase after administrations of benzodiazepines, observed for the frontal beta (below), the power of rolandic beta decreases [61].

1.5.4.2 Frontal Beta Rhythm

Its maximum is located at frontal lobes, anterior to the precedent rhythm, and usually it's observed in power spectra as wide peaks of small amplitude in frontal regions, indicating a low amplitude and an irregular pattern. In contrast to the previous rhythm, frontal beta arises in cognitive tasks related to stimulus assessment and decision-making and is a reliable measure of frontal lobe activation during this condition. Therefore, frontal beta rhythm is not usually recorded in the resting state.

It has been shown a close relationship between the EEG power in beta band and frontal metabolic activity [62], suggesting that absolute values of modifications of EEG bands may be useful to estimate a cortical index of metabolic activation in specific tasks.

One of the most striking features of frontal beta rhythm is its sensibility to GABA agonists as barbiturates, non-barbiturate sedatives (e.g., chloral hydrate) and benzodiacepines, increasing the amplitude of frontal beta even more than 50% of the waking tracing [63].

Increase in beta activity, especially in regions out of the physiological frontal localization, corresponds to cortical over-activation and can be associated with epileptic activity, hallucinations observed in schizophrenia [64, 65], viral or toxic encephalopathies, and sometimes in attention deficit and hyperactivity (ADHD) population [66].

Networks of inhibitory interneurons (which explain the pharmacological effect described above) are involved in generating beta activity, and it has been speculated that this can be related to resetting the previous processing task.

1.5.5 Other Physiological Elements of the EEG

Besides the rhythms described above, the clinical EEG recording in clinical practice has other physiological elements which must be known by the physician. A detailed description is out of the scope of this succinct introduction. For a more detailed description, see references [48, 50].

1.6 Concluding Remarks

The EEG is a powerful diagnostic method of extraordinary importance and utility in clinical practice and research. The information obtained provides an image of the physiological and pathological state of the brain. In this chapter, we have briefly seen the neural systems that generate the different EEG signals, which allows us to establish a relationship of some specificity between the alteration of these bioelectrical signals and the neural substrate that generates them.

References

1. Lazarev VV. The relationship of theory and methodology in EEG studies of mental activity. Int J Psychophysiol. 2006;62:384–93.
2. Kropotov JD. Quantitative EEG, event-related potentials and neurotherapy. Oxford, UK: Elsevier; 2009; ISBN 978-0-12-374512-5.
3. Peled A. Brain profiling and clinical-neuroscience. Med Hypotheses. 2006;67(4):941–6. https://doi.org/10.1016/j.mehy.2006.03.029. Epub 2006 May 15. PMID: 16698189.
4. Pastor J, Vega-Zelaya L, Martin-Abad E. Necessity of quantitative EEG in daily clinical practice. In: Nakano H, editor. Electroencephalography. InTech; 2021. ISBN 978-1-83968-289-6.
5. Fingelkurts AA, Fingelkurts AA. Quantitative electroencephalogram (qEEG) as a natural and non-invasive window into living brain and mind in the functional continuum of healthy and pathological conditions. Appl Sci. 2022;12:9560. https://doi.org/10.3390/app12199560.
6. Berger H. Uber das elektroenzephalogramm des Menschen. Arch Psychiatr Nervenk. 1929;87:527–70.
7. Nunez PL, Srinivasan R. Electric fields of the brain. In: The Neurophysics of EEG. 2nd ed. New York: Oxford University Press; 2006. ISBN-13 978-0-19-505038-7.
8. Chaibi S, Lajnef T, Ghrob A, Samet M, Kachouri A. A robustness comparison of two algorithms used for EEG spike detection. Open Biomed Eng J. 2015;9:151–6. https://doi.org/10.2174/1874120701509010151. PMID: 26312076; PMCID: PMC4541300.
9. Wendt SL, Welinder P, Sorensen HB, Peppard PE, Jennum P, Perona P, Mignot E, Warby SC. Inter-expert and intra-expert reliability in sleep spindle scoring. Clin Neurophysiol. 2015;126(8):1548–56. https://doi.org/10.1016/j.clinph.2014.10.158.
10. Halford JJ, Shiau D, Desrochers JA, Kolls BJ, Dean BC, Waters CG, Azar NJ, Haas KF, Kutluay E, Martz GU, Sinha SR, Kern RT, Kelly KM, Sackellares JC, LaRoche SM. Inter-rater agreement on identification of electrographic seizures and periodic discharges in ICU EEG recordings. Clin Neurophysiol. 2015;126(9):1661–9. https://doi.org/10.1016/j.clinph.2014.11.008.
11. Halford JJ, Arain A, Kalamangalam GP, LaRoche SM, Leonardo B, Basha M, Azar NJ, Kutluay E, Martz GU, Bethany WJ, Waters CG, Dean BC. Characteristics of EEG interpreters associated with higher interrater agreement. J Clin Neurophysiol. 2017;34(2):168–73. https://doi.org/10.1097/WNP.0000000000000344.
12. Halford JJ, Westover MB, LaRoche SM, Macken MP, Kutluay E, Edwards JC, Bonilha L, Kalamangalam GP, Ding K, Hopp JL, Arain A, Dawson RA, Martz GU, Wolf BJ, Waters CG, Dean BC. Interictal epileptiform discharge detection in EEG in different practice settings. J Clin Neurophysiol. 2018;35(5):375–80. https://doi.org/10.1097/WNP.0000000000000492.
13. Jing J, Herlopian A, Karakis I, Ng M, Halford JJ, Lam A, Maus D, Chan F, Dolatshahi M, Muniz C, Chu C, Sacca V, Pathmanathan J, Ge W, Sun H, Dauwels J, Cole AJ, Hoch DB, Cash SS, Westover MB. Interrater reliability of experts in identifying interictal epileptiform discharges in electroencephalograms. JAMA Neurol. 2019;77(1):49–57. https://doi.org/10.1001/jamaneurol.2019.3531.
14. Hirsch LJ, LaRoche SM, Gaspard N, Gerard E, Svoronos A, Herman ST, Mani R, Arif H, Jette N, Minazad Y, Kerrigan JF, Vespa P, Hantus S, Claassen J, Young GB, So E, Kaplan PW, Nuwer MR, Fountain NB, Drislane FW. American clinical neurophysiology society's standardized critical care EEG terminology: 2012 version. J Clin Neurophysiol. 2013;30(1):1–27. https://doi.org/10.1097/WNP.0b013e3182784729. PMID: 23377439.
15. Herman ST, Abend NS, Bleck TP, Chapman KE, Drislane FW, Emerson RG, Gerard EE, Hahn CD, Husain AM, Kaplan PW, LaRoche SM, Nuwer MR, Quigg M, Riviello JJ, Schmitt SE, Simmons LA, Tsuchida TN, Hirsch LJ, Critical Care Continuous EEG Task Force of the American Clinical Neurophysiology Society. Consensus statement on continuous EEG in critically ill adults and children, part I: indications. J Clin Neurophysiol. 2015;32(2):87–95. https://doi.org/10.1097/WNP.0000000000000166.
16. Admiraal MM, van Rootselaar AF, Horn J. International consensus on EEG reactivity testing after cardiac arrest: towards standardization. Resuscitation. 2018;131:36–41. https://doi.org/10.1016/j.resuscitation.2018.07.025.

17. Claassen J, Taccone FS, Horn P, Holtkamp M, Stocchetti N, Oddo M, Neurointensive Care Section of the European Society of Intensive Care Medicine. Recommendations on the use of EEG monitoring in critically ill patients: consensus statement from the neurointensive care section of the ESICM. Intensive Care Med. 2013;39(8):1337–51. https://doi.org/10.1007/s00134-013-2938-4.
18. Robertson JA, Thomas AW, Prato FS, Johansson M, Nittby H. Simultaneous fMRI and EEG during the multi-source interference task. PLoS One. 2014;9(12):e114599. https://doi.org/10.1371/journal.pone.0114599.
19. Klein C, Hänggi J, Luechinger R, Jäncke L. MRI with and without a high-density EEG cap—what makes the difference? NeuroImage. 2015;106:189–97. https://doi.org/10.1016/j.neuroimage.2014.11.053.
20. Omidvarnia A, Kowalczyk MA, Pedersen M, Jackson GD. Towards fast and reliable simultaneous EEG-fMRI analysis of epilepsy with automatic spike detection. Clin Neurophysiol. 2019;130(3):368–78. https://doi.org/10.1016/j.clinph.2018.11.024.
21. Vega-Zelaya L, Garnés-Camarena O, Sánz-García A, Ortega GJ, Pastor J. Mathematical foundations of quantified electroencephalography. Clinical Advances in Neurophysiology. Chapter 3. Avid Science. 2016 February 10.
22. Hirsch LJ. Continuous EEG monitoring in the intensive care unit: an overview. J Clin Neurophysiol. 2004;21:332–40.
23. Kennedy JD, Gerard EE. Continuous EEG monitoring in the intensive care unit. Curr Neurol Neurosci Rep. 2012;12:419–28.
24. Ney JP, van der Goes DN, Nuwer MR, Nelson L, Eccher MA. Continuous and routine EEG in intensive care: utilization and outcomes, United States 2005-2009. Neurology. 2013;81:2002–8.
25. Sutter R, Stevens RD, Kaplan PW. Continuous electroencephalographic monitoring in critically ill patients: indications, limitations and strategies. Crit Care Med. 2013;41:1124–32.
26. Sackellares JC, Shiau DS, Halford JJ, LaRoche SM, Kelly KM. Quantitative EEG analysis for automated detection of nonconvulsive seizures in intensive care units. Epilepsy Behav. 2011;22:S69–73.
27. Dericioglu N, Yetim E, Bas DF, Bilgen N, Caglar G, Arsava EM, Topcuoglu MA. Non-expert use of quantitative EEG displays for seizure identification in the adult neuro-intensive care unit. Epilepsy Res. 2015;109:48–56.
28. Haider HA, Esteller R, Hahn CD, Westover MB, Halford JJ, Lee JW, Shafi MM, Gaspard N, Herman ST, Gerard EE, Hirsch LJ, Ehrenberg JA, LaRoche SM, Critical Care EEG Monitoring Research Consortium. Sensitivity of quantitative EEG for seizure identification in the intensive care unit. Neurology. 2016;87(9):935–44. https://doi.org/10.1212/WNL.0000000000003034. Epub 2016 Jul 27. PMID: 27466474; PMCID: PMC5035158.
29. Hodgkin AL, Huxley AF. A quantitative description of membrane current and its application to conduction and excitation in nerve. J Physiol. 1952;117(4):500–44. https://doi.org/10.1113/jphysiol.1952.sp004764. PMID: 12991237; PMCID: PMC1392413.
30. Junge D. Ionic properties and active membranes. In: Nerve and muscle excitation. 3th ed. Massachusetts: Sinauer Associates, Inc; 1992. p. 32–57.
31. Koester J, Siegelbaum SA. Membrane potential and the passive electric properties of the neuron. In: Kandel ER, Schwartz JH, Jessell TM, Siegelbaum SA, Hudspeth J, editors. Principles of neural science. 5th ed. New York: Elsevier; 2013. ISBN 978-0-07-139011-8.
32. Shepherd GM. Electrotonic properties of axons and dendrites. In: Squire B, McConnell R, Spitzer Z, editors. Fundamental Neuroscience. Boston: Academic Press; 2003.
33. Johnston D, Wu SM. Functional properties of dendrites. Foundations of cellular neurophysiology. MIT Press; 2001. p. 55–120.
34. Plonsey R, Barr RC. Bioelectricity. a quantitative approach. 3rd ed. New York: Springer; 2007. ISBN 978-0-387-48864-6.
35. Amaral DG. The functional Organization of Perception and Movement. In: Kandel ER, Schwartz JH, Jessell TM, Siegelbaum SA, Hudspeth J, editors. Principles of neural science. 5th ed. New York: Elsevier; 2013. ISBN 978-0-07-139011-8.
36. Olson CR, Colby CL. The Organization of Cognition. In: Kandel ER, Schwartz JH, Jessell TM, Siegelbaum SA, Hudspeth J, editors. Principles of neural science. 5th ed. New York: Elsevier; 2013. ISBN 978-0-07-139011-8.
37. Casanova MF, Casanova EL. The modular organization of the cerebral cortex: evolutionary significance and possible links to neurodevelopmental conditions. J Comp Neurol. 2019;527(10):1720–30. https://doi.org/10.1002/cne.24554. Epub 2018 Nov 15. PMID: 30303529; PMCID: PMC6784310.
38. Nieuwenhuys R, Voogd J, van Huijzen C. Telencephalon: Neocortex. In: The human central nervous system. 4rd ed. Berlin: Springer-Verlag. ISBN 978-3-540-34684-5.
39. Vega-Zelaya L, Pastor J. The network systems underlying emotions: the rational foundation of deep brain stimulation psychosurgery. Brain Sci. 2023;13:943. https://doi.org/10.3390/brainsci13060943.
40. Richerson GB, Aston-Jones G, Saper CB. The modulatory functions of the brain stem. In: Kandel ER, Schwartz JH, Jessell TM, Siegelbaum SA, Hudspeth J, editors. Principles of neural science. 5th ed. New York: Elsevier; 2013. ISBN 978-0-07-139011-8.
41. Scarabino T, Salvolini U, Di Salle F, Duvernoy H, Rabischong P. Atlas of morphology and functional anatomy of the brain. Berlin Heidelberg New York: Springer; 2003. ISBN-10 3-540-29628-X.
42. Buzsáki G, Anastassiou CA, Nie C. The origin of extracellular fields and currents—EEG, ECoG, LFP and spikes. Nat Rev Neurosci. 2012;13:407–20.

43. Häusser M. The Hodgkin-Huxley theory of the action potential. Nat Neurosci. 2000;Suppl 11:1165. https://doi.org/10.1038/81426.
44. Pastor J. Biophysical foundations of the neuronal activity. Rev Neurol. 2000;30(8):741–55. https://doi.org/10.33588/rn.3008.99539
45. Hughes JR, John ER. Conventional and quantitative electroencephalography in psychiatry. J Neuropsychiatry Clin Neurosci. 1999;11:190–208.
46. Kondacs A, Szabo M. Long-term intra-individual variability of the background EEG in normals. Clin Neurophysiol. 1999;110:1708–16.
47. Buzsáki G. Rhythms of the brain. Oxford University Press.; ISBN 9780199863716; 2006. https://doi.org/10.1093/acprof:oso/9780195301069.001.0001.
48. Krishnan V, Chang BS, Schomer DL. Normal EEG in wakefulness and sleep: adults and elederly. In: Shomer DL, Lopes da Silva F, editors. Niedermeyer's Electroencephalography: basic principles, clinical applications and related fields. 7th ed. Oxford University Press; 2018. ISBN 9780190228484.
49. Acevedo J-M, Yelvington K, Tatum WO. Chapter 9 Normal EEG variants. In: Levin KH, Chauvel P, editors. Handbook of clinical neurology, vol. 160. Elsevier; 2019. p. 143–60. ISSN 0072-9752, ISBN 9780444640321, https://doi.org/10.1016/B978-0-444-64032-1.00009-6.
50. Tatum WO, Husain AM, Benbadis SR, Kaplan PW. Handbook of EEG interpretation. Demos; 2008. ISBN-10: 1-933864-11-7.
51. Arellano AP, Schwab RS, Casby JU. Sonic activation. Electroencephalogr Clin Neurophysiol. 1950;2(2):215–7. https://doi.org/10.1016/0013-4694(50)90041-1. PMID: 15421288.
52. Asada H, Fukuda Y, Tsunoda S, Yamaguchi M, Tonoike M. Frontal midline theta rhythms reflect alternative activation of prefrontal cortex and anterior cingulate cortex in humans. Neurosci Lett. 1999;274(1):29–32. https://doi.org/10.1016/s0304-3940(99)00679-5. PMID: 10530512.
53. Sasaki K, Tsujimoto T, Nishikawa S, Nishitani N, Ishihara T. Frontal mental theta wave recorded simultaneously with magnetoencephalography and electroencephalography. Neurosci Res. 1996;26(1):79–81. https://doi.org/10.1016/0168-0102(96)01082-6. PMID: 8895895.
54. Hughes SW, Lörincz M, Cope DW, Blethyn KL, Kékesi KA, et al. Synchronized oscillations at alpha and theta frequencies in the lateral geniculate nucleus. Neuron. 2004;42:253–88. [crossref].
55. Hughes SW, Lörincz ML, Blethyn K, Kékesi KA, Juhász G, et al. Thalamic gap junctions control local neuronal synchrony and influence macroscopic oscillation amplitude during EEG alpha rhythms. Front Psychol. 2011;2:193. [crossref].
56. Pastor J, Vega-Zelaya L. Normative structure of resting-state EEG in bipolar derivations for daily clinical practice: a pilot study. Brain Sci. 2023;13:167. https://doi.org/10.3390/brainsci13020167.
57. Lörincz ML, Kékesi KA, Juhász G, Crunelli V, Hughes SW. Temporal framing of thalamic relay-mode firing by phasic inhibition during the alpha rhythm. Neuron. 2009;63:683–96.
58. Halgren M, Ulbert I, Batuji H, Fabó D, Eröss L, et al. The generation and propagation of the human alpha rhythm. PNAS. 2019;116:23772–82.
59. Kropotov JD, Grin-Yatsenko VA, Ponomarev VA, Chutko LS, Yakovenko EA, Nikishena IS. ERPs correlates of EEG relative beta training in ADHD children. Int J Psychophysiol. 2005;55(1):23–34. https://doi.org/10.1016/j.ijpsycho.2004.05.011. PMID: 15598513.
60. Hari R, Salmelin R. Human cortical oscillations: a neuromagnetic view through the skull. Trends Neurosci. 1997;20(1):44–9. https://doi.org/10.1016/S0166-2236(96)10065-5. PMID: 9004419.
61. Jensen O, Goel P, Kopell N, Pohja M, Hari R, Ermentrout B. On the human sensorimotor-cortex beta rhythm: sources and modeling. NeuroImage. 2005;26(2):347–55. https://doi.org/10.1016/j.neuroimage.2005.02.008. Epub 2005 Apr 7. PMID: 15907295.
62. Wang X, Li M, Xie J, Chen D, Geng X, Sun S, Liu B, Wang M. Beta band modulation by dopamine D2 receptors in the primary motor cortex and pedunculopontine nucleus in a rat model of Parkinson's disease. Brain Res Bull. 2022;181:121–8. https://doi.org/10.1016/j.brainresbull.2022.01.012. Epub 2022 Jan 22. PMID: 35077843.
63. Sutter R, Pang T, Kaplan PW. EEG in metabolic disorders, intoxications and epileptic encephalopathies. In: Shomer DL, Lopes da Silva F, editors. Niedermeyer's electroencephalography: basic principles, clinical applications and related fields. 7th ed. New York: Oxford University Press; 2018. ISBN 9780190228484.
64. Ropohl A, Sperling W, Elstner S, Tomandl B, Reulbach U, Kaltenhäuser M, Kornhuber J, Maihöfner C. Cortical activity associated with auditory hallucinations. Neuroreport. 2004;15(3):523–6. https://doi.org/10.1097/00001756-200403010-00028. PMID: 15094516.
65. Gong R, Roth RW, Chang AJ, Sinha N, Parashos A, Davis KA, Kuzniecky R, Bonilha L, Gleichgerrcht E. EEG ictal power dynamics, function-structure associations, and epilepsy surgical outcomes. Neurology. 2024;102(12):e209451. https://doi.org/10.1212/WNL.0000000000209451. Epub 2024 May 31. PMID: 38820468.
66. Gola M, Magnuski M, Szumska I, Wróbel A. EEG beta band activity is related to attention and attentional deficits in the visual performance of elderly subjects. Int J Psychophysiol. 2013;89(3):334–41. https://doi.org/10.1016/j.ijpsycho.2013.05.007. Epub 2013 May 18. PMID: 23688673.

Electromagnetism and Biophysics of the EEG

2

Misunderstanding of electric fields and potentials often leads to fallacious and, in some cases, even absurd physiologic interpretations.

Nunez, Srinivasan, Electric Fields of the Brain, 2nd Ed, 2006

2.1 Introduction

The above quote from Nunez and Srinivasan [1] should be inscribed in the frontispiece of every department of Clinical Neurophysiology. Frequently many physicians use the EEG in a superficial manner, like a kind of stethoscope,[1] without the prevention that EEG is not only based in the brain physiology (what presumably they know) but in Electromagnetic theory (EM) and, more precisely, in the sub-domain of Electrostatic theory (ET). And, without this field of knowledge, the information obtained, at the best, is a small part of the total theoretical amount of data, when not directly erroneous.

Therefore, it is difficult to deeply understand EEG without the right understanding of ET. Obviously, a detailed description of this theory is out of the scope of this work, but we shall provide the necessary bulk of theory to properly understand the features of EEG recordings and its numerical treatment.

Nevertheless, we must keep in mind that ET is not exactly the scenario we need to understand EEG. Indeed, the real field of study is that of bioelectricity, which is a part of biophysics [2]. ET studies slow-moving or stationary electric charges under not-time-varying or slow-changing fields. Typically, therefore, its field of application is electrical phenomena in materials that do not conduct current, called insulators or dielectrics. However, bioelectricity deals with bodies capable of generating propagated currents in conductive volumes. In this case, the variation of the electric fields is slow enough to consider the electrostatic approximation valid. This is why the ET formalism can be applied with great benefit to the study of bioelectricity. Obviously, some concepts are different, but knowledge of ET is extraordinarily useful for understanding the processes that generate the EEG. Please, see Chap. 14 for a detailed desciption of vector analysis.

[1] Of course, I do not underestimate the in-depth knowledge required for the use of the stethoscope by cardiologists or pneumologists, for example. I mean that it is an extraordinarily used medical instrument and that many doctors (like me) use it superficially for basic cardiac, pulmonary, or abdominal auscultation.

J. Pastor, *Quantitative EEG Practical Guide*, https://doi.org/10.1007/978-3-032-05457-9_2

2.2 The Electric Field and Its Relationship with the Scalar Potential

The key concept of the theory is the presence of electric fields. In the nineteenth century, Michael Faraday (1791–1863), a genius with an elementary mathematical formation, had one of the deepest insights in Physics. In fact, he thought that the so-called distant action, the relationship between two separated bodies (a concept that horrified previous scientists), can be explained by the existence of a field. This would be a real modification of the surrounding space induced by an electric charge. This field only interacted with other charged bodies, and it would be responsible for the effect induced at a distance from the source. The presence of the electric field ($\vec{E}$) can only be deduced from the effect induced in a second charged body (usually called test charge) and, therefore, strictly speaking, cannot be measured (only the force induced onto the probe).

The force (in N) measured between two charged bodies is empirically described by Coulomb's law [3],

$$\vec{F}(\vec{r}) = \frac{1}{4\pi\varepsilon_0} \frac{qq'}{R^2} \hat{R} \tag{2.1}$$

where R is the relative position between the bodies with unitary vector $\hat{R}$, q' is the source charge,[2] q the test charge (both in C), and $\varepsilon_0 = 8.854 \times 10^{-12}$ F/m is a constant called vacuum permittivity.

The relationship between the force $\vec{F}$ generated by q and $\vec{E}$ is given by

$$\vec{F}(\vec{r}) = q\vec{E}(\vec{r}) \tag{2.2}$$

And combining with Coulomb's law we obtain the next definition for the electric field

$$\vec{E}(\vec{r}) = \frac{1}{4\pi\varepsilon_0} \frac{q'}{R^2} \hat{R} \tag{2.3}$$

In this expression, we have made explicit the vector nature of the magnitude. This will be very important in EEG analysis. Also, please note we have started from point charge as source and, obviously, the $\vec{E}$ is measured outside the placement of this source. In real systems, instead of punctual charges we have a set of charges (dq, differential of charge; e.g., ions) in a volume (dV, differential of volume) usually so close and uniformly distributed that we can use the concept of volume charge density ρ (C/m^3), defined as

$$\rho = \frac{dq}{dV} \tag{2.4}$$

where dV is the differential of volume. From this expression we can obtain the differential of charge as $dq = \rho dV$. We can use this magnitude to compute the $\vec{E}$ generated by a macroscopic charge distributed into a volume V' (Fig. 2.1), by means of this expression,

$$\vec{E}(\vec{r}) = \frac{1}{4\pi\varepsilon_0} \int_{V'} \frac{\rho(\vec{r}')dV'}{R^2} \hat{R} \tag{2.5}$$

Please refer to Chap. 16 to a deeper discussion about volume integration.

The main problem with this definition is the vector nature, which obligates to compute for all the space dimensions, which usually extraordinarily complicates the calculations. To facilitate the management of these systems we can introduce a related magnitude, the potential (ϕ) what has the feature of being scalar, not vector. The relationship between both concepts is given by

$$\vec{E}(\vec{r},t) = -\nabla\phi(\vec{r},t) \tag{2.6}$$

We have explicitly stated the time dependence for this expression including the term t. This general expression uses the concept of gradient (see Chap. 14) and can be used in three-dimensional problems. However, when one-dimensional approximation is taken, the following form can be used:

$$\vec{E}(r,t) = -\frac{d\phi(r,t)}{dr} \hat{r} \tag{2.7}$$

We can solve this differential equation[3] (see Chap. 17), obtaining this definition for the potential,

[2]Henceforth, all the magnitudes referring the source will be denoted with an apostrophe.

[3]Pay attention that the coordinate system used in this expression is spherical (r, θ, φ) not Cartesian (x, y, z). However, this fact do not change the main concepts.

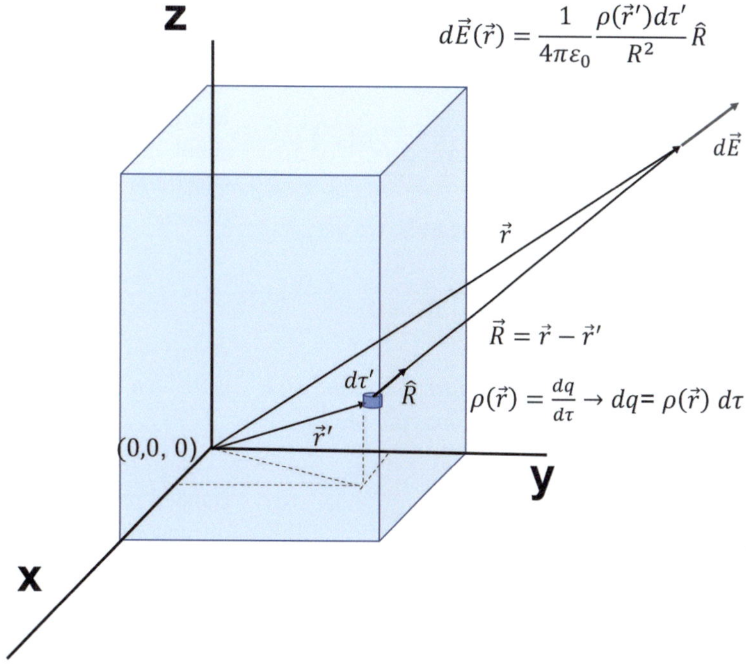

Fig. 2.1 Electric field induced in a point by a volumetric density of charge. The $d\vec{E}$ originated by a differential volume density of charge $d\tau'$ (another notation for dV', with the same meaning) placed at point $\vec{r}'$ is showed. The total $\vec{E}$ at point $\vec{r}$ can be computed after integration over the whole volume where charge exists

$$\phi(r) = \frac{q}{4\pi\varepsilon_0 R} + k \quad (2.8)$$

Please pay attention that we have supposed that the placement of the charge ($\vec{r}'$) can be different from origin, and, therefore, we have $\vec{R} = \vec{r} - \vec{r}'$. Since the units of the electric field are [*V*/*m*], the potential will be [*V*]. Note that there is a constant term (*k*), which indicates that the potential does not have an absolute value, except for $x = \infty$, $\varnothing = 0$. Therefore, we can only know potential differences between points, but not the absolute potentials. From a physical point of view, the scalar potential can be considered as the energy required to move a charge in a uniform electric field, thus, it makes sense to talk about the energy per unit charge, in fact $V = \frac{J}{C}$.

For a tridimensional system we must consider the $\rho(\vec{r}')$ instead of a punctual charge. In this case, to calculate the potential we must integrate over all the volume,

$$\phi(\vec{r}) = \frac{1}{4\pi\varepsilon_0}\int_{V'}\frac{\rho(\vec{r}')dV'}{R} \quad (2.9)$$

We have seen (see Chap. 14) that the curl of a gradient is zero. Therefore, taking Eq. 2.6, we have

$$\nabla \times -\nabla\phi = 0 \Rightarrow \nabla \times \vec{E} = 0 \quad (2.10)$$

It is legitimate to ask why new magnitudes should be introduced in EM. The reason is as follows. As we have indicated above, the analysis of vector quantities is more complex than the analysis of scalar magnitudes. Therefore, if we can do the basic calculations using a scalar quantity and then convert it to the electric field using the gradient (Eq. 2.6), we will have greatly simplified the resolution of many problems [4].

The magnitude of the electric field passing through a closed surface is generated by the amount of whole charge enclosed by that surface (Q_{enclosed}). This statement constitutes Gauss's theorem, which is of great importance in numerous problems of EM. Its mathematical statement is as follows [4]:

$$\oint_{S'}\vec{E}(\vec{r}) \circ d\vec{a} = \frac{1}{\varepsilon_0}\int_{V'}\rho(\vec{r}')dV' = \frac{Q_{\text{enclosed}}}{\varepsilon_0} \quad (2.11)$$

It is important to find the differential expression for this theorem. To do this, we make use of the Gauss-Ostrogadsky theorem (Chap. 16), so we obtain

$$\oint_{S'} \vec{E}(\vec{r}) \circ d\vec{a} = \int_{V'} \left(\nabla \circ \vec{E}(\vec{r})\right) dV' = \int_{V'} \frac{\rho(\vec{r}')}{\varepsilon_0} dV'$$

and grouping both volume integrals, we get

$$\int_{V'} \left[\left(\nabla \circ \vec{E}\right) - \frac{\rho(\vec{r}')}{\varepsilon_0} \right] dV' = 0$$

from which we obtain the differential expression of Gauss's law, which is one of Maxwell's equations

$$\nabla \circ \vec{E} = \frac{\rho(\vec{r}')}{\varepsilon_0} \tag{2.12}$$

In many cases, we cannot use Eq. 2.8 to determine the potential. To solve this problem, we combine Eqs. 2.6 and 2.10, obtaining Poisson's equation

$$\nabla^2 \phi = -\frac{\rho}{\varepsilon_0} \tag{2.13}$$

2.3 Multipolar Expansion of the Scalar Potential

A very important feature of potential is that its expression depends on the complexity of the sources. Suppose a charge enclosed in an irregular body. At distances very close to the charge, the morphology of the body must be considered for an adequate calculation. However, at increasing distances, we can dispense with the smaller features and consider only more general properties. This is achieved by developing the general expression of the potential through an expansion in monopoles, dipoles, quadrupoles, octopoles, etc. [5], whose formula is

$$\phi(r) = \frac{Q}{4\pi\varepsilon_0 r} + \frac{\vec{p} \bullet \vec{r}}{4\pi\varepsilon_0 r^3} + \frac{1}{4\pi\varepsilon_0 r^5} \frac{1}{2} \sum_{j=x,y,z} \sum_{k=x,y,z} jkQ_{jk} + \ldots$$

In EEG we are only interested in the first two terms of expansion, which are the monopolar ($\varnothing_M$) and the dipolar ($\varnothing_D$) terms. Terms of higher order than 2 are important when the spatial distribution of charges is relevant and this occurs at small distances from the sources. The expressions and meanings from the first terms are as follows:

$$\phi_M(r) = \frac{Q}{4\pi\varepsilon_0 r} \tag{2.14}$$

where Q represents the monopolar moment and is defined as $Q = \sum_i q_i$. The magnitude decreases inversely with the distance.

Nevertheless, in electrically excitable tissue, isolated monopoles are not found. Instead, pairs of monopoles of equal and opposite signs, close together, are the fundamental unit source in EEG. And this source is the dipolar moment, defined as

$$\vec{p}(r) = \sum_i q_i \vec{r_i} \tag{2.15}$$

that is, the sum of the products of the charges by their position. The units of the dipole moment, therefore, will be [m C]. The expression for the potential generated by a dipole term will be,

$$\phi_D(r) = \frac{\vec{p} \bullet \vec{r}}{4\pi\varepsilon_0 r^3} \tag{2.16}$$

Moreover, computing the dot product of the numerator (see Chap. 14), the expression will be,

$$\phi_D(r,\theta) = \frac{p\cos\theta}{4\pi\varepsilon_0 r^2} \tag{2.17}$$

This expression is very important and tells us that the dipole potential depends directly on the cosine of the angle between the dipole and the recording electrode and, in turn, depends inversely on the square of the distance (Fig. 2.2).

Even more useful is this modification of the formula, for calculating the potential generated by a current dipole, not a voltage dipole

$$\phi_D(r,\theta) = \frac{Id\cos\theta}{4\pi\sigma r^2} \tag{2.18}$$

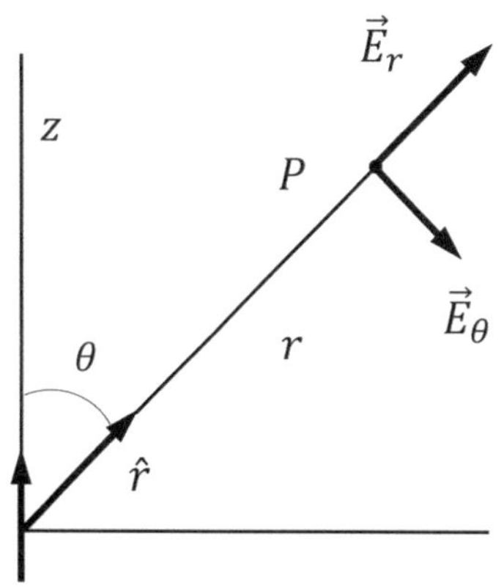

Fig. 2.2 Electric field induced by a vertical dipole located at the origin of coordinates

where I is the current and d the distance separating the sources/sinks generating the dipole and σ the conductivity.

Although this formula does not exactly obtain the same result of two opposite monopoles, when the source–field distance is five to ten times the monopole separation, the dipole is a good approximation to the magnitude and to the pattern of potentials computed as two monopoles [2].

We can compute the electric field generated by a dipole from equation Eq. 2.17 and considering that the gradient must be expressed in spherical coordinates (see Chap. 14, Eq. 14.15)

$$\vec{E}(r,\theta,\varphi) = -\frac{\partial \phi_D}{\partial r}\hat{r} - \frac{1}{r}\frac{\partial \phi_D}{\partial \theta}\hat{\theta} - \frac{1}{r\sin\theta}\frac{\partial \phi_D}{\partial \varphi}\hat{\varphi}$$

Firstly, observe that E_φ~0, indicating a cylindrical symmetry along the z-axis. Then we can compute the other components as

$$E_r = -\frac{\partial}{\partial r}\left(\frac{Id\cos\theta}{4\pi\sigma r^2}\right) = \left(\frac{Id}{4\pi\sigma}\right)\frac{2\cos\theta}{r^3} \quad (2.19)$$

$$E_\theta = -\frac{1}{r}\frac{\partial}{\partial \theta}\left(\frac{p\cos\theta}{4\pi\sigma r^2}\right) = \left(\frac{p}{4\pi\sigma}\right)\frac{\sin\theta}{r^3} \quad (2.20)$$

From Eq. 2.11 we can compute the equipotential surfaces, which corresponds to $\varnothing_D$ = const,

$$r^2 = \left(\frac{Id}{4\pi\sigma\varnothing_D}\right)\cos\theta = C_D\cos\theta \quad (2.21)$$

The $\vec{E}$ lines are perpendicular to the equipotential surfaces. These lines are tangent to the direction of $\vec{E}$ at every point. Let's have a small displacement along the line of $\vec{E}$ (line of force) named as $d\vec{s}_{lf}$. Therefore, we have $d\vec{s}_{lf} = k\vec{E}$. We can use this expression to find the differential equation to describe this line. We can obtain the expression of $d\vec{s}_{lf}$ in spherical coordinates from

$$d\vec{s}_{lf} = dr\hat{r} + rd\theta\hat{\theta} = E_r\hat{r} + E_\theta\hat{\theta}$$

Therefore, we obtain $dr = kE_r$ and $rd\theta = kE_\theta$
So, solving for k, we have,

$$k = dr / E_r = rd\theta / E_\theta$$

And re-ordering the terms and using equations Eq. 2.33 and Eq. 2.34

$$\frac{dr}{rd\theta} = \frac{E_r}{E_\theta} = \frac{2\cos\theta}{\sin\theta} \rightarrow \frac{d\ln r}{d\theta} = \frac{2d(\ln\sin\theta)}{d\theta}$$

We can integrate this equation

$$\int d\ln r = 2\int d(\ln\sin\theta) \rightarrow \ln r = \ln\sin^2\theta + \ln K_D = \ln(K_D\sin^2\theta)$$

where K_D is a constant. Finally, the expression searched is

$$r = K_D\sin^2\theta \quad (2.22)$$

The multipole expansion of the potential not only allows us to eliminate some terms that contribute only minimally to the potential, given the dimensions of the system analyzed. It also has a huge advantage over Eq. 2.8. In the former, the scalar field defining the distance is $\vec{R} = \vec{r} - \vec{r}'$, which depends on both $\vec{r}$ and $\vec{r}'$. However, the multipole expansion allows us to eliminate the dependence on the position of the sources, so that what matters is the distance to the field point; that is, the value of the term considered depends only on $\vec{r}$, but not on the position of the sources ($\vec{r}'$).

2.4 Conductors

A conductor is a body in which electric charges are free to move under the effect of an electric field. The movement of these charges is what generates the electric current I. This magnitude is defined as the amount of charge per time [6]

$$I = \frac{dq}{dt} \tag{2.23}$$

and its units are C/s [A]. This charge can freely move under the effect of $\vec{E}$, and therefore is labeled as free charge (and denoted with the subscript *f*). In a volume of tissue, we can define the magnitude volume charge density (ρ_f)

$$\rho_f = \frac{dq_f}{dV} \tag{2.24}$$

Free charges in a volume, under the presence of a $\vec{E}$, give rise to a density of current [$\vec{J}$ in A/m^2], defined as

$$\vec{J} = \sigma \vec{E} \tag{2.25}$$

This is the generalized Ohm's law. The symbol σ denotes conductivity [S/m]. It's interesting to observe that such conductivity as dielectric constants of linear media generally vary with field frequency [7].

Usually, Ohm's law is expressed for unidimensional systems and considering σ as a constant. In this case, the formula becomes as

$$I = \frac{V}{R} \tag{2.26}$$

where $R = \sigma^{-1}$, is called resistance[4] [Ω].

Therefore, the resistance is the difficulty to spread current through the tissue. It's a very important property and, in a first approximation, for a piece of tissue with thickness *d* and cross-sectional area *A*, resistance is given by

$$R = \eta \frac{d}{A} \tag{2.27}$$

where η is a specific property of the tissue and is called resistivity [Ω·m]. This expression indicates that the current will flow worst for pieces of tissue long and narrow.

One of the oldest uses of conductors in electrostatics is the storage of electrical charge. Systems with this property are called capacitors (usually pair of conductors able to accumulate electric charge) and the quantitative measure of this capacity is called capacitance [8], defined as

[4]Please pay attention that *R* in this formula refers to resistance and not magnitude of the relative position vector.

$$C = \frac{Q}{V} \tag{2.28}$$

In addition, it is the ratio between the accumulated charge (*Q*) and the voltage (*V*) needed to displace that charge. Therefore, the units, called Farads, are defined as *F* = *C*/*V*. It's important to realize that the capacitance of a system strongly depends on the morphology. Neural structures exhibit high values of capacitance, mostly arising from cell membranes. The presence of this capacitance markedly affects the tissue's natural electrical behavior. Therefore, although capacitance effects are very important for cellular neurophysiology, in EEG large-scale capacitive effects are negligible in neocortex, and the main effects are mainly resistive [9].

2.5 Dielectrics

In biological and non-biological materials there are numerous regions where $\vec{E}$ cannot give rise to a *current* (at least carried out by ions), because the charges are bounded and they cannot freely move. These regions are called dielectrics. The charge density is given by ρ_b, where the subscript *b* denotes bounded. Let's see the origin of this charge.

The average effect of microscopic dipoles can be defined by a macroscopic magnitude, defined as polarization ($\vec{P}$, in C/m^2), defined as dipole moment per unit volume of tissue,

$$\vec{P}(\vec{r},t) = \frac{d\vec{p}(\vec{r},t)}{dV} \tag{2.29}$$

Therefore, the whole dipolar moment generated by a volume of polarized matter will be

$$\vec{p}(\vec{r}) = \int_{V'} \vec{P}(\vec{r})\,dV' \tag{2.30}$$

From this equation we can define the two sources of this potential, the volumetric density of bounded charge, distributed over all the volume (ρ_b, [C/m^3]) and the surface density of bounded charge (σ_b, [C/m^2]), over the boundary, defined as,

$$\rho_b = -\nabla' \bullet \vec{P} \tag{2.31}$$

$$\sigma_b = \vec{P} \bullet \hat{n}' = P_n \quad (2.32)$$

where $\hat{n}'$ is the normal to the surface unitary vector. The electric field inside a dielectric is not zero, as it is in conductors. Let us assume that the electrical properties of an infinitesimal volume (density, charge, etc.) are homogeneous. Let us assume a local microscopic electric field [5], which we can call $\vec{e}$. This electric field should have the same properties as the field studied in vacuum (see Eqs. 2.9 and 2.11). That is, we expect the equalities to be fulfilled

$$\nabla \circ \langle \vec{e} \rangle = \frac{\rho_b}{\varepsilon_0} \quad (2.33)$$

$$\nabla \times \langle \vec{e} \rangle = 0 \quad (2.34)$$

where $\langle \vec{e} \rangle$ means the average over a volume. So, from this approach, we can accept that $\langle \vec{e} \rangle = \vec{E}$. That is, the average properties of the microscopic field, when considered for sufficiently large volumes, are equal to the properties of the field in vacuum, which allows us to apply the formalism developed so far for fields in the presence of matter.

The neutral material is formed by the same amount of positive and negative charges. But, under the influence of an external $\vec{E}_0$ the positive and negative charges will move in the opposite directions (Fig. 2.3). Even if the charges are separated, on the surfaces of volume elements of width Δx, such as that shown in the figure, there will be a number of negative charges equal to the number of opposite charges on the other side of the surface, so the total charge will be zero and, consequently, there will be no field. However, on the real surfaces of the body, the charges will be uncompensated, so there will be a number of uncompensated charges with charges of opposite sign. This means that two opposite surfaces with net charge would have been formed, which creates a surface charge density. This charge separation induces an induced electric field ($\vec{E}_{in}$) in the opposite direction to the external one.

In the brain there is a mixture of conductive and dielectric regions, e.g., some regions formed by an enough density of ions will behave like a

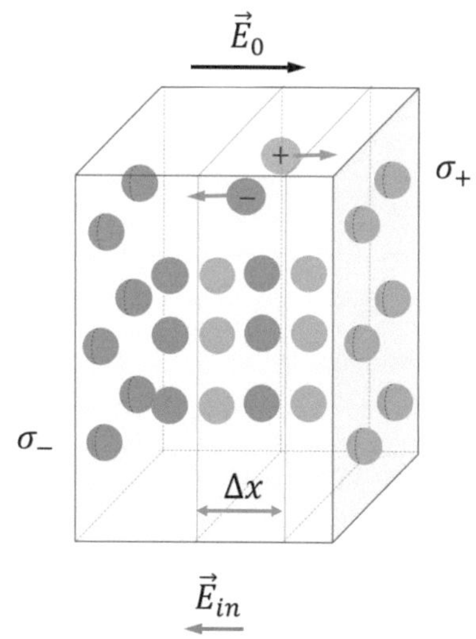

Fig. 2.3 Effect of an electric field ($\vec{E}_0$) on a non-conducting tissue. The partial separation of charges generates surface charge densities (σ_+ and σ_-) on surfaces orthogonal to the field. These charge densities generate an induced field of lower intensity and opposite direction ($\vec{E}_{in}$)

conductor, given rise to currents in the presence of a $\vec{E}$ and others will behave mainly like dipoles, typically the cellular membranes and those regions where the ionic density will be too low. Nevertheless, the strict differentiation between both kinds of tissues is difficult and sometimes depends on the strength of $\vec{E}$ [9]

Therefore, we will have both bound (ρ_b) and free (ρ_f) charges. The total charge density (ρ_t) will be

$$\rho_t = \rho_f + \rho_b = \rho_f - \nabla \circ \vec{P} \quad (2.35)$$

where we have used the definition ρ_b Of from Eq. 2.31.

Then, operating in Eq. 2.12, and using Eq. 2.35, we get

$$\nabla \circ \left(\varepsilon_0 \vec{E} + \vec{P} \right) = \rho_f \quad (2.36)$$

This is a very important equation, as it tells us that there is a new field whose sources are the free and bounded charges. This field is called electric displacement (C/m²), defined as

$$\vec{D} = \varepsilon_0 \vec{E} + \vec{P} \quad (2.37)$$

2.6 Maxwell's Equations

These equations represent one of the highest intellectual achievements of the Science. We will discuss only the version of equations for macroscopic fields [10].

Previous to the analysis of the equations, we must first define two fields, the first is the displacement vector, discussed above,

$$\vec{D} = \varepsilon \vec{E} \tag{2.38}$$

This is the constitutive equation for electric fields in dielectric media, where ε is the permittivity of the medium (F/m) and indicates the capacity of the medium to store charge and $\vec{E}$ is the electric field. The vacuum permittivity $\varepsilon_0 = 8.84 \times 10^{-12}$ s/Ω·m is a constant. The ratio between the two values is called the relative dielectric constant and is defined as $\kappa = \varepsilon/\varepsilon_0$ (unitless).

The other field is the magnetic field ($\vec{H}$), which is related to the magnetic induction by means of the expression,

$$\vec{H} = \vec{B}/\mu \tag{2.39}$$

where μ is the magnetic permeability of the medium.

Maxwell's equations are

$$\nabla \bullet \vec{D} = \rho \tag{2.40}$$

$$\nabla \bullet \vec{B} = 0 \tag{2.41}$$

$$\nabla \times \vec{E} = -\frac{\partial \vec{B}}{\partial t} \tag{2.42}$$

$$\nabla \times \vec{B} = \mu \vec{J} + \frac{\partial \vec{D}}{\partial t} \tag{2.43}$$

Equation 2.40 corresponds to Gauss' law and the symbol ρ is the charge density, being $\vec{J}$ in Eq. 2.43 indicates the current density.

To these equations, we must add the law of conservation of current, which is

$$\nabla \bullet \vec{J} + \frac{\partial \rho}{\partial t} = 0 \tag{2.44}$$

These equations explain all EM phenomena, including the occurrence of electromagnetic waves (comprising light), which requires the reciprocal interaction of electric and magnetic fields through very fast time variations (of the order of kHz), defined in Eqs. 2.41 and 2.42. However, in the case of the nervous system, the fastest recorded activity is well below this frequency, implying that EM waves will not be generated by brain activity. This greatly simplifies the analysis of the system and allows Maxwell's equations to be modified for static conditions [11], being the modified equations,

$$\nabla \times \vec{E} = 0 \tag{2.45}$$

$$\nabla \times \vec{B} = \mu \vec{J} \tag{2.46}$$

The set of equations thus modified is adequate for the study of electrostatic phenomena in the Nervous System.

2.7 The Dipolar Model in the EEG

The dipole model is the best theoretical approach we have to the EEG generation (Fig. 2.4a), although there are several questions not well-explained. However, as a first approximation it is good enough to explain most of the features of EEG. In the cortex, the sources separation can be about 2 mm, therefore a distance of 1–2 cm would be adequate for recording the potentials generated and, therefore, use of the dipolar model. Obviously, this distance is into the order of magnitude of EEG recordings. In the brain, however, there cannot be, strictly speaking, stationary and static electrical charges. What we have, in reality, on the surface of the cell membrane are sources and sinks. These regions are generated by the presence of transmembrane currents that are generated by the movement of ions. The expressions that we have seen (Eq. 2.8) do not take into account currents as generators of potential, but rather charges. Nevertheless, Thévenin's theorem allows us to use a similar formalism with currents, instead of charges [12]. So, the voltage generated by a current (I) is

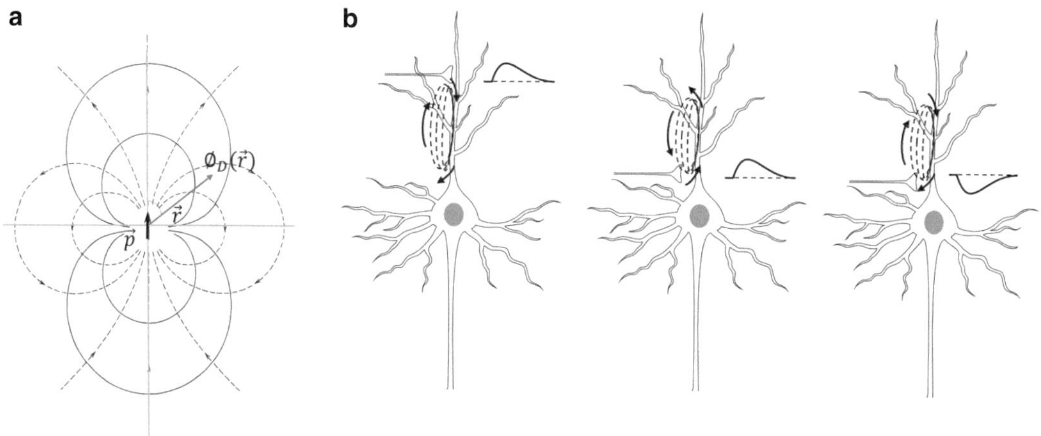

Fig. 2.4 Cortical dipole model. (**a**) Equipotential surfaces (generated from Eq. 2.21) from a dipole located at the origin of coordinates. The field lines (dashed) are orthogonal to the potential surfaces and are given by 2.22. (**b**) Model of the generation of dipoles in apical dendrites. EPSP at layer II (left) and IV (middle) originates opposite dipoles. However, an IPSP at layer IV (right) originates a dipole similar to left

$$\phi(r) = \frac{I}{4\pi\sigma r} \tag{2.47}$$

where σ is the conductivity (Ω^{-1}), and, consequently, the equivalent dipolar current (equivalent to 2.11) will be

$$\phi_D(r,\theta) = \frac{Id\cos\theta}{4\pi\sigma r^2} \tag{2.48}$$

In this case, the dimensions of the dipolar moment are given by $[p] = Am$, using the adequate units.

To understand the contribution of cortical neurons to the EEG, consider the inward flow of current produced by an excitatory post-synaptic potential (EPSP) or the outward current induced by an inhibitory post-synaptic potential (IPSP) on the apical dendrite of a cortical pyramidal neuron. Current flows into the dendrite at the site of generation of the EPSP, creating a current sink. It must complete a loop by flowing down the dendrite and back out across the membrane at other sites, creating a current source. Both sink and source can be approximate to a current dipole [13, 14]. Obviously, an IPSP will generate an opposite dipole, because the current at the synapse is a source, not a sink.

In the cortex, excitatory inputs from the contralateral hemisphere or other cortical regions contact the pyramidal neurons mostly on distal parts of the dendrite, in layers 2 and 3, whereas ThC inputs terminate in layer 4 [15]. The activity measured at a surface EEG electrode will have opposite polarities for these two inputs (Fig. 2.4b).

The potential due to sources generated to a cortical column can be represented by the dipole term of the multipole expansion. The magnitude at a defined point will be the average of the weighted sum of dipole contributions from all the columns. However, this is a simplistic approximation. Studies in animal models, determining the current sources density (CSD), have demonstrated the existence of other current sources besides dipolar sources. Specifically, monopolar and quadrupolar sources have been identified [16]. Considering that the latter have a distance dependence inversely proportional to r^4, their contribution to the potential recorded in scalp is not very important. On the contrary, this same argument, but in the opposite direction, would give greater relevance than previously assumed for monopolar sources, which could be generated by diffusion currents, much slower than ohmic currents [17]. However, the actual contribution of monopolar and quadrupolar sources in the genesis of scalp EEG is still under debate, although reports using monopolar sources in mesial regions have been described [18].

In addition to this, the cerebral cortex is formed of pyramidal cells located at different depths, other neuronal and glial types, with significant electrogenic capacity and numerous axonal tracts that, globally, complicate even more the correct understanding of the potentials generated in the brain [19] and recorded in scalp. Furthermore, some numerical models suggest that not only the local field potentials (LFP), but also the action potentials could contribute more than initially considered to the EEG recording [20, 21].

In summary, even taking into account all these warnings about the genesis of potentials, the dipolar model is a good approximation in clinical practice and will be the one used in this book.

The theory followed in this work, on the genesis of the activity recorded in the scalp EEG, is that the overall activity (including postsynaptic currents and probably action potentials) of a cortical column can be modelled as a unit dipole. Each of these dipoles will generate a dipolar electric field. The potential at any point will therefore be the vectorial sum of all the unit dipoles that present enough synchronization.

2.8 Currents in the Head

A homogeneous system is one that presents the same properties in all its parts, without any different elements being visible. On the other hand, a system will be isotropic at a point if it exhibits the same properties in all directions [22]. Let's have a look at this situation in more detail.

A homogeneous system implies that the properties will be the same at any point. Figure 2.5a shows that the properties, represented by the vectors (which can indicate the electric field, the concentration gradient, or any other vectorial magnitude), are the same at all the points. Since the sum of all the vectors is zero, this result is marked as a black dot. A homogeneous but anisotropic system is shown in Fig. 2.5b. The result of the vector sum is the same at all points, although each point has a well-defined direction (black vector). However, in Fig. 2.5c, the property varies from point to point, and the direction of the resulting vector is different for each point. It is therefore an inhomogeneous and anisotropic system.

Let's now assume that each of these systems is subject to an electric field indicated by the vertical vector, and that the relevant magnitude at each point is the conductivity. For the homogeneous and isotropic system (Fig. 2.5a, lower), the current density will not be modified and will be the same at each point. For the homogeneous and anisotropic system (Fig. 2.5b, lower), the current density direction is different from the electric field direction. And, finally, for the inhomogeneous and anisotropic system (Fig. 2.5c, lower), the current density is different from point to point from the electric field direction.

The relevant point is that conductivity is a constant in homogeneous and isotropic structures, but for other systems, conductivity is actually a tensor, so the volumetric current density will be given by this expression.

$$\begin{pmatrix} J_x \\ J_y \\ J_z \end{pmatrix} = \begin{bmatrix} \sigma_{xx} & \sigma_{xy} & \sigma_{xz} \\ \sigma_{yx} & \sigma_{yy} & \sigma_{yz} \\ \sigma_{zx} & \sigma_{zy} & \sigma_{zz} \end{bmatrix} \begin{pmatrix} E_x \\ E_y \\ E_z \end{pmatrix} \tag{2.49}$$

This is actually a system of three vector equations,

$$\begin{aligned} \vec{J}_x &= \sigma_{xx} E_x \vec{i} + \sigma_{xy} E_y \vec{j} + \sigma_{xz} E_z \vec{k} \\ \vec{J}_y &= \sigma_{yx} E_x \vec{i} + \sigma_{yy} E_y \vec{j} + \sigma_{yz} E_z \vec{k} \\ \vec{J}_z &= \sigma_{zx} E_x \vec{i} + \sigma_{zy} E_y \vec{j} + \sigma_{zz} E_z \vec{k} \end{aligned} \tag{2.50}$$

Therefore, in this case, although the electric field at each point is known, the volumetric current density can vary both in magnitude and direction. It is very important to properly understand this point [23]. The electric fields generated by sources and sinks in a tissue volume will propagate according to the laws of electromagnetism we have seen above.

Any field relative to a given surface can be decomposed into two components: normal and tangential to surface, i.e.,

$$\vec{E} = \vec{E}_n + \vec{E}_t \tag{2.51}$$

We must ask ourselves whether the change between tissue 1 and tissue 2 (e.g., between the cortex and white matter) involves any modifica-

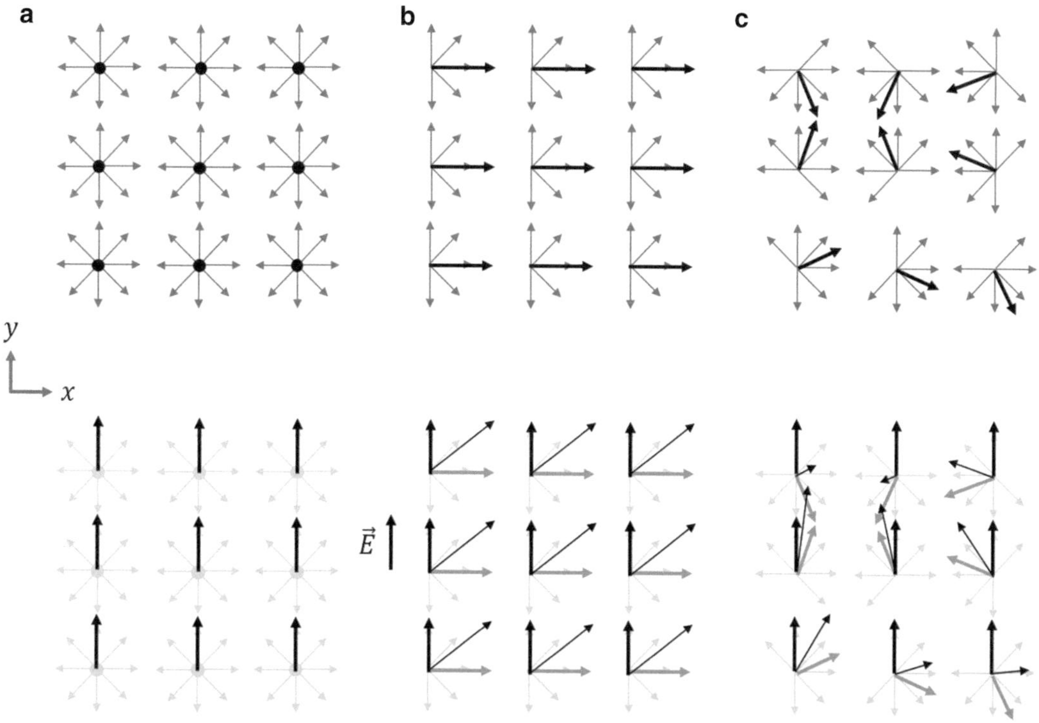

Fig. 2.5 Homogeneity and isotropy in two-dimensional systems and the effect onto an electric field. (**a**) Homogeneous and isotropic systems. (**b**) Homogeneous and anisotropic systems. (**c**) Inhomogeneous and anisotropic systems. The bottom row represents the same system with a vertical electric field imposed on it. The field is indicated by the vertical black vector at each point. The resulting current vector is represented by a thin black vector

tion in the direction or magnitude of the field [24] besides changes in conductivity tensor. The theory of boundary conditions at a conductive surface allows us to answer this question. The modification of the normal component will depend on the divergence of the field. To do this, the infinitesimal changes around the discontinuity surface (h) are computed by passing to the limit. The mathematical expression sought will be the following:

$$E_{n2} - E_{n1} = \lim_{h\to 0}\left[h\left(\nabla \circ \vec{E}\right)\right] \quad (2.52)$$

And, from Eq. 2.12, we have $\lim_{h\to 0}\left[h\left(\frac{\rho}{\varepsilon_0}\right)\right] = \sigma$,therefore

$$E_{n2} - E_{n1} = \sigma \quad (2.53)$$

This means that the difference in the modulus of the normal component in tissue 2 (E_{n2}) will be different from that in tissue 1 (E_{n1}) only if there is a surface charge density (σ) between them. If this density does not exist ($\sigma = 0$), both normal components will be equal $E_{n1} = E_{n2}$.

The modification of the tangential component depends on the rotational component, according to this expression

$$\hat{n}\times\left(\vec{E}_2 - \vec{E}_1\right) = \lim_{h\to 0}\left[h\left(\nabla\times\vec{E}\right)\right] \quad (2.54)$$

where $\hat{n}$ represents the unitary normal vector. And, from Eq. 2.10, $\lim_{h\to 0}\left[h(0)\right] = 0$, therefore

$$\vec{E}_{2t} = \vec{E}_{1t} \quad (2.55)$$

This means that the modifications of the current circuits do not depend primarily on changes in the field, but on changes in the conductivity tensor (Eq. 2.49) that depend on the local properties of the tissue [25, 26]. Obviously, the conductivity modulus varies tremendously between different tissues. Some of these conductivities are shown in Table 2.1.

Table 2.1 Typical conductivity of several tissues (modified from [9])

Tissue	Conductivity (mS/cm)
CSF	15.62
Spinal cord (longitudinal)	5.56
Spinal cord (transverse)	0.83
Cortex	4.35
White matter (average)	1.53
Bone	0.125–0.062

This table shows that the skull resistivity (the lowest conductivity, what is the same) is the highest that any intracranial current circuit has to deal with. The human skull consists of several bones joined by cranial sutures. These bones are made up of two parts: the outer and inner tables (cortical bone), which are compact, and the spongy central part (diploe or cancellous bone), which contains more blood vessels than outer and inner cortical. The cortical bone is the most resistant, with diploe having much greater conductivity. Therefore, the areas of the skull with the greatest spongy content will be the regions that best allow current to pass through.

In summary, the important effect derived from these reflections is that the modifications of the current circuits (conductivity differential, resistivity, anatomic differences) make it difficult to identify intracranial sources from extracranial voltages.

2.9 Measurements on the Scalp Surface

The measurements made in scalp are called potentials and their units are, therefore, μV. However, what we actually are measuring are currents. This explains why changes in conductivity conditions affect the recording so much, which would not occur if voltages were actually measured. The conditions involved in recording these currents are as follows:

1. Conductivity. According to Ohm's law (Eq. 2.26), the current generated by a given voltage depends inversely on the resistance. Therefore, doubling the resistance will halve the current. In conclusion, the current circuits will change depending on the conductivity of different tissues, which will vary not only depending on cellular composition but also on the preferential orientation of fibers (think of the internal capsule). Similarly, when these circuits reach the skull, they will encounter great resistance, so a large portion of these currents will not reach the surface. This explains why the presence of a cranial alteration (trauma, previous surgery, or anatomical anomaly) decreases local resistance and promotes "current leakage" compared to regions of the intact skull. This explains the breach rhythm [27].
2. The magnitude of the currents. Let's assume that each cortical column ($r = 1.5$ mm) generates an approximately constant amount of current. The total current will be the contribution of all these unit currents. To increase the magnitude of the current, given that we have assumed an unitary current per cortical column approximately constant, we need to increase the number of synchronous cortical columns within a short period of time. A cortical region will be synchronized when cortical columns produce activity in the same time period. Synchronization will be achieved through neural signals propagated from a defined region. Since the propagation speed is finite, $v = r/\Delta t$; the maximum signal propagation radius will be given by $r = v\Delta t$. Therefore, all those regions located beyond this radius will not generate their activity in the significant period and, therefore, will not contribute to the synchronized activity. An interesting topic not always covered in textbooks is the relationship between the synchronized cortical surface and the frequency of activity recorded in scalp. We do not attempt to develop an accurate model, nor will we use realistic numerical values. We merely aim to describe, in an intuitive way, why it is easier to record low-frequency cortical signals than those of faster bands. See Fig. 2.6a.

The total current (I) is given by the expression

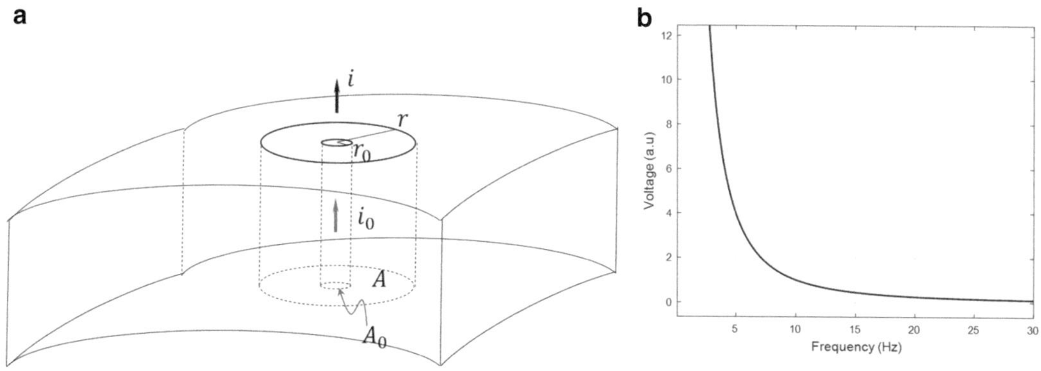

Fig. 2.6 Cortical generators and their scalp recording. (**a**) Model of the cerebral cortex with a cortical macrocolumn of radius r_0 and surface area (A_0), centered on a synchronized cortical surface A. The gray vector indicates the unitary current generated by the column, while the black vector represents the total current. (**b**) Qualitative representation of the relationship between the voltage reaching the surface electrodes and the frequency of the generating rhythm

$$I = NI_0 \tag{2.56}$$

where I_0 represents the unitary current generated by a cortical column and N is the number of columns discharging in the period Δt. The total area synchronized can be estimated from

$$A = \pi r^2 = \pi \left(v \Delta t \right)^2 \tag{2.57}$$

Considering the inverse relationship between frequency and period, $f = \pi / \Delta t$, we can substitute in the above equation, obtaining

$$A = \pi \left(\frac{v}{f} \right)^2 \tag{2.58}$$

This equation indicates that there is an inverse square relationship between the synchronized area and the frequency of activity. This is logical, since synchronizing a large region of cortex can only be achieved if the oscillation frequency is slow. Conversely, a fast frequency cannot cover a very large area since the propagation speed is (approximately) constant.

The radius of a cortical macrocolumn is approximately $r_0 = 1.5$ mm $\equiv 0.15$ cm. Therefore, the unit area occupied by a column will be

$$A_0 = \pi r_0^{\,2} \tag{2.59}$$

Therefore, the number of columns that will be in the synchronized area will be the ratio $N = A/A_0$, and substituting Eqs. 2.59 and 2.58

$$N = \frac{\pi \left(\frac{v}{f} \right)^2}{\pi r_0^{\,2}} = \frac{v^2}{f^2 r_0^{\,2}} \tag{2.60}$$

So, the total current generated by the region synchronized, from Eq. 2.56 and substituting Eq. 2.60, will be

$$I = NI_0 = \frac{v^2 I_0}{f^2 r_0^2} \tag{2.61}$$

Using now, in a clearly unrealistic way (but for the sake of simplicity), Ohm's law (Eq. 2.26), to estimate the voltage generated on the surface by this current, we will have that

$$V = \frac{v^2 I_0}{f^2 r_0^2} R_{\text{whole}} \tag{2.62}$$

where R_{whole} represents the resistance of all the structures from the cortical origin to the scalp electrodes. A functional description of the behavior of this expression in given in Fig. 2.6b.

It is important to emphasize that this is not intended to be a realistic model of the system, but rather to explain why low-frequency oscillations require larger cortical surfaces than higher-frequency oscillations, which are generated by synchronized surfaces of smaller area. This fact, in turn, explains why the currents generated by beta (and especially gamma) frequencies are very small and may not reach the scalp electrodes with a signal-to-noise ratio sufficient to be adequately recorded.

2.10 Concluding Remarks

The biophysical and neurophysiological properties of the cranio-brain system explain the characteristics of EEG recordings obtained from the scalp. Understanding these concepts is essential for correct interpretation of the study, which, of course, goes beyond assessing the presence or absence of epileptic spikes.

Concepts such as homogeneity, isotropy, dipolar sources, resistive tissue, capacitive properties, or conductivity are essential to understanding how currents generated in the cortex are distorted and partially blocked in the skull and, finally, recorded at scalp electrodes.

References

1. Nunez PL, Srinivasan R. Fallacies in EEG. In: Electric fields of the brain. The Neurophysics of EEG. 2nd ed. New York: Oxford University Press; 2006a. ISBN-13 978-0-19-505038-7.
2. Plonsey R, Barr RC. Sources and Fields. In: Bioelectricity. A quantitative approach. 3rd ed. New York: Springer; 2007. ISBN 978-0-387-48864-6.
3. Wagness RK. The Electric Field. In: Electromagnetic fields. 2nd ed. Danvers: Wiley & Sons; 1986a. ISBN 0-471-81-186-6.
4. Fitzpatrick R. The Electric Scalar Potential. Maxwell's equations and the principles of electromagnetism. Massachusetts: Infinity Science Press; 2008b. ISBN: 978-1-934015-20-9.
5. Wagness RK. Electric Multipoles. In: Electromagnetic fields. 2nd ed. Danvers: Wiley & Sons; 1986b. ISBN 0-471-81-186-6.
6. Wagness RK. Electric Currents. In: Electromagnetic fields. 2nd ed. Danvers: Wiley & Sons; 1986c. ISBN 0-471-81-186-6.
7. Nunez PL, Srinivasan R. Current sources in homogeneous and isotropic medium. In: Electric fields of the brain. The Neurophysics of EEG. 2nd ed. New York: Oxford University Press; 2006b. ISBN-13 978-0-19-505038-7.
8. Blomberg C. Electricity: the core of reductionism basis. In: Physics of Life. The Physicist's Road to Biology. Amsterdam: Elsevier; 2007. ISBN 978-0-444-52798-1.
9. Nunez PL, Srinivasan R. Electric fields and currents in biological tissue. In: Electric fields of the brain. The neurophysics of EEG. 2nd ed. New York: Oxford University Press; 2006c. ISBN-13 978-0-19-505038-7.
10. Wagness RK. Maxwell's Equations. In: Electromagnetic fields. 2nd ed. Danvers: Wiley & Sons; 1986d. ISBN 0-471-81-186-6.
11. Nunez PL, Srinivasan R. Appendices B. Quasi-static Reduction of Mawell's Equations. In: Electric fields of the brain. The Neurophysics of EEG. 2nd ed. New York: Oxford University Press; 2006d. ISBN-13 978-0-19-505038-7.
12. Nunez PL, Srinivasan R. An overview of the electromagnetic field. In: Electric fields of the brain. The Neurophysics of EEG. 2nd ed. New York: Oxford University Press; 2006e. ISBN-13 978-0-19-505038-7.
13. Koester J, Siegelbaum SA. Membrane potential and the passive electrical properties of the neuron. In: Kandel ER, Schwartz JH, Jessell TM, Siegelbaum SA, Hudspeth J, editors. Principles of Neural Science. 5th ed. New York: Elsevier; 2013. ISBN978-0-07-139011-8.
14. Michel CM, He B. EEG source localization. Handb Clin Neurol. 2019;160:85–101. https://doi.org/10.1016/B978-0-444-64032-1.00006-0.
15. Wadman WJ, Lopes da Silva FH. Biophysical aspects of EEG and MEG generation. In: Shomer DL, Lopes da Silva F, editors. Niedermeyer's electroencephalography: basic principles, clinical applications and related fields. 7th ed. Oxford University Press; 2018. ISBN 9780190228484.
16. Riera JJ, Ogawa T, Goto T, Sumiyoshi A, Nonaka H, Evans A, Miyakawa H, Kawashima R. Pitfalls in the dipolar model for the neocortical EEG sources. J Neurophysiol. 2012;108(4):956–75. https://doi.org/10.1152/jn.00098.2011. Epub 2012 Apr 25. PMID: 22539822.
17. Halnes G, Ostby I, Pettersen KH, Omholt SW, Einevoll GT. Electrodiffusive model for astrocytic and neuronal ion concentration dynamics. PLoS Comput Biol. 2013;9(12):e1003386. https://doi.org/10.1371/journal.pcbi.1003386. Epub 2013 Dec 19. PMID: 24367247; PMCID: PMC3868551.
18. Pastor J, de la Prida LM, Hernando V, Sola RG. Voltage sources in mesial temporal lobe epilepsy recorded with foramen ovale electrodes. Clin Neurophysiol. 2006;117(12):2604–14. https://doi.org/10.1016/j.clinph.2006.07.311. Epub 2006 Oct 6. PMID: 17029955.
19. Butson CR, Cooper SE, Henderson JM, McIntyre CC. Predicting the effects of deep brain stimulation with diffusion tensor based electric field models. Med Image Comput Comput Assist Interv. 2006;9(Pt 2):429–37. https://doi.org/10.1007/11866763_53.
20. Li Z, Cui D, Li X. Unbiased and robust quantification of synchronization between spikes and local field potential. J Neurosci Methods. 2016;269:33–8.

https://doi.org/10.1016/j.jneumeth.2016.05.004. Epub 2016 May 13.

21. Swindale NV, Rowat P, Krause M, Spacek MA, Mitelut C. Voltage distributions in extracellular brain recordings. J Neurophysiol. 2021;125(4):1408–24. https://doi.org/10.1152/jn.00633.2020. Epub 2021 Mar 10. PMID: 33689506.
22. Venkadesh S, Van Horn JD. Integrative Models of Brain Structure and Dynamics: Concepts, Challenges, and Methods. Front Neurosci. 2021;15:752332. https://doi.org/10.3389/fnins.2021.752332. PMID: 34776853; PMCID: PMC8585845.
23. Kauppinen RA, Thotland J, Pisharady PK, Lenglet C, Garwood M. White matter microstructure and longitudinal relaxation time anisotropy in human brain at 3 and 7 T. NMR Biomed. 2023;36(1):e4815. https://doi.org/10.1002/nbm.4815. Epub 2022 Sep 12. PMID: 35994269; PMCID: PMC9742158.
24. Wagness RK. Boundary Conditions at a surface of discontinuity. In: Electromagnetic fields. 2nd ed. Danvers: Wiley & Sons; 1986e. ISBN 0-471-81-186-6.
25. Forssell M, Goswami C, Krishnan A, Chamanzar M, Grover P. Effect of skull thickness and conductivity on current propagation for noninvasively injected currents. J Neural Eng. 2021;18(4). https://doi.org/10.1088/1741-2552/abebc3.
26. Acar ZA, Makeig S. Evaluation of skull conductivity using SCALE head tissue conductivity estimation using EEG. Annu Int Conf IEEE Eng Med Biol Soc. 2022;2022:4826–9. https://doi.org/10.1109/EMBC48229.2022.9872004.
27. Brigo F, Cicero R, Fiaschi A, Bongiovanni LG. The breach rhythm. Clin Neurophysiol. 2011;122(11):2116–20. https://doi.org/10.1016/j.clinph.2011.07.024. Epub 2011 Aug 26.

3 Analog and Digital Signals

3.1 Introduction

The introduction of computers has brought about the greatest conceptual and methodological revolution in the field of clinical neurophysiology, and especially in electroencephalography. Many of the mathematical tools developed before the advent of computers were based on continuous mathematics, which is solidly grounded in the work of mathematicians such as Weierstrass, Gauss, Euler, or Leibniz, to mention only a few of the most representative.

However, computers do not work with continuous values of physical or numerical quantities, but rather with discrete values. This necessitated the adaptation and, in many cases, reformulation in the form of discrete mathematics of what had proved very useful in continuous tools.

Although it is not part of our field of study, it is important to consider this example to appreciate the importance of the proper use of discrete mathematics. I shall refer to the ultraviolet catastrophe, also known as Rayleigh-Jeans catastrophe. It was a failure of classical EM to explain the electromagnetic emission of a body in thermal equilibrium with its environment. According to the predictions of classical EM, an ideal black body in thermal equilibrium should emit energy in all frequency ranges; the higher the frequency, the greater the energy. This result was reached at using the hypothesis that the energy distribution was continuous. However, in 1900, Max Planck replaced integrals with summations, after assuming that energy was distributed discretely, that is, quantized. This change made it possible to find the correct expression for black body radiation, eliminating the previous catastrophe and giving rise to the birth of Quantum Mechanics [1].

In this chapter, we will justify the procedures that allow us to move from a continuous measurement to a discrete one. Although we will not justify the methods with mathematical exactitude, we will provide an idea that will help us understand the differences between the two types of mathematical analysis.

3.2 Discrete and Continuous Mathematics

A quantitative variable can be of two types:

1. Continuous: a variable that can take any value within a given interval, meaning it can take on an infinite number of values whiting a given range. An example would be the amplitude of a waveform. Mathematically

$$f:[a,b] \to \mathbb{R}, a,b \in \mathbb{R}; a \le x \le b \text{ and } f(x) = y \tag{3.1}$$

2. Discrete: a variable that can only take on a given number of values, such as the number of EEG channels, so

$$f:[p,q] \to \mathbb{R},\ p,q \in \mathbb{N}; p \le n \le q \text{ and } f(n) = y \tag{3.2}$$

J. Pastor, *Quantitative EEG Practical Guide*, https://doi.org/10.1007/978-3-032-05457-9_3

Discrete mathematics is the branch of mathematics that studies discrete groups. A group is discrete when it either has a finite number of elements or is countable, meaning that a correspondence can be established between this group and a subgroup of natural numbers *N*.

In contrast to continuous mathematics, which studies concepts such as continuity and continuous change, discrete mathematics studies structures whose elements can be counted one by one separately and is, therefore, the kind of mathematics used in computation [2].

The key point in discrete mathematics is the impossibility to handle the ideas of proximity or limit and smoothness in curves, as is possible in analysis.

A function defined on an interval of integers is called a succession that can either be finite or infinite. Such a discrete function is explicitly defined by a list (if its domain is finite), or by a formula for its *n*-term, or it can also be given implicitly by a recurrence relation or difference equation. Difference equations are similar to differential equations (see Chap. 17) but replace derivatives by differences between adjacent terms and can be used to approximate differential equations.

The use of finite increments implies that it is essential to determine and bound the error or difference between the real function and its digitalized equivalent or the deviation from the differential equations of the results in difference equations [3].

3.3 Analog-to-Digital Conversion

Digital signal processing necessarily involves the conversion of analog-to-digital signals. An analog signal is continuous over a time interval. Voltage signals can be represented as $V(t)$, indicating that this signal is a function of time. These signals are converted into a series of numbers that can be stored, represented, and mathematically analyzed.

Modern equipment in clinical neurophysiology consists of cheap analog-to-digital converters (ADC) with high sampling rate multiplexers. Briefly, this process samples a registered potential value defined during a very short time interval (sampling period, or T_s, usually measured in milliseconds or even microseconds -10^{-3} to 10^{-6} s). Thus, the continuous signal is converted into a sequence of discrete measured voltage values at constant intervals of time and amplitude. We won't go into detail about the electronics of EEGs, but we will briefly describe the mathematical basis that justifies these processes.

3.3.1 Measurement Chain

Measuring biopotentials requires electronic devices with defined characteristics. We will briefly outline the EEG measurement procedure. The first part of the acquisition is analog and allows obtaining a signal with suitable characteristics for subsequent digitalization by the process of analog-to-digital (ADC or A/D) conversion.

Biopotential's measurement is performed through electrodes. Their characteristics and particularities would be a topic of their own, but we don't have time right now. These electrodes are connected to a preamplifier or head stage. This device has a high input impedance, which is essential to prevent current leakage from the system being recorded. EEG devices have an input impedance ranging from 10 to more than 100 MΩ [4]. The preamplifier is connected to an amplifier to increase the size of the signal, which is usually very small. Increasing the signal's magnitude also increases the noise level. Therefore, analog (or digital) filters are necessary to limit this noise. Typically, a series of analog or hardware filters are used, which limit the signal's frequency range. These filters generally allow signals with frequencies below 1000 Hz to pass through.[1] Above this, the maximum frequency is limited by the anti-aliasing filter, which prevents high-frequency signals from having to be digitized. In addition to these filters, a notch filter is commonly used, which eliminates the alternating

[1] In research equipment or v-EEG for preoperative assessment in epilepsy, the frequency may be much higher.

current frequency (50 Hz in most of Europe, although in England or the USA, as well as other countries, the frequency is 60 Hz).

This amplified and filtered signal reaches a special type of electronic circuit called a sample and hold (S/H) circuit, which is capable of retaining the potential value at a defined point in time. These voltages are maintained throughout the sampling period and constitute the actual signal to be digitized. The amplifier's input range is generally around 1–10 V. This input voltage amplitude must be adjusted to the input range once the signal is amplified. Let's assume that the signal amplification of the preamplifier and amplifier are 100x respectively, so the total amplification will be 10,000x. If the input range is 1 V, the input range for the ACD will be 1 V/10,000 = 10^{-4} V, or 100 μV. One of the most important properties of an A/D converter is its number of bits (n). This number indicates the number of steps into which we will divide the input range once it has been amplified. The number of steps is calculated as a base 2 power function. Let's assume the converter has 12 bits, so the number of steps will be $2^{12} = 4096$. So, the resolution in amplitude will be 10^{-4} V/4096 = 2.44×10^{-8} V, which is equivalent to 0.244 nV.

Amplitude quantization is not the only way we'll convert our analog signal into a digital signal. It will also be digitized or quantized over time. To do this, S/H measurements will be taken at specific, and generally equal, time intervals. This interval is called the sampling interval (T_s). Magnitudes with the subscript s (for sampling) indicate discrete measurements. Magnitudes without this subscript correspond to analog measurements.

This is a good time to review Chap. 21. Trigonometry to refresh the relationships between frequencies, periods, or harmonic functions.

3.3.2 Dirac Delta Function

The mathematical processing of the digitized signal requires the introduction of a powerful mathematical tool, used in various fields of physics and mathematics, such as the Dirac delta function (δ), defined by its assigned properties [5]

$$\delta(x) = 0; x \neq 0 \qquad f(0) = \int_{-\infty}^{\infty} f(x)\delta(x)dx \tag{3.3}$$

As a special case we have

$$\int_{-\infty}^{\infty} \delta(x)dx = 1$$

From this equation, we can observe that $\delta(x)$ must be an infinitely high, infinitely thin spike at $x = 0$. The problem is that no such function exists. This problem can be treated rigorously by considering the integral as the limit of a series of functions, which is called a distribution theory formalized in the late 1940s by Laurent Schwartz. A rigorous approach to this theory is beyond the scope of this work, and therefore, we will assume the properties of Eq. 3.3 and continue to delve deeper into the theory of signal digitization.

The Dirac delta is an even function, that is $\delta(x) = \delta(-x)$. A more important property is the so-called sifting property

$$f(n) = \int_{-\infty}^{\infty} f(t)\delta(t-n)dt \tag{3.4}$$

Using this property, we can sift out specific values of a continuous function $x(t)$ at given values of n.

The Dirac delta can be used to formalize discrete series from an analog or continuous series. However, to do so, we need a discrete definition of the delta function. Suppose a rectangular function, centered at $x = 0$, whose area is 1. A function like this can be seen in Fig. 3.1a. For the discrete time, we can define the Dirac delta function or unit impulse, as

$$\delta(n) = 0; n \neq 0 \qquad 1 = \sum_{n=-\infty}^{\infty} \delta(n) \tag{3.5}$$

We can represent the sampling procedure as a continuous time function $x(t)$ that is sampled at regular periods T_s, and that is considered zero in between the sampling times. For each of the sampled periods, a Dirac delta shifted exactly to that period will be considered, according to Eq. 3.4, namely $\delta(t - nT_s)$, weighted by the value of $x(t)$ at $t = nT_s$. According to this reasoning, we

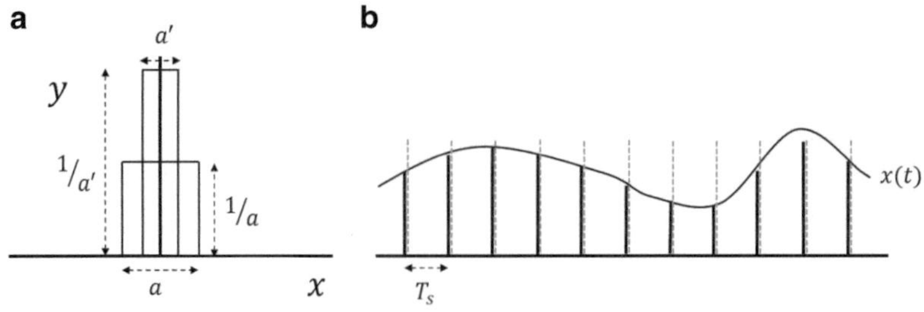

Fig. 3.1 Graphical representation of the discrete-time Dirac delta. (**a**). The unit impulse can be considered a square wave with duration a and amplitude $1/a$ in which $a \to 0$. Reminder that in continuous time, the amplitude of the unit impulse is ∞, whereas the amplitude is 1 in the discrete time version. (**b**) Sampling a continuous function $x(t)$ by multiplication with the Dirac unit impulse series (dashed grey lines) generates discrete time function $x_s(n)$ (thick vertical bars)

make it believable that we can represent the sampled equivalent ($x_s(n)$) of continuous time function as

$$\begin{aligned} x_s(nT_s) &= \sum_{n=-\infty}^{\infty} x(nT_s)\delta(t-nT_s) \\ &= x(t)\sum_{n=-\infty}^{\infty}\delta(t-nT_s) \end{aligned} \tag{3.6}$$

3.3.3 The Nyquist Theorem

The Nyquist theorem [6] is a fundamental theorem of information theory. It shows that if the following two conditions are met: (i) the signal is band-limited, and (ii) the sampling rate is greater than twice its bandwidth, the complete information of the original analog signal is described by the entire series of samples resulting from the sampling process. It applies to both analog and digital signals, which is the field of application that concerns us. Despite the name, it seems that Harry Nyquist, while working for Bell Laboratories, stated it as a hypothesis and that it was later Shannon who demonstrated it.

Nevertheless, it may be obvious that the sample rate at which $x_s(n)$ is obtained must be sufficient to represent the change in the continuous signal $x(t)$. The Nyquist theorem shows that the minimum sampling frequency (f_{Nyquist}) that must be used in a band-limited signal, whose maximum frequency is $f_{\max}$, to preserve all the spectral structure, is at least twice that maximum frequency [7], that is

$$f_{\text{Nyquist}} \geq 2f_{\max} \tag{3.7}$$

Strictly speaking, this inequality is not an equation, but we number it consecutively to maintain proper indexing.

Although we will not discuss the Nyquist theorem in detail, we will briefly address it herein [8]. Let's assume a periodic signal at a frequency around 10 Hz (Fig. 3.2a). According to Eq. 3.7, the minimum sampling frequency to preserve frequency information would be >20 Hz. However, for the digitized signal to be similar to the real signal, the sampling frequency must not only be the f_{Nyquist} frequency, but it must be several times higher (at least 4 times higher). In Fig. 3.2b, the sampling frequency is 100 Hz ($f_{\text{Nyquist}} \times 4$) and it can be seen that the digitized signal is similar to the original. When this sampling frequency is reduced, but is still greater than f_{Nyquist}, the frequency composition is preserved, but it can be seen that the morphology of the signal is severely altered (Fig. 3.2c). However, when the sampling frequency is lower than f_{Nyquist}, it results in a digitized signal with a lower frequency than the original signal (Fig. 3.2d) which completely modifies the information of the reconstructed signal.

Since the frequency content of scalp EEG recordings is not very high (at best, up to 60 Hz),

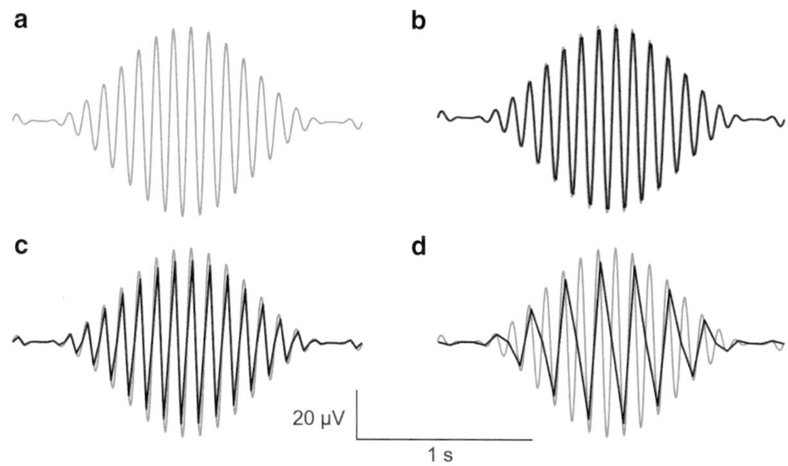

Fig. 3.2 Sampling frequency and information preservation. (**a**) Continuous original signal. (**b**) Reconstructed signal with a sampling frequency of 100 Hz, (**c**) 30 Hz, and (**d**) 15 Hz. The digitized signals are shown in black, and the superimposed original signal is shown in gray

the $f_{Nyquist}$ is not usually a problem in practice. However, when performing intracranial studies, this is a point that must be taken into account.

In conclusion, the analysis of analog-to-digital signal conversion is a field of extraordinary breadth and, of course, cannot be rigorously addressed in this work. However, these few ideas are sufficient to understand the fundamentals on which most (if not all) EEG equipment and the analysis methods used are based.

3.4 Concluding Remarks

There are important differences between continuous mathematical concepts and their discrete equivalents. These differences, in general, do not have a major impact on clinical practice, since any errors in numerical analysis due to discretization are smaller than errors due to noise in the recording. However, it is useful to be aware of these differences because, in some cases, proper handling of the sampling frequency or the Nyquist theorem is important for good recording. This is especially true when using numerical analysis packages and, even more so, if you plan to create a script for the analysis.

References

1. Nauenberg M. Max Planck and the birth of the quantum hypothesis. Am J Phys. 2016;84:709–20. https://doi.org/10.1119/1.4955146.
2. Rosen KH. Discrete mathematics: and its applications. 7th ed. New York: McGraw-Hill; 2007. ISBN 978-0-07-338309-5.
3. Gerald CF, Wheatley PO. Applied numerical analysis. 7th ed. New York: Addison Wesley; 2004. ISBN 0-321-13304-8.
4. Górecka J, Makiewicz P. The dependence of electrode impedance on the number of performed EEG examinations. Sensors (Basel). 2019;19(11):2608. https://doi.org/10.3390/s19112608. PMID: 31181738; PMCID: PMC6603518.
5. Arfken GB, Weber HJ. Vector analysis. In: Mathematical methods for physicists. 6th ed. Amsterdam: Elsevier Academic Press; 2005. ISBN-13: 978-0-12-059876-2.
6. Maciejewski MW, Mobli M, Schuyler AD, Stern AS, Hoch JC. Data sampling in multidimensional NMR: fundamentals and strategies. Top Curr Chem. 2012;316:49–77. https://doi.org/10.1007/128_2011_185. Erratum in: Top Curr Chem. 2012;316:149. PMID: 21773916; PMCID: PMC6078388.
7. Adeli H, Gosh-Dastidar S. Time and frequency domain analyses. In: Automated EEG-based diagnosis of neurological disorders. Inventing the future in neurology. CRC Press; 2010. ISBN 978-1-4398-1531-1.
8. Van Drongelen W. Data acquisition. In: Signal processing for neuroscientists. Amsterdam: Elsevier; 2007. ISBN 13:978-0-12-370867-0.

4 Analysis in the Frequency Domain

4.1 Introduction

Neurophysiological recordings in general and, in particular EEG, are always obtained as time-dependent functions, i.e., in the time domain (see Chap. 16). This somewhat cryptic expression refers to the mathematical formulation of a function for which the domain is the set of values that can take the function. Therefore, the useful information is obtained as change in voltages along the time, e.g., $V(t)$.

However, the time-domain analysis does not allow us to access much neurophysiological information, such as the frequencies composition in the signal. Therefore, to assess this structure, the analysis should be moved to the frequency domain [1]. This means that all the information contained into the function $V(t)$ must be represented in a function which independent variables would not be the time, but the frequency.

The most useful ways to obtain this pass, from the time to the frequency domain in neurophysiology, are the Fourier transform and the power spectrum derived. This is the topic we will discuss in detail because it is the main tool used in our method. There are, also, other analytical methods in the frequency domain and, by this reason, we shall discuss briefly. Between these methods, we have the compressed spectral array (CSA) and the wavelet transformations.

For a most rigorous treatment of the chapter, see [2–4].

4.2 The Fourier Transform

Before beginning, we will define a periodic function. This function is whatsoever that it acquires the same value when a defined interval of time, called period (T, s), is reached, which is explicitly defined as

$$f(t) = f(t + nT); n = 1,2,\ldots \quad (4.1)$$

The function value is equal at the $t + T$, $t + 2T$, $t + 3T$, etc. The trigonometric functions sine, cosine, and their multiple combinations are typical examples of such functions (see Chap. 16, Sect. 16.5).

The interesting aspect to be noted here is that periodic functions can be easiest treated algebraically by means of an expansion in a series of sines and cosines, called a Fourier series. Its use in Mathematics and Physics is impressive, involving waves (the use we are interested for), phenomena related to harmonic functions or solving differential equations.

4.2.1 The Wave Equation

One of the most important equations in the history of science is the one that describes the motion of a vibrating string or the movement of waves, whether in a material medium, such as water, or in a vacuum, such as EM radiation. This equation is the wave equation. The number of fields of Physics affected by this equation

J. Pastor, *Quantitative EEG Practical Guide*, https://doi.org/10.1007/978-3-032-05457-9_4

includes EM, thermodynamics, or quantum mechanics. Obviously, this is not the time or place to deal with these implications, but it is worth introducing this equation, because it is the origin of the fundamental theory of our work.

The unidimensional wave equation is

$$\frac{\partial^2 \psi}{\partial x^2} = \frac{1}{v^2}\frac{\partial^2 \psi}{\partial t^2} \quad (4.2)$$

where $\psi(x, t)$ is the displacement of a point, which depends on both its position and time and v is the speed of wave propagation (see Chap. 17).

The general solution for this equation can be written as [5]. See also the paragraph 17.6

$$\psi(x,t) = \int_{-\infty}^{\infty} F(\omega) e^{\left[i\frac{\omega}{\upsilon}(x+\upsilon t)\right]} d\omega + \int_{-\infty}^{\infty} G(\omega) e^{\left[i\frac{\omega}{\upsilon}(x-\upsilon t)\right]} d\omega \quad (4.3)$$

where $F(\omega)$ and $G(\omega)$ are arbitrary functions.

The method found to obtain a general periodic function as a linear combination of sine and cosine functions is known as Fourier's theorem, developed in 1822, precisely, to solve the differential equation of heat propagation (Jean-Baptiste Joseph Fourier, 1768–1830).

4.2.2 The Fourier Series for a Period of 2π

Fourier's theorem states that any periodic and continuous function $f(t)$ (of period 2π) that complies with specific mathematical conditions—which we will not describe in detail here[1]—can be written as a series of an infinite sum of weighted sine and cosine functions (see [6] for a detailed description). This result is profound and awesome, especially because this approach can be used to reconstruct functions similar to sines or cosines (Fig. 4.1a), as well as functions that bear no resemblance to these periodic functions (Fig. 4.1b) or even to random functions, such as noise (Fig. 4.1c).

In the figure, we show a few components of the Fourier series (1, 5, 10, 15, 20, and 30). It is evident that the frequencies (except for the $N = 1$ component in Fig. 4.1b, in fact $N = 2$) are the same in each term, but the amplitudes are very different. The 30th term (the last one shown) is not necessarily the last one needed to make the series almost exact to the original function, since this resemblance will require adding more or fewer terms depending on the complexity of the original function. This fact is very important for understanding the meaning of the transformed function. We will go into more detail on this important point below, as it explains the meaning of these power spectra.

Before formally showing the Fourier series, we have to show a fact of extraordinary importance about the orthogonality of functions. Without going into detail, we state the following results:

$$\int_{-\pi}^{\pi} \sin nt \sin mt dt = \pi\delta_{nm}$$

$$\int_{-\pi}^{\pi} \cos nt \cos mt dt = \pi\delta_{nm} \quad (4.4)$$

$$\int_{-\pi}^{\pi} \cos nt \sin mt dt = 0$$

Where $\delta_{nm} = \begin{cases} 1, if\, m = n \\ 0, if\, m \neq n \end{cases}$, known as the Kronecker delta.

When it is verified that $-\pi \leq t \leq \pi$, the Fourier series can be formally stated as follows:

$$f(t) = \frac{a_0}{2} + \sum_{n=1}^{\infty} a_n \cos nt + \sum_{n=1}^{\infty} b_n \sin nt \quad (4.5)$$

We can determine the value of the coefficients a_n and b_n using the relations of Eq. 4.4. To do this, multiply both sides of the Eq. 4.5 by a trigonometric function, for example sinpt and integrate

$$\int_{-\pi}^{\pi} f(t)\sin pt = \frac{a_0}{2}\int_{-\pi}^{\pi} \sin pt dt + \int_{-\pi}^{\pi} \sum_{n=1}^{\infty} a_n \cos nt \sin pt dt$$
$$+ \int_{-\pi}^{\pi} \sum_{n=1}^{\infty} b_n \sin nt \sin pt dt$$

The first integral of the right side is null. The second is also null by the orthogonality (Eq. 4.4) between cos nt and sin pt functions for any pair n,

[1] Briefly, the conditions imposed to $f(t)$ are that this function have only a finite number of extreme values in the interval $[-\pi, \pi]$ and a finite number of finite discontinuities. Both conditions are met in non-artifacted biological functions. These conditions are known as the Dirichlet conditions. Please, note that the choice of a symmetric interval, like this, is not essential.

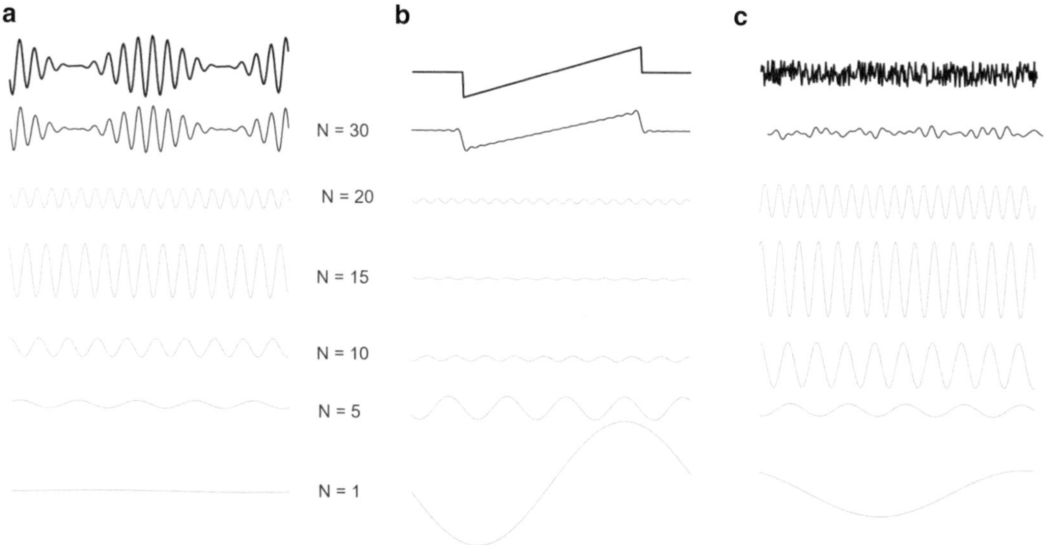

Fig. 4.1 Fourier series for different functions, represented in black, are shown in the top row. (**a**) A predominantly sinusoidal and evidently periodic function. (**b**) A linear function with discontinuity points. (**c**) Gaussian random noise. From bottom to top, some definite terms of the Fourier series needed to reconstruct the original signal. The last term shown is the series for $N = 30$ is not the 30th term, but the sum of all the terms. In the case of a, this term is practically identical to the function. In b, the similarity is very significant, but it is not exactly the same. In c, the sum of 30 terms is not sufficient to completely reconstruct the signal. The functions have been represented in a period $t \in [-\pi, \pi]$

p. However, let's see what happens with the third integral

$$\int_{-\pi}^{\pi} \sum_{n=1}^{\infty} b_n \sin nt \sin pt\,dt = \sum_{n=1}^{\infty} b_n \int_{-\pi}^{\pi} \sin nt \sin pt\,dt$$

The last integral vanishes, except when $n = p$, so, finally we have

$$\int_{-\pi}^{\pi} f(t) \sin pt = b_p \pi$$

The same procedure can be used to obtain a_p. Obviously, the notation n or p is not relevant and only indicates a natural number. Therefore, we can write the coefficients of the Fourier series as

$$a_0 = \frac{1}{\pi} \int_{-\pi}^{\pi} f(t)\,dt \tag{4.6}$$

$$a_n = \frac{1}{\pi} \int_{-\pi}^{\pi} f(t) \cos nt\,dt \tag{4.7}$$

$$b_n = \frac{1}{\pi} \int_{-\pi}^{\pi} f(t) \sin nt\,dt \tag{4.8}$$

The power spectrum is the representation of the sum of the squared coefficients as a function of the n-term, that is $a_n^2 + b_n^2$. See below for a detailed justification.

But how does the Fourier series provide information about the harmonic components of a signal? Let's look at the basics of spectral analysis. The number of terms required for the Fourier series to reliably reconstruct the original function $f(x)$ will depend on the complexity of the original signal, so that a highly periodic signal with trigonometric morphology will need a single term for its reconstruction. We will evaluate the degree of adequacy of the signal reconstructed by the Fourier series ($F_n(x)$), which depends on the number of terms, n, and the original function ($f(x)$), using the expression $r = f(x)/F_n(x)$ and its dynamics informs about the convergence between both functions along the terms. All the algebraic functions are numerically developed, with a sampling frequency of 256 Hz and are represented and analyzed for a time period between $-\pi$ and π.

The trigonometric function $f(t) = \sin(3t)$ is shown in Fig. 4.2a. The only component in its spectrum is at 3 Hz (third row). It can be seen (lower row) that the first 18 spectral components do not contribute anything to the function. In fact, 23 terms are needed for $r = 0.99443$. Figure 4.2b shows the function $f(t) = 0.5\sin(4.7t) + \sin(5t) + 0.5\sin(5.3t)$, the sum of three functions with close frequencies and different amplitudes. The spectrum of this function has several components, with a maximum at 5 Hz, but with more than 10 non-zero terms. Overall, 139 terms were needed to obtain $r = 0.999196$. However, it can be seen (lower row) that the contribution between terms >60 is very small and a sufficiently good representation would be obtained by truncating the series from here on (not shown, but the simulation in this case produces $r = 0.98579$). Figure 4.2c represents a waveform with characteristics similar to a slow wave, simulated by $f(t) = 4e^{-(t+0.3)^2/0.2} - e^{-(t+0.3)^2/1.5}$. The spectrum of this function has 5 components, with a maximum at 1.11 Hz. Above 1.59 Hz there are no significant terms, only requiring 11 terms to obtain $r = 0.99995$. A minimal change in the function ($f(t) = 4e^{-(t+0.3)^2/0.02} - e^{-(t+0.3)^2/1.5}$) causes the waveform to acquire a sharper morphology (Fig. 4.2d). To fit the series to the function, we need a larger number of terms (actually 34 instead of 11), with a spectrum that presents more fast components, reaching up to 4 Hz in this case and $r = 0.999997$. An even more complex function is shown in Fig. 4.2e, which is given by the expression

$$f(t) = 4e^{-(t+0.3)^2/0.02} - e^{-(t-0.3)^2/1.5} + 4e^{-(t+0.6)^2/0.2}.$$

It can be seen that to adequately represent this function, which is more complex than in D, more terms are needed in the spectrum, with a total of 47 terms in the series. Furthermore, the added terms have higher frequencies, although the fit is still very good, with $r = 0.99998$. Finally, in Fig. 4.2f, the presence of random noise has been introduced to simulate a more realistic waveform. In this case, the convergence of both functions occurs for a range of terms between 150 and 160, with the subtraction of both functions becoming divergent again (bottom row). However, although the fit in this case is not as good as in the previous ones, it can be seen that with 155 terms, a $r = 0.99658$. As expected, random noise varies the number of absolute terms and the number of spectral components, producing a significant modification with respect to the noise-free waveform.

4.2.3 The Fourier Series for a Period Different of 2π. Complex Fourier Series

When the period is different from 2π, it is necessary to slightly modify the development seen above, by means of a change of variable. Let us consider a function with a period T, that is, $-T/2 \le t \le T/2$. This interval can be increased or decreased by introducing the new variable $z(t) = 2\pi t/T$, that $-\pi/2 \le z \le \pi/2$ (Fig. 4.3).

The Fourier series, for the new variable, will be

$$f(z) = \frac{a_0}{2} + \sum_{n=1}^{\infty} a_n \cos\frac{2\pi nt}{T} + \sum_{n=1}^{\infty} b_n \sin\frac{2\pi nt}{T}$$

The angular frequency is defined as $\omega = 2\pi\nu = 2\pi/T$, which is measured in radians/s (rad/s). We have used the inverse relationship between ν (Hz) and T (s^{-1}) in this expression. For every component of the frequency, n, we modify the angular frequency in this way $\omega_n = 2\pi n/T$. Thus, the above equation will become

$$f(z) = \frac{a_0}{2} + \sum_{n=1}^{\infty} a_n \cos\omega_n t + \sum_{n=1}^{\infty} b_n \sin\omega_n t \quad (4.9)$$

With this variable, the coefficients, a_n and b_n, can be defined as follows (a_0 is a_n when $n = 0$):

$$a_0 = \frac{2}{T}\int_{-T/2}^{T/2} f(z)\,dz$$

$$a_n = \frac{2}{T}\int_{-T/2}^{T/2} f(z)\cos\omega_n t\,dt \quad (4.10)$$

$$b_n = \frac{2}{T}\int_{-T/2}^{T/2} f(z)\cos\omega_n t\,dt$$

It can be useful to obtain a more compact form of the series Eq. 4.9. From Chap. 21, we can remember the relationships between hyperbolic and non-hyperbolic trigonometric functions

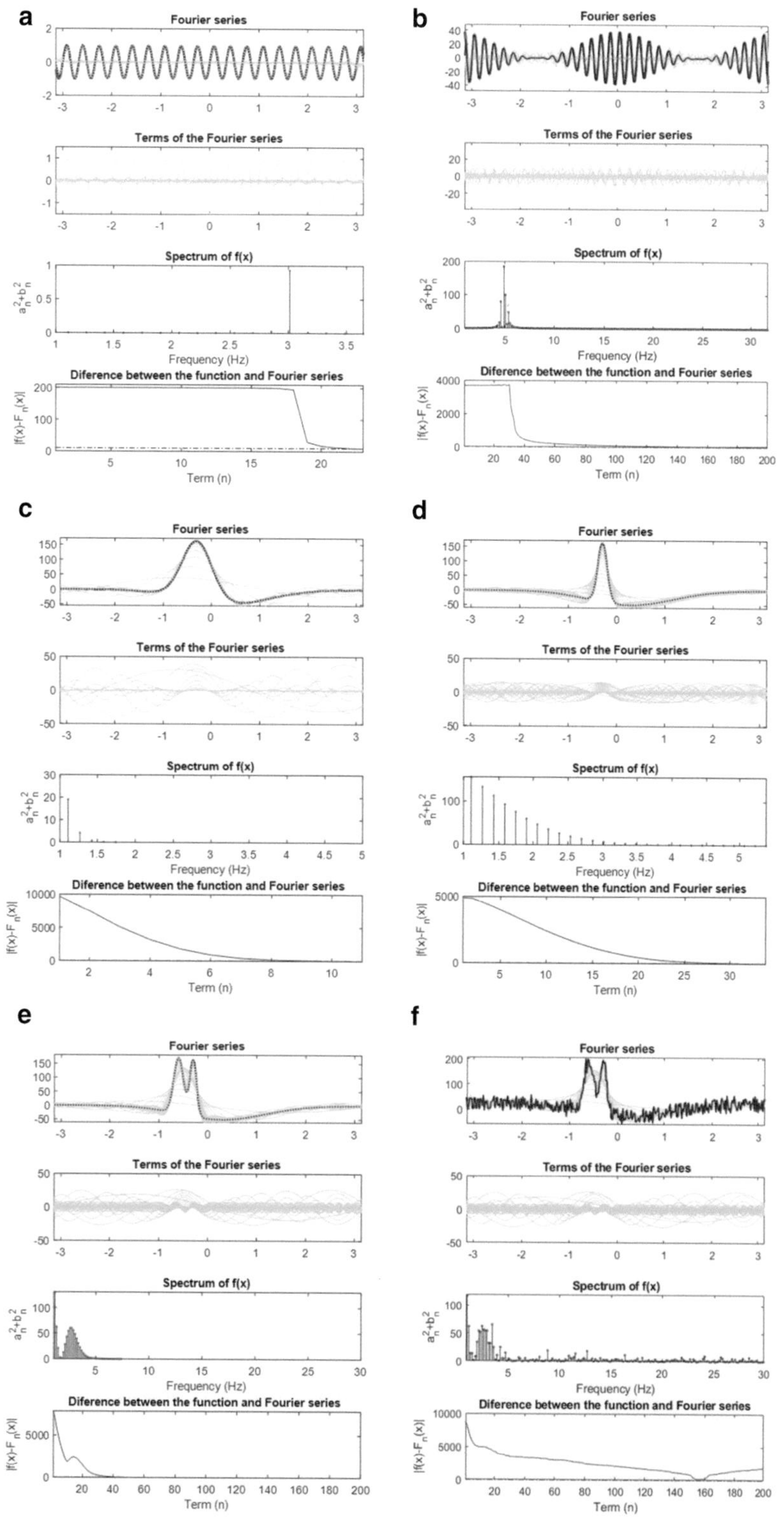

Fig. 4.2 Meaning of the power spectrum (see the text for more details). (**a**) Sine function. (**b**) Sum of three sine functions of similar frequency. (**c**) Waveform with slow-wave morphology. (**d**) Waveform with sharper morphology. (**e**) Complex waveform. (**f**) The same waveform as in E but with added Gaussian noise. For each part, from top to bottom, the following panels are shown: first row, superposition of the represented function (black), of the successive terms of the Fourier series (grey) and the final series (dashed red line), second row, superposition of the individual Fourier terms ($a_n + b_n$), third row, power spectrum, and fourth row, absolute value of the difference between the function and the Fourier series as a function of n

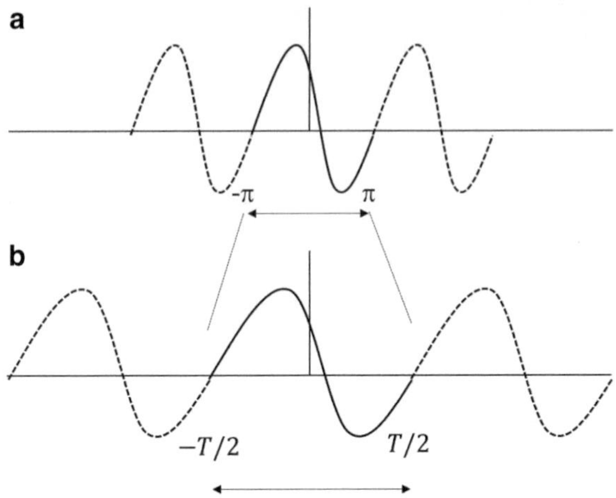

Fig. 4.3 Modification of the period for the use of the Fourier series in functions with arbitrary periods. (**a**) Periodic function with period 2π. (**b**) Function with similar morphology, but with a longer period T

$$\cosh ix = \frac{e^{ix}+e^{-ix}}{2} = \cos x$$

$$\sinh ix = \frac{e^{ix}-e^{-ix}}{2} = i\sin x \qquad (4.11)$$

To obtain these equalities, we have used Euler's identity. We can substitute Eq. 4.11 in Eq. 4.9, obtaining

$$f(z) = \frac{a_0}{2} + \sum_{n=1}^{\infty} a_n \frac{e^{i\omega_n t}+e^{-i\omega_n t}}{2} - i\sum_{n=1}^{\infty} b_n \frac{e^{i\omega_n t}-e^{-i\omega_n t}}{2}$$

Operating and grouping terms

$$f(z) = \frac{a_0}{2} + \frac{1}{2}\sum_{n=1}^{\infty}\left(a_n e^{i\omega_n t} + a_n e^{-i\omega_n t}\right) - i\frac{1}{2}\sum_{n=1}^{\infty}\left(b_n e^{i\omega_n t} - b_n e^{-i\omega_n t}\right)$$

$$f(z) = \frac{a_0}{2} + \frac{1}{2}\sum_{n=1}^{\infty}\left(a_n - ib_n\right)e^{i\omega_n t} + i\frac{1}{2}\sum_{n=1}^{\infty}\left(a_n + b_n\right)e^{-i\omega_n t} \qquad (4.12)$$

We can define

$$c_n = \frac{1}{2}\left(a_n - ib_n\right) c_{-n} = \frac{1}{2}\left(a_n + ib_n\right) c_0 = \frac{1}{2}a_0 \qquad (4.13)$$

And, replacing at 4.12, we obtain the expression

$$f(z) = \frac{a_0}{2} + \sum_{n=1}^{\infty} c_n e^{i\omega_n t} + \sum_{n=1}^{\infty} c_{-n} e^{-i\omega_n t} \qquad (4.14)$$

Nevertheless, we can observe that $\sum_{n=1}^{\infty} c_{-n} e^{-i\omega_n t} = \sum_{n=-\infty}^{-1} c_n e^{i\omega_n t}.$

So, the precedent series Eq. 4.14 can be rewritten in a more compact and easiest form as follows:

$$f(t) = \sum_{n=-\infty}^{\infty} c_n e^{i\omega_n t} \qquad (4.15)$$

In this equation, we have again used the variable t, instead of z, only for clarity. But this does not mean that they are different functions, since, in fact, it has always been taken for granted that $z = z(t)$.

And, combining Eq. 4.10 and Eq. 4.13, we get

$$c_n = \frac{1}{T}\int_{-T/2}^{T/2} f(\tau)\cos\omega_n\tau\,d\tau - i\frac{1}{T}\int_{-T/2}^{T/2} f(\tau)\sin\omega_n\tau\,d\tau = \frac{1}{T}\int_{-T/2}^{T/2} f(\tau)\left(\cos\omega_n\tau - i\sin\omega_n\tau\right)d\tau = \frac{1}{T}\int_{-T/2}^{T/2} f(\tau)e^{-i\omega_n\tau}\,d\tau \qquad (4.16)$$

This expression is quite interesting, because it allows us to increase the period to infinity, what is needed to use the Fourier series for any arbitrary function and not only for periodic functions [4].

We must briefly comment on the Gibbs phenomenon [2]. This is an overshoot observed at discontinuities (Fig. 4.4) and is not limited to the Fourier series. The difference between the overshoot and

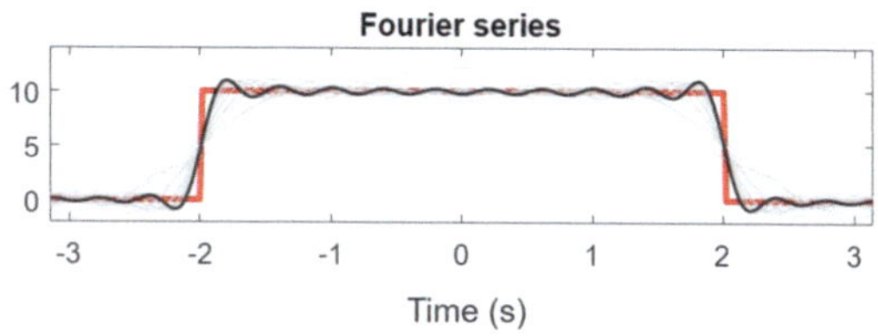

Fig. 4.4 The addition of Fourier series terms produces a damped oscillation at discontinuities, called the Gibbs phenomenon. The red line shows the function $f(t)=\begin{cases}0, & t<2\\ 10, & 2\le t\le 2\\ 0, & t>2\end{cases}$, defined at $t\in[-\pi,\pi]$. The thin gray lines show the different Fourier series terms, and the black line shows the result of the series, truncated for n = 15. The Gibbs phenomenon is especially evident near discontinuities

the true signal is inversely proportional to the number of terms used in computation. However, this phenomenon does not vanish completely and we can observe it at any discontinuity (e.g., spikes).

4.2.4 The Fourier Transform for Non-periodic Functions

We can think of a non-periodic function as a periodic function whose period is infinite, i.e., $T\to\infty$ (Fig. 4.5). However, we cannot use the equation Eq. 4.16 to find this limit, because the period is in the denominator and therefore $\lim\limits_{T\to\infty}\frac{1}{T}=0.$ Therefore, to make the step to the limit, we have to eliminate this difficulty. To do this, let us suppose that between two consecutive angular frequencies, the frequency interval ($\Delta\omega_n$) is

$$\omega_{n+1}-\omega_n \cong \Delta\omega_n = 2\pi(n+1)/T-2\pi n/T=2\pi/T$$

From this equality, we have $1/T$ = $\Delta\omega_n/2\pi$, what we can use in Eq. 4.15, to get

$$c_n=\frac{\Delta\omega_n}{2\pi}\int_{-T/2}^{T/2} f(t)e^{-i\omega_n t}dt \qquad (4.17)$$

The factor that caused the difficulty has disappeared, so we can now make the step to the limit. We introduce, first of all, the equation Eq. 4.16 in Eq. 4.15

$$f(t)=\sum_{n=-\infty}^{\infty}\frac{\Delta\omega_n}{2\pi}\int_{-T/2}^{T/2} f(\tau)e^{-i\omega_n\tau}d\tau e^{i\omega_n\tau} \qquad (4.18)$$

operating

$$f(t)=\frac{1}{2\pi}\sum_{n=-\infty}^{\infty}\Delta\omega_n\int_{-T/2}^{T/2} f(\tau)e^{-i\omega_n\tau}d\tau e^{i\omega_n\tau}$$

Then

$$\lim_{\substack{T\to\infty\\ \Delta\omega_n\to 0}}\left(\frac{1}{2\pi}\sum_{n=-\infty}^{\infty}\Delta\omega_n\int_{-T/2}^{T/2} f(\tau)e^{-i\omega_n\tau}d\tau e^{i\omega_n\tau}\right)$$

$$=\frac{1}{2\pi}\int_{-\infty}^{\infty}d\omega e^{i\omega t}\int_{-\infty}^{\infty} f(\tau)e^{-i\omega\tau}d\tau$$

Then, we can write

$$\mathcal{F}(i\omega)=\int_{-\infty}^{\infty} f(t)e^{-i\omega t}dt \qquad (4.19)$$

$$f(t)=\mathcal{F}^{-1}(\omega)=\frac{1}{2\pi}\int_{-\infty}^{\infty}\mathcal{F}(i\omega)e^{i\omega t}d\omega \qquad (4.20)$$

The first equation is the continuous Fourier transform (CFT) and the second is the inverse continuous Fourier transform (ICFT).

However, the EEG signal is a time-varying signal, sometimes transient (e.g., spikes or bursts) and typically nonstationary. In practice, we not only need to know the frequency components but we also want to know the dynamics. Time-frequency analysis is especially suitable for addressing such problems [7]. The simplest method uses a short-time Fourier transform (STFT) to increase the time resolution, given by this expression

$$\mathcal{F}_{ST}(i\omega)=\int_{-\infty}^{\infty} f(t)w(t)e^{-i\omega t}dt \qquad (4.21)$$

where $w(t)$ is the window function. This equation is called Gabor transform and the Fourier transform-based time-dependent spectrum is called periodogram. In fact, this is the approach we used in our method and the expressions $\mathcal{F}(i\omega)$ and $\mathcal{F}_{ST}(i\omega)$, from now on, must be considered as interchangeable. The window used can be different but we use the simplest one, named rectangular, what is defined as

$$w(t)=\begin{cases}1, t\in[0,T]\\ 0, \text{in other case}\end{cases}$$

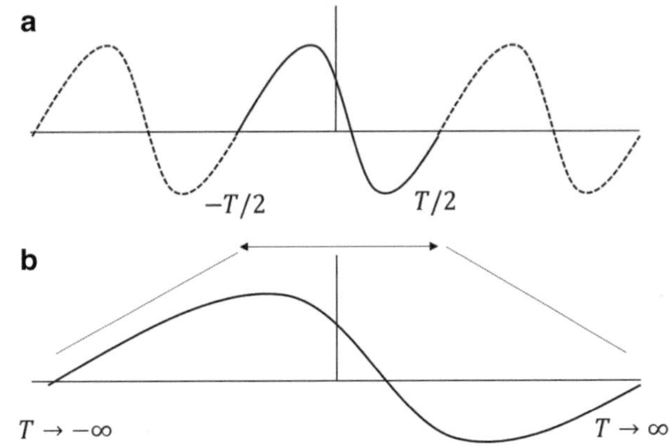

Fig. 4.5 Modification of the period from an arbitrary period to non-periodic function. (**a**) Periodic function with period T. (**b**) Function with similar morphology, but non-periodic

There are other windows with different morphologies, such as the Bartlett (or triangular) type, the Hamming type, or the Hann type. However, when analyzing a transient signal, such as a seizure or a noise burst, the rectangular window may be most appropriate.

The use of this strategy, however, has effects that modify the continuous Fourier transform.

We have explicitly written the expression for $\mathcal{F}(i\omega))$ as a function of the angular frequency and the imaginary term i to show that it is a complex value. Therefore, we cannot use this expression directly in our analysis, but we must convert it into a real value.

4.2.5 Discrete Fourier Transform and Fast Fourier Transform

The CFT is a very powerful expression for finding the transformation of algebraic functions. However, the EEG voltage time series are finite series of real numbers that cannot be integrated with Eqs. 4.19 or 4.20. In real-time series, both the time and frequency domains must be finite. Therefore, we need to find a discrete version of the CFT to be applied to these types of signals.

Let us assume a real function T length, sampled at ν_s (in Hz). So, the time increment (accuracy or precision) is Δt (in s). Both parameters determine the range and precision of the discrete Fourier transform (DFT).

The analog signal shall be denoted as $f(t)$, where $t \in [0, T]$, $t \in R$ and the digitized signal will be $f(t_n)$, where $n = 1, 2 \ldots N$, $n \in N$

The total sampled elements of the digitized series is N, and is related with T by this expression

$$T = N\Delta t \tag{4.22}$$

The frequency accuracy, which represents the frequency interval between two consecutive frequency values, is,

$$\Delta \nu = \frac{1}{T} \tag{4.23}$$

And, the maximum frequency possible for these conditions will be

$$\nu_{\max} = \frac{1}{\Delta t} \tag{4.24}$$

We can reformulate the two last equations in angular frequency terms (in rad/s), and using 4.22, to obtain

$$\Delta \omega = \frac{2\pi}{N\Delta t} \tag{4.25}$$

And, the range of angular frequency

$$\Omega = \omega_{\max} = N\Delta \omega = \frac{2\pi}{\Delta t} \tag{4.26}$$

Obviously, the number of elements for time and frequency domains must be equal. Time can only be positive and its range is $[0, T]$, while frequency has both positive and negative (imaginary) parts. The frequency limit cannot be $[-\infty,\infty]$, but $[-\Omega/2,\Omega/2]$. With these two series, the temporal one and the frequency one, time n-element and a k-frequency element, are defined as follows:

$$t_n = n\Delta t \tag{4.27}$$

$$\omega_k = k\Delta\omega = \frac{2\pi k}{N\Delta t} \tag{4.28}$$

Then, the discrete form of Eq. 4.19, sampled over a finite interval including N samples, for a k-frequency can be written as

$$\mathcal{F}_a(i\omega_k) = \sum_{n=1}^{N} f(t_n)e^{-i\omega_k t_n}\Delta t$$

And, operating

$$\mathcal{F}_a(i\omega_k) = \Delta t\sum_{n=1}^{N} f(t_n)e^{-i\frac{2\pi}{N}nk}$$

And, finally, considering a small value for Δt, we can remove from the equation, obtaining

$$\mathcal{F}_a(i\omega_k) = \sum_{n=1}^{N} f(t_n)e^{-i\frac{2\pi}{N}nk} \tag{4.29}$$

What is the discrete Fourier transform (DFT).

It is common that in engineering and signal analysis texts, the following equivalences are used:

$\mathcal{F}_a(i\omega_k) = X(k); f(t_n) = x(n)$ and $e^{-i\frac{2\pi}{N}nk} = W_N^{kn}$

so the above equation is written, more compactly, as

$$X(k) = \sum_{n=1}^{N} x(n)W_N^{kn} \tag{4.30}$$

What is the standard definition of DFT.

In turn, the inverse discrete Fourier transform (IDFT) can be obtained from the ICFT (Eq. 4.20). To do that, we proceed as above,

$$f_a(t_n) = \mathcal{F}_a^{-1}(i\omega_k) = \frac{1}{2\pi}\sum_{k=-N/2}^{\frac{N}{2}-1} \mathcal{F}_a(i\omega_k)e^{i\omega_k t_n}\Delta\omega$$

Please pay attention that the upper limit is $N/2-1$, not $N/2$. We can change the summation from $-N/2$ to $N/2-1$, to 1 to N without affecting the result and, using the Eqs. 4.25, 4.27, and 4.28, we have

$$\begin{aligned} f_a(t_n) &= \frac{1}{2\pi}\sum_{k=1}^{N} \mathcal{F}_a(i\omega_k)e^{i\omega_k t_n}\frac{2\pi}{N\Delta t} \\ &= \frac{1}{N\Delta t}\sum_{k=1}^{N} \mathcal{F}_a(i\omega_k)e^{i\frac{2\pi}{N}nk} \end{aligned} \tag{4.31}$$

This is the IDFT approximation. However, to unify notation, we can use the same changes in notation made for DFT and smuggle Δt back, obtaining the usual standard expression

$$x(n) = \frac{1}{N}\sum_{k=1}^{N} X(i\omega)W_N^{-kn} \tag{4.32}$$

The weighting factor introduced as W_N in the preceding expressions and known as twiddle factor plays an important role, mainly in the development of the optimized algorithm known as fast Fourier transform (FFT). The importance of this factor relies in its periodicity [8].

We are not going to go into detail about the FFT method, but simply indicate why it is faster and, above all, why it requires well-defined numbers of elements for its implementation. The standard computation of the DFT of an N length time series requires N^2 multiplications. The two fundamental aspects of the algorithm are:

1. The summation can be separated into even and odd components.
2. The periodic nature of the twiddle factor.

Before starting, we change the range of the summation between 0 and N-1, instead of the original between 1 and N, because it is simpler, although this fact does not modify the results. Let's suppose that $N = r_1r_2$, where $r_1, r_2 \in N$. Then, let's express the indices in this way

$k = k_1r_1 + k_0$; $k_0 = 0, 1, \ldots r_1 - 1$; $k_1 = 0, 1, \ldots r_2 - 1$, then $k = k(k_0, k_1)$

$n = n_1 r_2 + n_0$; $n_0 = 0, 1, \ldots r_2 - 1$; $n_1 = 0, 1, \ldots r_1 - 1$, then $n = n(n_0, n_1)$

Using these expressions in Eq. 4.30

$$\begin{aligned} X(k_0,k_1) &= \sum_{n=0}^{N-1} x(n_0,n_1) W_N^{k(n_1 r_2 + n_0)} \\ &= \sum_{n_0=0}^{r_2-1} \sum_{n_1=0}^{r_1-1} x(n_0,n_1) W_N^{k n_1 r_2} W_N^{k n_0} \end{aligned} \tag{4.33}$$

Since

$$W_N^{k n_1 r_2} = W_N^{(k_1 r_1 + k_0) n_1 r_2} = W_N^{k_1 r_1 n_1 r_2} W_N^{k_0 n_1 r_2} = W_N^{k_0 n_1 r_2} \tag{4.34}$$

Because $W_N^{k_1 r_1 n_1 r_2} = W_N^{k_1 n_1 N} = 1$

And using Eq. 4.32 in Eq. 4.31, we obtain

$$\begin{aligned} X(k_0,k_1) &= \sum_{n_0=0}^{r_2-1} \sum_{n_1=0}^{r_1-1} x(n_0,n_1) W_N^{k_0 n_1 r_2} W_N^{k n_0} \\ &= \sum_{n_1=0}^{r_1-1} x(n_0,n_1) W_N^{k_0 n_1 r_2} \sum_{n_0=0}^{r_2-1} W_N^{k n_0} \end{aligned} \tag{4.35}$$

The inner sum over n_1 defines the new array

$$x_1(n_0,k_0) = \sum_{n_1=0}^{r_1-1} x(n_0,n_1) W_N^{k_0 n_1 r_2} \tag{4.36}$$

Rewritten the result as

$$X(k_0,k_1) = \sum_{n_0=0}^{r_2-1} x_1(n_0,k_0) W_N^{k n_0} \tag{4.37}$$

There are N elements in the array x_1, each requiring $r_1$operations, giving a total of Nr_1 operations. Similarly, it takes Nr_2 operations to calculate X. Therefore, this two-step algorithm requires a total $T = Nr_1 + Nr_2 = N(r_1 + r_2)$ operations. Therefore, if we apply this procedure m times, we need a total amount of operations of

$$T = N(r_1 + r_2 + \ldots + r_m) = N \sum_{i=1}^{m} r_i \tag{4.38}$$

What are m operations. Then if we have all the r_i, $i = 1, 2, \ldots m$ equal, we obtain $N = rr \ldots r = r^m$ and we can obtain m from the logarithmic operation

$$m = \log_r N$$

Then, using this value with Eq. 4.38, we obtain the total number of operations needed as

$$T(r) = rN \log_r N$$

This argument justifies why the number of elements in the time series must be an integer exponent with base 2. Furthermore, we can still reduce the total number of operations if we take into account the periodicity of the twiddle factor. It's straightforward to check the following identity

$$e^{-i\frac{2\pi}{N}(k+N)} = e^{-i\frac{2\pi}{N}k} e^{-i\frac{2\pi}{N}N} = e^{-i\frac{2\pi}{N}k} e^{-i2\pi} = e^{-i\frac{2\pi}{N}k}$$

Therefore, the twiddle factor for element k returns to the same value N elements later, and this allows us to need an even smaller number of operations to obtain the DFT, according to this expression

$$T(r) = \frac{N}{2} \log_r N \tag{4.39}$$

Let see the influence of the FFT algorithm with respect to the raw analysis of the DFT. If we have a 5 s period sampled at 512 Hz (2^9), the number of elements shall be $N = 512 \times 5 = 2560$. Using DFT, we have $T_{\text{DFT}} = 2560^2 = 6553600$. However, using Eq. 4.37, $T_{\text{FFT}} = \frac{2560}{2} \log_2 2560 = 14490$. Therefore, the ratio between both numbers of operations is $\frac{T_{\text{DFT}}}{T_{\text{FFT}}} = \frac{6553600}{14490} > 452$

Thus, the FFT has reduced the number of necessary operations by up to a factor of 500, allowing spectral analysis to be performed in a very short time.

4.2.6 Spectral Analysis

The most common approach to characterize the Fourier transform is to present the power spectrum S of the given signal [4]. We know that each frequency component is given by the components obtained in Eq. 4.13, but, in those equations we have the imaginary term. To obtain a real value for each frequency component, we multiply c_n by its complex conjugate c_n^* (see Chap. 18).

We have the product (after eliminating the term ½)

$$c_n \times c_n^* = (a_n - ib_n) \times (a_n + ib_n) = a_n^2 + b_n^2 \tag{4.40}$$

Repeating the process for all frequencies ω_n, we obtain the power spectrum

$$S(\omega) = \frac{X(k)X^*(k)}{N} \tag{4.41}$$

The power spectrum is normalized by the number of data points from which it was obtained (N). This normalization is derived from Parseval's theorem, which states that the energy of the signal in the time domain is equal to the energy of the transformed signal in the frequency domain.

Since the Fourier transform has a range between $-\infty$ and $+\infty$ (Eq. 4.18), which appears when constructing the complex expression of the Fourier series (Eq. 4.14), negative frequencies also exist and therefore $S(\omega)$ is symmetric about the midpoint. Obviously, the negative frequencies have not real meaning and can be ignored.

Therefore, the spectral function on the y-axis shows the absolute value of the power (Eq. 4.21), measured in units of $\mu V^2/Hz$, as a function of each of the frequencies analyzed (x-axis). In practice, only the positive part of the frequency represents all information.

A related method is the amplitude spectrum, defined as

$$AS(\omega) = 2\sqrt{\frac{X(k)X^*(k)}{N^2}} \tag{4.42}$$

A third commonly used measurement is the phase spectrum, which depicts the phase (ϕ) versus frequency, where phase can be calculated as

$$\phi = \tan^{-1}\frac{Im(X)}{Re(X)} \tag{4.43}$$

where Im(X) and Re(X) indicate the imaginary and real parts of X. No normalization is required for the phase.

The information provided by S and AS is considerably similar, although AS is used in filter analysis rather than in qEEG. The phase spectrum is used in some tools that measure synchronization between pairs of signals.

It is very important to remember that there is an inverse relationship between period analyzed time series and resolution of the spectra. Suppose a time series of duration T and sampling frequency ν (Δt^{-1}). According to Eq. 4.25, we can obtain the value of frequency step $\Delta\omega$, which will be the maximum frequency. However, for this case, we must remember that the spectrum is a series with a positive and a negative part, which is specular and therefore does not provide useful information. This makes the frequency range over the positive part of the spectrum only half (Fig. 4.6a). This is why increasing the analyzed period reduces the frequency step in the power spectrum. In the Fig. 4.6b we show the difference of power spectra from the same channel obtained at 1 s length period (black line) and at 6 s length period (grey line). The last spectrum shows sharper spectral components.

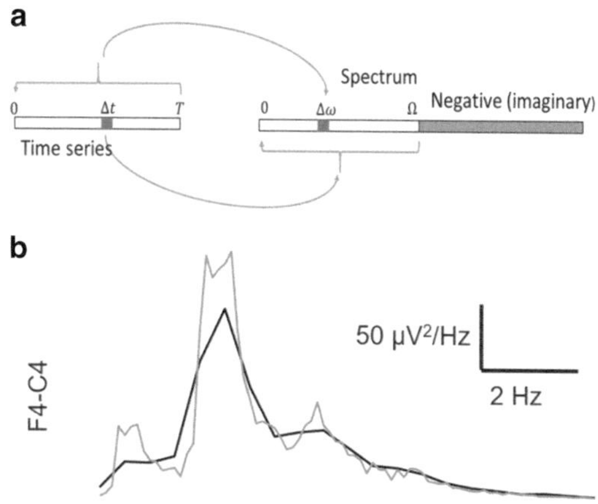

Fig. 4.6 Overview of the relationship between the properties of the time series (total period and sampling frequency) and the features of the spectra. (**a**) Example of time series showing specifically the step of frequency and total range of frequency and their relationships with power spectrum properties. In the last case, we must remember that only one half of spectrum (the positive) is useful, therefore, the range over the positive part of spectrum is half of the total. (**b**) Power spectra obtained with 1 s length (black) and 6 s length periods (grey)

However, this type of analysis suffers from various problems, such as fluctuations caused by inconsistencies in the statistical estimation of the periodogram [9] or the fact that this analysis applies mainly to stationary signals (whose probability density distribution does not change over time [10]), which is not the case for most of the

EEG signals. Therefore, to extend the power of spectral analysis signals to non-stationary or short-term signals, the analysis of time-frequency has been introduced, which consists of presenting functions that depend on both the frequency and time.

4.3 Density Spectral Array

Numerous techniques are available for time-frequency analysis (for a detailed description, see [9]), but a detailed analysis of these techniques is beyond the scope of this work. However, we will focus on two that are most commonly used: the Density Spectral Array (DSA) and wavelet transform (WT). We will start with the analysis of DSA.

The DSA (see Fig. 4.4a, also sometimes known as color DSA or CDSA) represents the change in the frequency composition of a signal over time. It is estimated using spectral analysis methods, such as classical nonparametric methods (FFT), model-based methods (autoregressive or moving average), or time-frequency methods, such as the SFT or WT [11–13]. In this type of graphic, one axis (horizontal in the example shown) represents the frequency, whereas the other (vertical) axis represents time. The recording is analyzed in fixed time windows from every periodogram. The spectral composition for each frequency and time is represented by a color code: red indicates the maximum power, and blue indicates the minimum power for the example shown (codes may vary in other cases). This analytical tool is useful, especially for neurophysiologists, allowing the easy identification of complex spectral changes over time at a single glance.

Figure 4.7 shows the two-channel DSA during an ECoG recording. The spectral difference between the two channels is evident, with a much larger frequency range in the upper channel. Furthermore, in this channel, the presence of epileptogenic activity shifts the spectral composition toward higher frequencies, while in periods without spikes, the frequency range is much narrower. This makes the frequency profile much more oscillatory or irregular than in the lower channel, where the frequency profile is more uniform.

These differences between DSA for different channels are very useful to identify the presence of epileptogenic activity in ECoG and although in recordings in ICU.

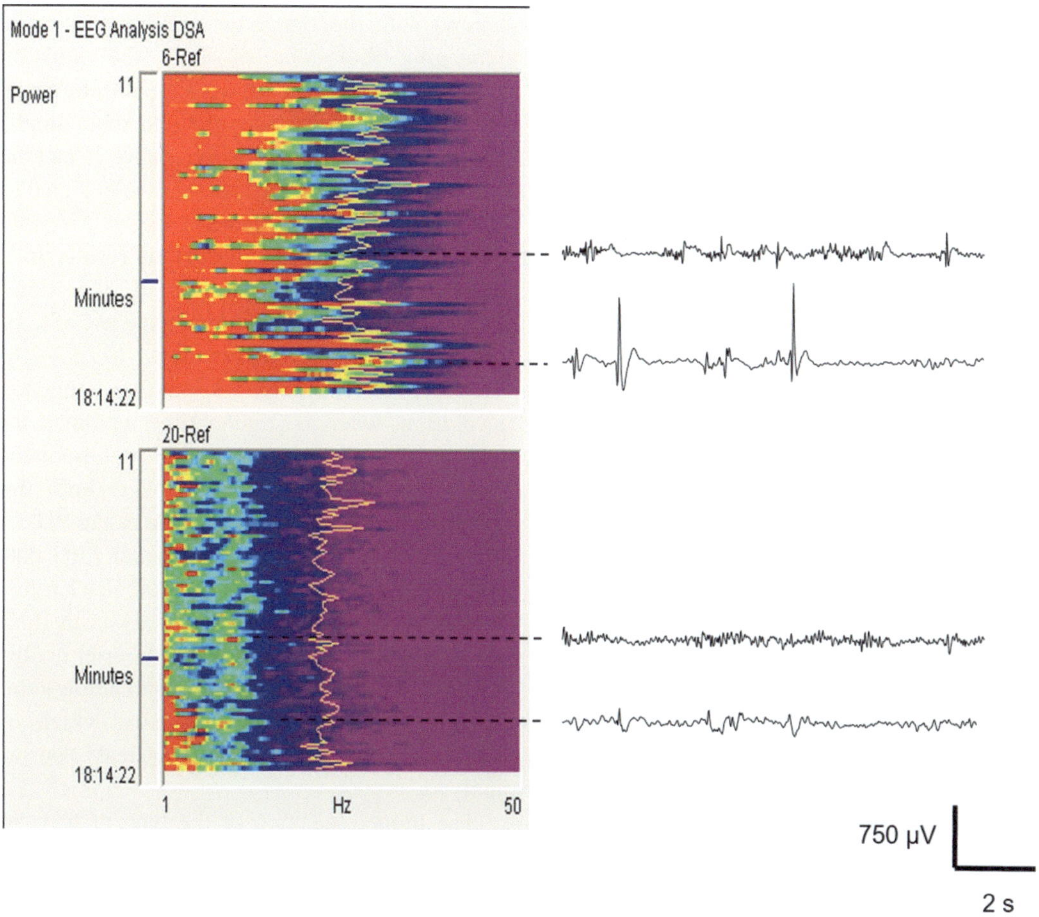

Fig. 4.7 Two-channel DSA during an ECoG recording. The top channel shows a channel with epileptogenic activity, while the bottom shows the presence of a lesion, with loss of bioelectrical activity due to the presence of an underlying tumor. For each channel, two different recording periods and the associated spectral moment are shown

4.4 Welch Method

The Welch method enhances resolution and reduces errors in results compared to the average period method. To do that, the method allows overlap between data segments and the rectangular window is changed by the Hamming window function ($\nu(n)$), which ameliorates the potential distortion [14].

Let's suppose we have $\{x_k(n)\}$, $k = 1, 2, \ldots S$ data segments, M length each. The overlapping coefficient is usually 50% (M/2). We compute the general average of window function of $\left|v_{(n)}^2\right|$ from

$$P = \frac{1}{M}\sum_{n=1}^{M}\left|v_{(n)}^2\right|^2 \tag{4.44}$$

Then, we estimate the spectrum for each k-segment

$$P_k(f) = \frac{1}{M}\frac{1}{P}\left|\sum_{n=1}^{M} v(n)x_k(n)e^{-i2\pi nf}\right|^2 \tag{4.45}$$

And, finally, the Welch spectrum estimate is given by

$$P_w(f) = \frac{1}{S}\sum_{k=1}^{S} P_k(f) \tag{4.46}$$

where f represents the sampling frequency.

In fact, the Welch is one of the most widely used power spectrum analysis methods because it reduces the influence of boundary effects and provides a more stable power spectral results than FFT/short-time Fourier transform (STFT) methods [15, 16].

4.5 Wavelet Analysis

The other widely used tool is the wavelet analysis (WT). Wavelet analysis is a generalization of the short-term Fourier transform that allows for basic functions that are more general than a sine or cosine wave.

The general expression for a WT is as follows [17]:

$$W(a,b) = \frac{1}{\sqrt{a}} \int_{-\infty}^{\infty} f(t) \Psi^{*} \left(\frac{t-b}{a} \right) dt \quad (4.47)$$

where Ψ^* represents the type of wavelet used for the transformation, and the asterisk indicates its complex conjugate. In addition, two parameters are used to define the transformation, such as the scale factor, a (which is >0), and b, the positional parameter. The main property of this transformation is that it is linear and does not depend on the translation or magnification. The operation within the integral is the convolution between the function $f(t)$ and the wavelet. It is the product of two functions, so to obtain a single numerical value, the product is integrated. For a detailed explanation of the convolution, see Chap. 19.

The form or type of wavelet varies [4], but the most frequent forms are the Haar, Daubeuchies, Morlet, and Mexican hat forms. As stated above, a convolution is performed (see below) with the signal for each wavelet scale (changing the value of a in the equation Eq. 4.46). For subsequent scales, different values are color-coded according to their amplitude. Thus, by changing the scale, it can be repeatedly applied to the same wavelet signal to cover large frequency ranges.

Therefore, this transformation results in a more flexible representation of the time-frequency relationship, allowing the use of temporal analysis variable windows [9, 17–20], such that large windows allow a finer frequency resolution, whereas smaller time windows are used to increase the resolution of high frequencies. This flexibility, which covers both the lower and higher frequencies, makes the WT an ideal tool for analyzing paroxysmal EEG patterns, such as ictal and interictal discharges. Therefore, a multi-resolution analysis method avoids fixed frequency classical Fourier analysis and similar methods [21] and allows the analysis of non-stationary signals, which in principle is prohibited in classical Fourier analysis.

The proper selection of the type of wavelet and the number of levels to be tested is very important for the decomposition of the signal. For a suitable choice, the dominant frequency of the signal must be considered.

Figure 4.8 shows the time-frequency diagrams for the right hemisphere and midline channels of two waveforms, displayed on the Fp2-F4 and Fz-Cz graphs. These waveforms are K complexes obtained during NREM N2 sleep (left column) and during basal wakefulness (right column). It is evident that the spatiotemporal composition of both waveforms is the same for all the channels analyzed.

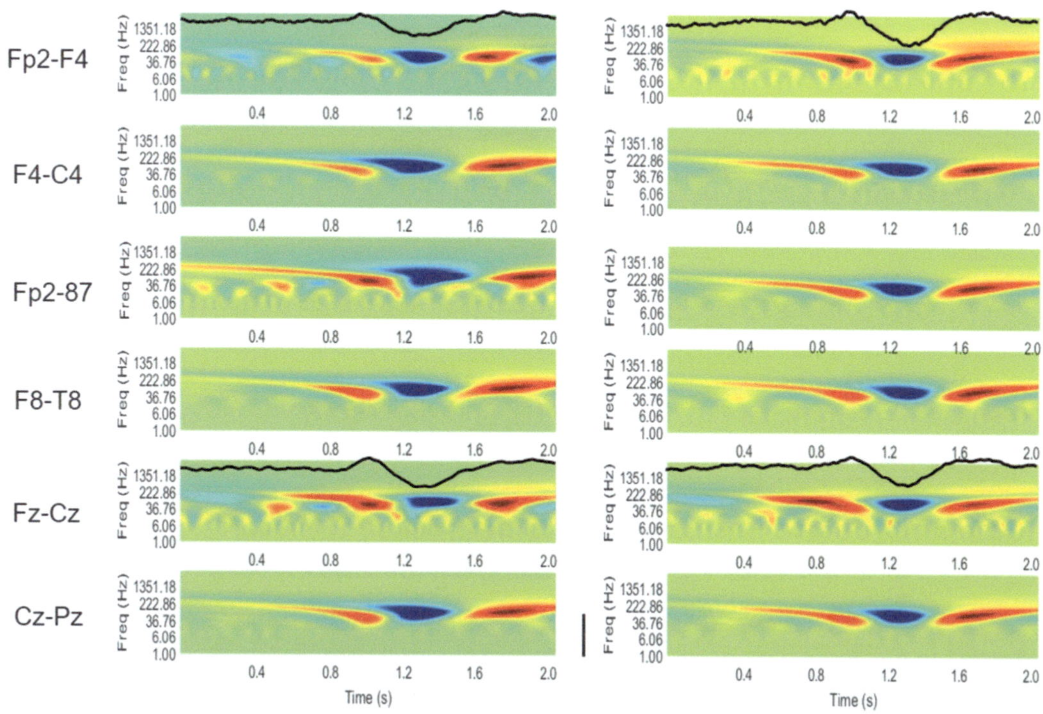

Fig. 4.8 Time-frequency wavelet analysis of a K-complex recorded during sleep (left column) and a graphoelement with a similar morphology recorded during wakefulness. Both are shown as a black line. The vertical calibration bar at the bottom of the figure indicates 100 μV

4.6 Concluding Remarks

Frequency-domain analysis tools are probably the most widely used in qEEG and, certainly, in those packages used in clinical practice. Therefore, we have dedicated this chapter to an in-depth analysis of the Fourier transform. There are other tools, many of them based on it, although other non-Fourier-based tools are increasingly applied to EEG.

Wavelet analysis is another widely used tool in time-frequency analysis. The graphical information it provides is probably more comprehensive than that of Fourier analysis. However, its computational requirements make it difficult to apply to multi-channel, long-duration recordings.

References

1. Vega-Zelaya L, Garnés-Camarena O, Sánz-García A, Ortega GJ, Pastor J. Mathematical foundations of quantified electroencephalography. Clinical Advances in Neurophysiology. Chapter 3. Avid Science. February 10, 2016.
2. Arfken GB, Weber HJ. Fourier series. Mathematical methods for physicists. 6th ed. Amsterdam: Elsevier Academic Press; 2005. ISBN 978-0-12-059876-2.
3. Sanei S, Chambers JA. Signal transforms and joint time–frequency analysis. In: EEG Signal Processing. London: Wiley&Son Ltd; 2007. ISBN-13 978-0-470-02581-9.
4. Van Drongelen W. Continuous, discrete and Fast Fourier transform. In: Signal processing for neuroscientists. Amsterdam: Elsevier; 2007. ISBN 13:978-0-12-370867-0.
5. Edwards CH, Penney DE. Linear Equations of Higher Order. Elementary differential equations. 6th ed. London: Prentice Hall; 2008. ISBN 0-13-239730-7.

6. Gabbiani F, Cox SJ. Mathematics for Neuroscience. Amsterdam: Elsevier; 2010. ISBN 978-0080890494.
7. Thakor NV, Shanbao T. Advances in quantitative electroencephalogram analysis methods. Annu. Rev. Biomed. Eng. 2004;6:453–95. https://doi.org/10.1146/annurev.bioeng.5.040202.121601.
8. Cooley JW, Tukey JW. An algorithm for machine calculation of complex Fourier series. Math Computation. 1965;19:297–301.
9. Hadjiyannakis K, Ogilvie RD, Alloway CE, Shapiro C. FFT analysis of EEG during stage 2-to-REM transitions in narcoleptic patients and normal sleepers. Electroencephalogr Clin Neurophysiol. 1997;103(5):543–53. https://doi.org/10.1016/s0013-4694(97)00064-3.
10. Sun J, Cao R, Zhou M, Hussain W, Wang B, Xue J, Xiang J. A hybrid deep neural network for classification of schizophrenia using EEG Data. Sci Rep. 2021;11(1):4706. https://doi.org/10.1038/s41598-021-83350-6. PMID: 33633134; PMCID: PMC7907145.
11. Wang Y, Sokhadze EM, El-Baz AS, Li X, Sears L, Casanova MF, Tasman A. Relative power of specific EEG bands and their ratios during neurofeedback training in children with Autism Spectrum Disorder. Front Hum Neurosci. 2016;9:723. https://doi.org/10.3389/fnhum.2015.00723. PMID: 26834615; PMCID: PMC4712412.
12. Perumalsamy V, Sankaranarayanan S, Rajamony S. Sleep spindles detection from human sleep EEG signals using autoregressive (AR) model: a surrogate data approach. J Biomed Sci Eng. 2009;02:294–303. https://doi.org/10.4236/jbise.2009.25044.
13. Saidatul A, Paulraj MP, Yaacob SB, Yusnita MA. Analysis of EEG signals during relaxation and mental stress condition using AR modelling techniques. In: IEEE International Conference on Control System, Computing and Engineering. Penang, Malaysia; 2011. p. 477–81. https://doi.org/10.1109/ICCSCE.2011.6190573.
14. Günes S, Polat K, Yosunkaya S. Efficient sleep stage recognition system based on EEG signal sing k-means clustering based feature weighting. Expert Syst Appl. 2010;37(12):7922–8. https://doi.org/10.1016/j.eswa.2010.04.043.
15. Zhang H, Zhou QQ, Chen H, Hu XQ, Li WG, Bai Y, Han JX, Wang Y, Liang ZH, Chen D, Cong FY, Yan JQ, Li XL. The applied principles of EEG analysis methods in neuroscience and clinical neurology. Mil Med Res. 2023;10(1):67. https://doi.org/10.1186/s40779-023-00502-7. PMID: 38115158; PMCID: PMC10729551.
16. Allen J. Short term spectral analysis, synthesis, and modification by discrete Fourier transform. IEEE Transactions on Acoustics, Speech, and Signal Processing. 1977;25(3):235–8. https://doi.org/10.1109/TASSP.1977.1162950.
17. Akin M. Comparison of wavelet transform and FFT methods in the analysis of EEG signals. J Med Syst. 2002;26(3):241–7. https://doi.org/10.1023/a:1015075101937.
18. Lawhern V, Hairston WD, McDowell K, Westerfield M, Robbins K. Detection and classification of subject-generated artifacts in EEG signals using autoregressive models. J Neurosci Methods. 2012;208(2):181–9. https://doi.org/10.1016/j.jneumeth.2012.05.017. Epub 2012 May 23.
19. Mousavi SR, Niknazar M, Vahdat BV. Epileptic seizure detection using AR model on EEG signals. In: Cairo International Biomedical Engineering Conference. Cairo, Egypt; 2008. p. 1–4. https://doi.org/10.1109/CIBEC.2008.4786067.
20. Sanei S, Chambers JA. Wavelet transform. In: EEG Signal Processing. London: Wiley&Son Ltd.; 2007. ISBN-13 978-0-470-02581-9.
21. Wijaya SK, Badri C, Misbach J, Soemardi TP, Sutarmo V. Electroencephalography (EEG) for detecting acute ischemic stroke. In: 4th International Conference on Instrumentation, Communications, Information Technology, and Biomedical Engineering (ICICI-BME). Bandung, Indonesia; 2015. p. 42–8. https://doi.org/10.1109/ICICI-BME.2015.7401312.

5 Measurements of Synchronization

5.1 Introduction

The word synchronization comes from two Greek words, χρονοζ (chronos) and συν (same), which means "sharing the same time"; therefore, a synchronized event is always composed by the temporal coincidence of two or more actions. Nevertheless, it is usually understood that the existence of an underlying mechanism is responsible for synchronization itself. In this sense, synchronization is assumed differently from chance, because no deterministic causal effect exists in the last one. On the other hand, the use of the concept of synchronization is generally associated with a mechanism, maybe unknown, that makes possible the temporal coincidence [1].

In general, synchronization between two systems is defined as the adjustment of their internal rhythms due to an existing (weak) interaction between them [2]. Therefore, the following ingredients are essential for synchronization:

1. There must exist two or more self-sustained oscillators, i.e., systems capable of generating their own rhythms.
2. The systems adjust their own rhythms due to a (weak) interaction between them, and
3. The adjustment of rhythms occurs in a certain range of systems' mismatch; in particular, if the frequency of one oscillator is slowly varied, the second system follows this variation.

The presence of self-sustained oscillators is needed for synchronization. This requirement is essential to differentiate actual synchronization from other phenomena, for instance, resonance. Two isolated different self-sustained oscillators, with different intrinsic frequencies, can oscillate at the same frequency when they interact with each other. Note that synchronization refers to a dynamical process instead of a stationary state. This means that the interaction allows (generally small) variations of the intrinsic rhythms, but always try to reach the common frequency.

Nowadays, in fact, there is no more a single synchronization phenomenon, but instead, the traditional term has been split into several, more specific terms to characterize the numerous forms of the underlying mechanism and also the different kinds of synchronized objects. New terms as lag or full synchronization are slowly emerging. Its detailed description is out of the scope of this chapter. The reader can be referred to [1] for a detailed description.

There are several methods for evaluating synchronization. It is often difficult to justify the choice of a specific method. Furthermore, each group often uses idiosyncratic variations in their implementation, making it difficult to compare results.

The relevant point here is that it is generally accepted that the brain could dynamically coordinate the flow of information by changing the

J. Pastor, *Quantitative EEG Practical Guide*, https://doi.org/10.1007/978-3-032-05457-9_5

strength, pattern, or frequency with which different brain areas participate in oscillatory synchrony. This hypothesis has resulted in an increase in the methods to evaluate synchrony.

Three types of connectivity can be distinguished [3]:

1. Anatomical, which shows the existence of anatomical pathways between two regions. Their presence does not presuppose the directionality or intensity of the connections.
2. Functional, defined as statistical dependencies among remote neurophysiological events.
3. Effective, defined as the simplest possible circuit diagram that would replicate the observed timing relationships between the recorded regions. Therefore, effective connectivity is dynamic (activity-dependent), and depends on a model of interactions or coupling.

It is common to identify synchronization between two regions, without attempting to study the direction of that communication. However, other tools analyze functional connectivity in detail. The use of complex tools and/or those requiring high-quality recording is difficult to implement in daily clinical practice, so it is usually sufficient to establish the existence of synchronization between regions and measure its intensity.

Methods for assessing functionality are divided into two classes, initially, depending on whether or not they allow for the determination of connectivity direction. That is, there are directed methods, which establish the direction of connectivity, and non-directed methods, which do not. Furthermore, each of these types is subdivided into other categories, depending on whether or not they use the assumption of a model. Furthermore, each of these divisions can be analyzed in both the time and frequency domains.

In the next table, we can see some of the most representative methods and the cleavage properties of classification (Table 5.1).

Connectivity analysis includes multiple types, such as signal morphology-based, signal phase-based, statistics-based, and information-based analyses [4].

Table 5.1 Classification of the most used methods for quantifying connectivity

Class	Model-based?	Time domain	Frequency domain
Non-directed	Yes	Correlation	Coherence Phase synchronization
	No	Mutual information	
Directed	Yes	Cross-correlation	Phase slope index
	Yes	Granger causality	
	No	Transfer entropy	

5.2 Correlation

Correlation coefficient (r_{xy}) measures the similarity between the signals x and y by calculating the covariance of signals. The simplest non-directed measure of linear dependence is Pearson's correlation coefficient, defined as the covariance (s_{jk}), normalized by the standard deviation of both vectors (s_j, s_k) [5]. The r is as follows: [6]

$$r_{xy} = \frac{\Sigma(x_i - \bar{x})(y_i - \bar{y})}{\sqrt{\Sigma(x_i - \bar{x})^2 (y_i - \bar{y})^2}} \tag{5.1}$$

5.3 Cross-correlation

It is a metric of the similarity between two signals x and y, frequently used to find relevant characteristics in one signal by comparing it with another one [7, 8]. Its value is obtained from the convolution of two functions, one of them displaced an interval of time τ.

$$C_{xy}(\tau) = \int_{t_0}^{t_1} x(t)\, y(t-\tau)\, dt \tag{5.2}$$

Cross-correlation identifies time-dependent synchronization features. Fig. 5.1, left column, shows two synchronous waveforms (which could correspond to autocorrelation), and some examples of the shift of $y(t - \tau)$ around $x(t)$ are shown. Figure 5.1c, left column, shows how the cross-

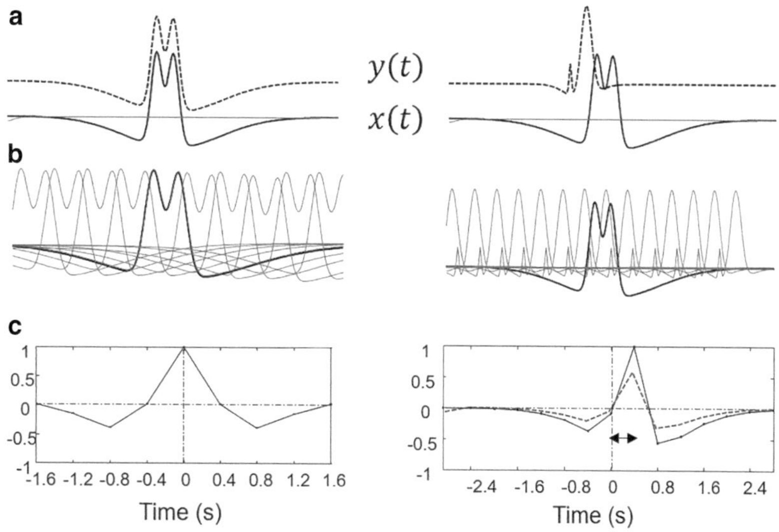

Fig. 5.1 Cross-correlation for two equal and synchronous signals (left column) and two different and asynchronous signals (right column). (**a**) Signals *x* and *y* are not superimposed for clarity. (**b**) Different $y(t\text{-}\tau)$ signals (grey) and the $x(t)$ signal (black). (**c**) Correlation of both signals as a function of τ

correlation graph is symmetrical with max lag equal to zero.

However, the right-hand column shows two waveforms that are neither equal nor synchronous. In section C of this same column, we can see how the cross-correlation graph is asymmetrical, and the maximum lag is shifted from zero (double horizontal arrow). The correlation between the waveforms shown in C is normalized, meaning that the maximum correlation will not necessarily be equal to one. In fact, the dashed gray line in section C on the right shows the actual correlation as a function of τ.

Therefore, although it is not a strictly directional measure, cross-correlation can be used to estimate the direction of interaction between two signals.

5.4 The Coherence, Coherency and the Phase Difference (PD) Spectrum

The coherence (coh) is the frequency domain equivalent to the time domain cross-correlation function [9]. Recall that the FFT is a complex series (see Chap. 4). Therefore, for the signals *x* and *y*, we will have the transforms

$$X(\omega) = x_1 + ix_2 \text{ and } Y(\omega) = y_1 + iy_2 \quad (5.3)$$

To obtain a useful magnitude, we need the values to be real, which we achieve through the concept of complex conjugate (see Chap. 18). The power spectra are obtained from these expressions [10]

$$\begin{aligned} S_{xx}(\omega) &= X(\omega)X^*(\omega) = (x_1 + ix_2)(x_1 - ix_2) = x_1^2 + x_2^2 \\ S_{yy}(\omega) &= Y(\omega)Y^*(\omega) = (y_1 + iy_2)(y_1 - iy_2) = y_1^2 + y_2^2 \end{aligned} \quad (5.4)$$

We can obtain another complex magnitude, which is called the cross-spectrum $S_{xy}(\omega)$

$$\begin{aligned} S_{xy}(\omega) &= X(\omega)Y^*(\omega) = (x_1 + ix_2)(y_1 - iy_2) \\ &= (x_1y_1 + x_2y_2) - i(x_1y_2 + x_2y_1) \end{aligned} \quad (5.5)$$

To obtain a real value, we can find the square of the modulus of a complex number (see Chap. 18)

$$\begin{aligned} \left|S_{xy}(\omega)\right|^2 = S_{xy}(\omega)S_{xy}^*(\omega) &= (x_1y_1 + x_2y_2)^2 \\ &\quad - (x_1y_2 + x_2y_1)^2 \end{aligned} \quad (5.6)$$

Therefore, coherence is defined as

$$\mathrm{coh}_{xy}(\omega) = \frac{\left|S_{xy}(\omega)\right|^2}{S_{xx}(\omega)S_{yy}(\omega)} \quad (5.7)$$

However, if we substitute the values obtained in equations Eq. 5.2 and Eq. 5.4, we will always obtain a value equal to one, which is useless. Therefore, in practice, coh is calculated by averaging over n epochs ($\langle \cdots \rangle_n$), so the previous equation is modified in this way

$$\mathrm{coh}_{xy}(\omega) = \frac{\left|S_{xy}(\omega)_n\right|^2}{S_{xx}(\omega)_n S_{yy}(\omega)_n} \quad (5.8)$$

It is important to remember that the average must be obtained before the absolute value, because otherwise the average value will be one [10].

Coherence is very sensitive to the phase changes of the signal but is little affected by energy changes [4]. Its squared value quantifies, as a function of frequency, the amount of variance in one of the signals that can be explained by the other signal, or vice versa, in analogy to the squared correlation coefficient in the time domain [11].

Complex numbers can also be represented in polar form. In the case of the signals analyzed, their expression would be

$$X(\omega) = A_1 e^{i\varphi_1} \text{ and } Y(\omega) = A_2 e^{i\varphi_2} \quad (5.9)$$

Therefore, the cross-spectrum would be

$$S_{xy}(\omega) = A_1 e^{i\varphi_1} A_2 e^{-i\varphi_2} = A_1 A_2 e^{i(\varphi_1 - \varphi_2)} \quad (5.10)$$

If we do not take the absolute value for this magnitude, we will have a complex value in which the phase is the difference between the phases of the original signals. This magnitude is called

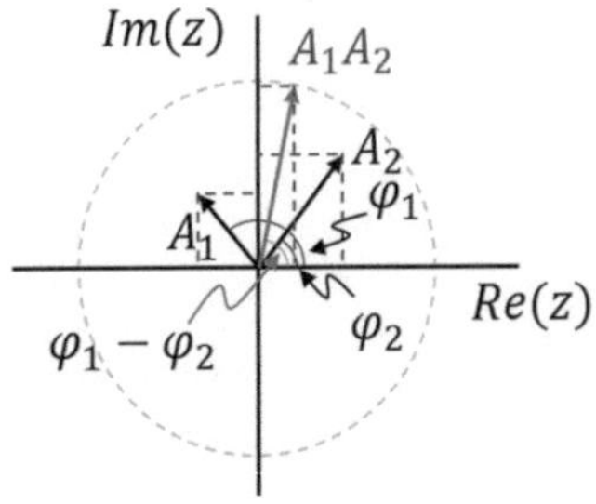

Fig. 5.2 Computing of phase lag from a pair of time series

coherency [11]. Computing the phase difference over a range of frequencies may give an estimate of the time delay between signals (Fig. 5.2). Then, the rhythmic interaction can be described as a predominantly unidirectional time-lagged linear interaction. By repeating the process for a large number of pairs and frequencies, we can obtain the difference spectrum slope. This interpretation, however, is only acceptable under the assumptions that the interaction is predominantly unidirectional and that it is a linear process [12].

5.5 Phase Lag Value (PLV) and Phase Lag Index (PLI)

Both metrics can be applied to assess the strength of phase synchronization [13, 14]. The instantaneous phase of is $x(t)$ generated according to this expression CONTINUAR

$$\varphi_x(t) = \tan^{-1}\frac{\tilde{x}(t)}{x(t)} \quad (5.11)$$

where $\tilde{x}(t)$ is the Hilbert transform, defined as follows:

$$\tilde{x}(t) = \frac{1}{\pi}\mathrm{PV}\int_{-\infty}^{\infty}\frac{x(t)}{t-\tau}dt \quad (5.12)$$

PV refers to the Cauchy principal value, a method for assigning values to improper integrals. The relative phase between signals can be computed as [15]

$$\Delta\varphi(t) = \arg\left(\frac{\left(x(t)+i\tilde{x}(t)\right)\left(y(t)+iy(t)\right)^*}{\left|\left(x(t)+i\tilde{x}(t)\right)\right|\left|y(t)+iy(t)\right|}\right) \quad (5.13)$$

The PLV is defined as

$$\mathrm{PLV}(t) = \left|E\left[e^{i\Delta\varphi(t)}\right]\right| \quad (5.14)$$

Where $E(.)$ Indicates the expected value (see Chap. 22).

PLV is a metric of the strength of synchronization, with a value of zero for lack of synchronization and one for a perfect synchronization.

PLI characterizes the asymmetry in the phase difference distribution between both signals and is defined as

$$\text{PLI}(t) = \left| E\left[sign\left(\Delta\varphi(t)\right)\right]\right| \quad (5.15)$$

5.6 Mutual Information (MI)

Represents how one signal provides information for another signal [6, 16]. For both signals $x(t)$ and $y(t)$, we can obtain the probability density functions, $P(x)$ and $P(y)$. The entropy is defined as follows:

$$H(x) = -\sum_{i=1}^{N} P(x_i)\log P(x_i) \text{and } H(y) = -\sum_{i=1}^{N} P(y_i)\log P(y_i) \quad (5.16)$$

We can define the conditional and the joint entropy, respectively, following these formulas:

$$H(x \mid y) = -E_x\left[E_y\left[\log P(y \mid x)\right]\right] \quad (5.17)$$

$$H(x,y) = -E_x\left[E_y\left[\log P(x,y)\right]\right] \quad (5.18)$$

Finally, the MI can be calculated as follows:

$$\text{MI}(x,y) = H(x) + H(y) - H(x,y) = H(y) - H(y|x) \quad (5.19)$$

Interestingly, MI can detect linear and nonlinear correlations between the signals. The main problem is that it requires mass data, which is not always available in EEG recordings.

5.7 Granger Causality (GC)

This is an auto-regressive model based on random time-series data, which can estimate the effective interactions and the directionality [17–19]. Unfortunately, GC can only provide information on the linear features of the signal and cannot offer information about nonlinear situations [4].

5.8 Some Pitfalls of Functional Connectivity

The knot of the problem is to determine whether the connectivity estimate reflects a genuine effect in terms of interaction between different brain regions or is an artefact [11].

Synchronization analysis in conventional EEG is far from a univocal method and is not free from interpretation problems. Therefore, the presence of spurious functional connectivity can result from the use of a common reference electrode. The use of a reference montage, whether average, midline, or a single cephalic electrode (either Cz or mastoid) can induce spurious synchronizations that do not correspond to brain activity, but rather to the effect that the same electrode is influencing the activity measured in several channels.

The best solution to this problem is to use bipolar leads to eliminate the effect of the common reference electrode [20]. This solution is sufficient to eliminate the fact that any device's recording is done with a hardware reference setup. However, the hardware reference is equally present in all channels and, therefore, has no specific effect on one or a group of them, so it can be removed.

The other major drawback to determining synchronization is volume conduction. This effect is due to the propagation of currents through the tissues surrounding the source of neural activity [21]. These currents can propagate to distant regions and affect numerous electrodes simultaneously. In the worst case, this can generate spurious synchronization. An important property of this volume conduction is that its effects are immediate; that is, the influence on spatially distant electrodes occurs at the same time and is therefore maximum at zero lag [21, 22].

The two strategies that we can use in the field of EEG are (i) trying to separate the measured signal into its underlying sources and (ii) using a connectivity metric that privileges interaction in phases other than 0 or 180°, such as the imaginary part of coherence, PLI, or the phase slope index.

5.9 Concluding Remarks

Synchronization measurements are of great importance for assessing the physiological or pathological state of the brain. Visual analysis of the EEG allows for the identification of numerous voltage-related magnitudes, but it is much more difficult to determine the degree of synchronization between separate regions. Therefore, these measurements can be crucial for determining clinical states.

There are multiple metrics for assessing connectivity, but their use is not perfectly defined, depending on the preferences of different authors. In our case, we opted those simple and robust tools that can be used in patients in whom it is assumed that the recordings do present optimal quality.

References

1. Pastor J, Sola RG, Ortega GJ. Hyper-synchronization, de-synchronization, synchronization and seizures. In: Epileptic seizures. InTech; 2012. p. 117–44. ISBN 978-953-307-737-6.
2. Pikovsky A, Rosenblum M, Kurths J. Synchronization: a universal concept in nonlinear sciences. Cambridge Nonlinear Science Series; 2001.
3. Friston KJ. Functional and effective connectivity: a review. Brain Connect. 2011;1(1):13–36. https://doi.org/10.1089/brain.2011.0008.
4. Zhang H, Zhou QQ, Chen H, Hu XQ, Li WG, Bai Y, Han JX, Wang Y, Liang ZH, Chen D, Cong FY, Yan JQ, Li XL. The applied principles of EEG analysis methods in neuroscience and clinical neurology. Mil Med Res. 2023;10(1):67. https://doi.org/10.1186/s40779-023-00502-7. PMID: 38115158; PMCID: PMC10729551.
5. Peña D. Descripción de datos multivariantes. In: Análisis de datos multivariantes. Madrid: McGraw-Hill; 2002. ISBN 84-481-3610-1.
6. Ortega GJ, Menéndez de la Prida L, Sola RG, Pastor J. Synchronization clusters of interictal activity in the lateral temporal cortex of epileptic patients: intracranial analysis. Epilepsia. 2008;49(2):269–80. https://doi.org/10.1111/j.1528-1167.2007.01266.x.
7. Abdullah H, Maddage NC, Cosic I, Cvetkovic D. Cross-correlation of EEG frequency bands and heart rate variability for sleep apnoea classification. Med Biol Eng Comput. 2010;48(12):1261–9. https://doi.org/10.1007/s11517-010-0696-9. Epub 2010 Nov 3.
8. Kumagai Y, Arvaneh M, Okawa H, Wada T, Tanaka T. Classification of familiarity based on cross-correlation features between EEG and music. Annu Int Conf IEEE Eng Med Biol Soc. 2017;2017:2879–82. https://doi.org/10.1109/EMBC.2017.8037458.
9. Gabbiani F, Cox SJ. Power and Cross Spectra. In: Mathematics for Neuroscientist. London: Academic Press; 2010. ISBN 978-0-12-374882-9.
10. Van Drongelen W. LTI systems, convolution, correlation and coherence. In: Signal processing for neuroscientists. Amsterdam: Elsevier; 2007. ISBN 13:978-0-12-370867-0.
11. Bastos AM, Schoffelen JM. A tutorial review of functional connectivity analysis methods and their interpretational pitfalls. Front Syst Neurosci. 2016;9:175. https://doi.org/10.3389/fnsys.2015.00175. PMID: 26778976; PMCID: PMC4705224.
12. Friston KJ, Bastos A, Litvak V, Stephan KE, Fries P, Moran RJ. DCM for complex-valued data: cross-spectra, coherence and phase-delays. Neuroimage. 2012;59(1):439–55. https://doi.org/10.1016/j.neuroimage.2011.07.048. Epub 2011 Jul 28. PMID: 21820062; PMCID: PMC3200431.
13. van Mierlo P, Papadopoulou M, Carrette E, Boon P, Vandenberghe S, Vonck K, et al. Functional brain connectivity from EEG in epilepsy: seizure prediction and epileptogenic focus localization. Prog Neurobiol. 2014;121:19–35.
14. Vasudeva B, Tian R, Wu DH, James SA, Refai HH, Ding L, He F, Yang Y. Multi-phase locking value: a generalized method for determining instantaneous multi-frequency phase coupling. Biomed Signal Process Control. 2022;74:103492. https://doi.org/10.1016/j.bspc.2022.103492. Epub 2022 Jan 7. PMID: 35111233; PMCID: PMC8803274.
15. Aydore S, Pantazis D, Leahy RM. A note on the phase locking value and its properties. Neuroimage. 2013;74:231–44. https://doi.org/10.1016/j.neuroimage.2013.02.008. Epub 2013 Feb 19. PMID: 23435210; PMCID: PMC3674231.
16. Tafreshi TF, Daliri MR, Ghodousi M. Functional and effective connectivity based features of EEG signals for object recognition. Cogn Neurodyn. 2019;13(6):555–66. https://doi.org/10.1007/s11571-019-09556-7. Epub 2019 Oct 1. PMID: 31741692; PMCID: PMC6825113.
17. Seth AK, Barrett AB, Barnett L. Granger causality analysis in neuroscience and neuroimaging. J Neurosci. 2015;35(8):3293–7. https://doi.org/10.1523/JNEUROSCI.4399-14.2015. PMID: 25716830; PMCID: PMC4339347.
18. Sanz-García A, Pérez-Romero M, Pastor J, Sola RG, Vega-Zelaya L, Monasterio F, Torrecilla C, Vega G, Pulido P, Ortega G. Identifying causal relation-

ships between EEG activity and intracranial pressure changes in neurocritical care patients. J Neural Eng. 2018;15(6).
19. Liu S, Molenaar P. Testing for Granger causality in the frequency domain: a phase resampling method. Multivar Behav Res. 2016;51(1):53–66. https://doi.org/10.1080/00273171.2015.1100528.
20. Pastor J, Vega-Zelaya L. Normative structure of resting-state EEG in bipolar derivations for daily clinical practice: a pilot study. Brain Sci. 2023;13(2):167. https://doi.org/10.3390/brainsci13020167. PMID: 36831710; PMCID: PMC9953767.
21. Schoffelen JM, Gross J. Source connectivity analysis with MEG and EEG. Hum Brain Mapp. 2009;30(6):1857–65. https://doi.org/10.1002/hbm.20745. PMID: 19235884; PMCID: PMC6870611.
22. Winter WR, Nunez PL, Ding J, Srinivasan R. Comparison of the effect of volume conduction on EEG coherence with the effect of field spread on MEG coherence. Stat Med. 2007;26:3946–57. https://doi.org/10.1002/sim.2978.

6 A Proposal for an Analytic System: CRANE

6.1 Introduction

Despite a cheaper and well-established method to evaluate the bioelectrical brain function, there is a factual understanding that EEG is mainly suitable to epilepsy, and is also reputed as a subjective method, depending strongly on the interpreter [1–4]. There has been an attempt to reduce inter-rater variability among interpreters by the introduction of a consensus for EEG interpretation [5–9]. Although these consensus and classifications introduce some objectivity in EEG analysis and increase the interchangeability of information, the variables are mostly qualitative or binary (present/absent), and specificity is therefore not very high.

To minimize these flaws, in the last decades, it has been sought by clinicians, neuroscientists, mathematicians, physicists, and engineers (among other experts) a way to overcome these defects, increasing the true field of EEG application and reducing subjectivity in diagnosis. A straightforward direction has been to develop mathematical tools to increase the consistency and deepen the information extracted by neurophysiological recordings. Collectively, this approach is called quantitative EEG (qEEG [10]). There is no canonical formulation of toolbox needed to define a qEEG. Instead, we can say that every EEG recording that use any kind of post-hoc numerical method to obtain a result about frequency spectrum, synchronization, network dynamics, or anything else bioelectrical magnitude can be considered as qEEG. This approach has been rising its popularity progressively after the introduction of digitalization in electroencephalography.

Numerous methods have been used in qEEG, although scarcely in daily clinical practice (see Chap. 13 for a detailed discussion). There is, therefore, a huge gap between the promising (even spectacular) results obtained with qEEG and its practical usefulness. To the best knowledge of authors this issue has not been systematically addressed, although it has been said that electroencephalographers are poorly trusting in mathematical models [10, 11].

The degree of mathematical complexity and abstraction is very different too. Obviously, not all the mathematical models can be included in the same category and it is of extreme importance that mathematical solutions to be robust with respect to physiological assumptions [10]. Then, some procedures imply relatively simple and straightforward methods as FFT [12–16] and synchrony measurements [17, 18], methods that clinicians are familiar with. On the contrary, other numerical methods are more complex or mathematical approaches are far away from physiological assumptions [19–22] and, probably this takes neurophysiologists out of its comfort zone.

Numerical analysis in clinical practice seems a true necessity because it can offer a greater exactness in comparison to visual analysis, diminishing the degree of the physician's subjectivity and even allowing the possibility of

J. Pastor, *Quantitative EEG Practical Guide*, https://doi.org/10.1007/978-3-032-05457-9_6

other diagnoses not accessible by naked eye evaluation. However, despite this necessity, the commercial offer of comprehensive applications to perform qEEG is quite restricted.

A practical device to be used in the routine clinical practice must satisfy some features. Among these, we would outline:

- Relatively simple mathematics. The definition of what this would be will always be an open debate. Nevertheless, every physician should be comfortable with the level of high school mathematics. Too abstract and elaborate numerical methods, although mathematical and computational justified, can yield results difficult to understand (e.g., dimensionless, hypersphericity, multi-dimensional regions in configuration spaces, etc.). Although well correlated with classification methods, probably paradigms like this would be more fitted to highly trained professionals in mathematics (e.g., physicists, engineers, mathematicians, etc.) than clinical practitioners. This is the approach and goal used in this book. It should be desirable that every expert can understand and justify the main lines of the techniques he/she use.
- Friendly and straightforward interpretation. As stated above, highly abstract results can be difficult to understand and put into the context of the pathology of the patient. For example, what meaning does a change in the Lyapunov exponent have for a patient with a low level of consciousness? On the contrary, a decrease in the power spectrum of the alpha and beta bands, associated with an increase in delta power can be easily understood and explained in the context of what we know about encephalopathies. Evidently, the degree of abstractness and complexity would depend on the particular training of every professional.
- Easily associated with or correlated with the brain patho-physiology. In the line of the above discussion, it should be searched that any method used to quantify could be easily translated and explained in pathophysiological foundations. Equivalence of concepts between different fields of science can sometimes be difficult to reach but it's very important because it can clarify the meaning and even serve as a heuristic method. Therefore, any method should be founded (as much as possible) in the physiology of the CNS.
- Relatively unconstrained. Despite all the comments mentioned above, it should be important that any system like this, well-grounded in patho-physiology of the brain, founded in sound but reliable mathematics and producing results easy to interpret, could be flexible enough to address the specific necessities that can change from patient to patient.

Various numerical systems, toolboxes, or even more or less complex teams of clinicians and engineers can fulfill these features. With this picture in mind, we developed our own method that we have been systematically using since 2016. In our department, we were familiar with numerical methods long before. Nonetheless, we used these methods always from the scientific point of view [23–25]. To find a funny and easily remembered name, we called Cerebral Research and ANalytical EEG (CRANE) to the system.

Nevertheless, there are several software packages available for download to perform numerical EEG analysis. Only a few of the most popular are mentioned [26].

- CARTOOL is software designed for the spatiotemporal analysis of EEG.
- EEGLAB is a toolbox that enables numerous statistical and dynamic analyses.
- Fieldtrip is a general-purpose software that includes tools for connectivity, sources, and time-frequency analysis, among many others.
- One of the most widely used commercial packages is Persyst. See [27] for an excellent description of its use in ICU patients.
- Other packages are LORETA, EEGNET, eConnectome, Epilab, etc.

The method used daily in our hospital will be described below.

6.2 Overview of CRANE

We have used the assumption that EEG is founded in a homeostatic system to obtain the main variables of our method [28, 29]. This aspect is of extraordinary relevance, because its specific involvement in different pathologies will lead to specific changes in the different numerical variables obtained. The EEG is the multivariate spatio-temporal determination of the electrical potentials generated by the brain and recorded on the surface of the scalp. The oscillatory activity of the EEG, in clinical practice, is divided into four bands, depending on its oscillation frequency: delta (δ; 0–4 Hz), theta (θ; 4–8 Hz), alpha (α; 8–13 Hz), and beta (β; 13–30 Hz). Frequencies above 30 Hz, although very important in cognitive research, are not customarily used in clinical practice because technical requirements needed.

The complex neuroanatomical homeostatic system (see Chap. 2 for a detailed description) is probably genetically determined and regulates basal levels of local synchronization, global interactions between different regions, spectral composition, and periodic signal space sampling [30–33].

We have two branches of analysis: power spectra and synchronization and, for both of them, we have dynamical (i.e., varying along the time) and average measurements (i.e., mean spectra or mean graph of synchronization). The process is summarized in Fig. 6.1.

Process used for qEEG follows the next steps:

Different-length raw records can be exported from the EEG device to ASCII file. Usually, artefacts are excluded by export of several artefact-free chunks, later put together for analysis. We have shown that this process does not change the main properties analyzed.

Differential EEG montage is reconstructed. Topographic placement of channels is defined on the scalp as the midpoint between the electrode pairs defining the channel, e.g., the Fp1–F3 channel would be placed at the midpoint of the geodesic between Fp1 and F3 electrodes.

All recordings can be divided into different-length moving windows (1–10 s each) with different overlaps (between 0 and 50%, but usually 10%). The total length used during the FFT is directly related to frequency precision in the power spectrum (PS). Overlap is used to minimize the effect produced by windowing [34].

For each window (n) and frequency (k), we computed the FFT of the voltage ($V^m(n)$) obtained from every channel (m) to obtain the power spectrum ($S^m_{n,k}$, in μV²/Hz). We used this expression:

$$S^m_{n,k} = \sum_{n=0}^{N-1} V^m(n)\, e^{-i\frac{2\pi}{N}kn} ; m = Fp_1, F_3, \dots \quad (6.1)$$

We also computed Shannon's spectral entropy (SSE) according to

$$\mathrm{SSE}^m_k = -\sum_{k=0}^{F} p_k \log_2 p_k \quad (6.2)$$

where F is the maximum frequency computed and p_k is the probability density of S, obtained from the expression

$$p_k = \frac{S^m_{n,k}}{\sum_{k=0}^{F} S^m_{n,k} \Delta k} \quad (6.3)$$

We computed the area under the $S^m_{n,k}$ according to the classical segmentation of EEG bands. We used this expression:

$$A_j(k) = \sum_{k=\inf}^{\sup} S^m_n(k) \Delta k ; j = \delta, \theta, \alpha, \beta \quad (6.4)$$

The expression *sup* refers to the upper limit of every EEG band and where Δk is the increment of frequency. This expression is what we show as PS dynamics. Although the dynamics of the power bands can be represented in absolute values, we frequently use the logarithmic transform (logPS).

The absolute value of Pearson's correlation coefficient (r) was computed for every pair of channels (i,j) according to this expression:

$$r^k_{ij} = \frac{\sum_{k=1}^{N_{\text{window}}} \left(x_i(k) - \bar{x}_i\right) \sum_{k=1}^{N_{\text{window}}} \left(x_j(k) - \bar{x}_j\right)}{\sqrt{\sum_{k=1}^{N_{\text{window}}} \left(x_i(k) - \bar{x}_i\right)^2 \sum_{k=1}^{N_{\text{window}}} \left(x_j(k) - \bar{x}_j\right)^2}} \quad (6.5)$$

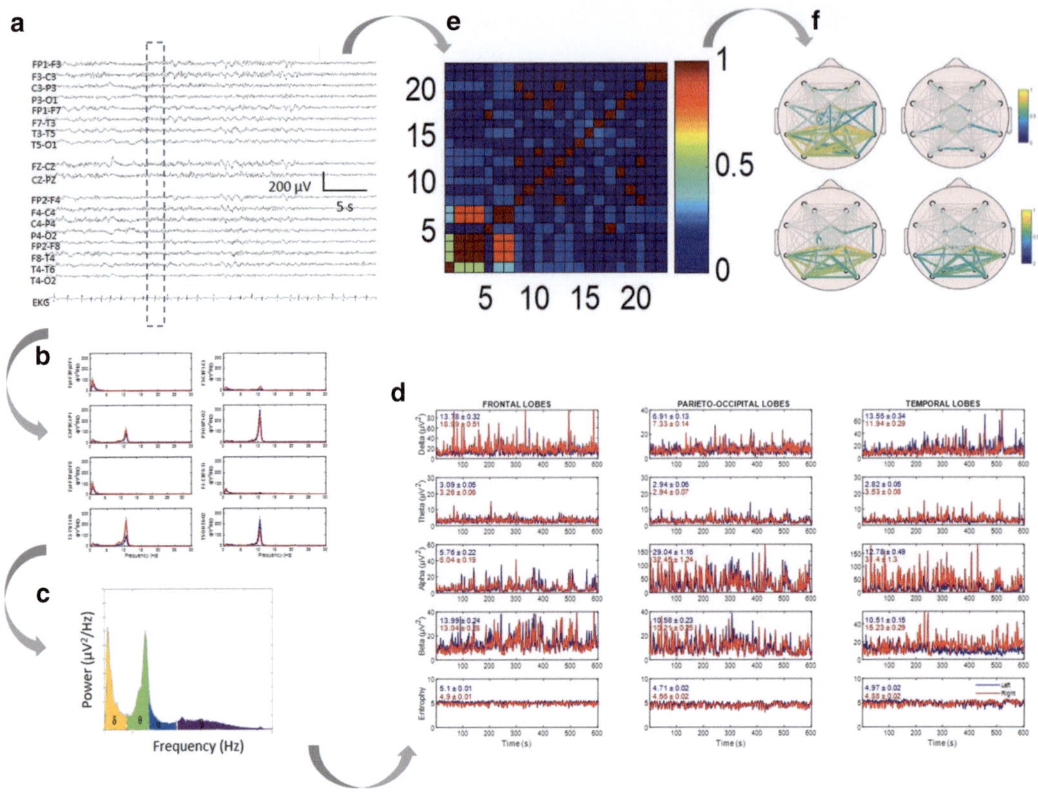

Fig. 6.1 Method of EEG quantification in two branches: power spectra (**b–d**) and synchronization (**e**, **f**). (**a**) Raw EEG tracing. The discontinuous rectangle shows the moving window used for analysis; (**b**) power spectra for each channel; (**c**) areas for delta, theta, alpha, and beta bands under the spectrum are highlighted in different colors; (**d**) dynamics of the four bands (and entropy in the lower row) for every lobe. Mean and SEM values for each tracing are displayed inside each graph. Red and blue lines indicate right and left hemispheres, respectively; (**e**) correlation matrix for the window; (**f**) mean correlation (left upper row), mean delta coherence (right upper row), mean alpha coherence (left lower row), and mean beta coherence (right lower row)

where N_{window} is the number of points included in a window (usually 512) and $\overline{x}_i, \overline{x}_j$ represent the mean of both channels.

The mean value of all windows was computed, obtaining the mean correlation matrix.

Pearson's correlation is the simplest linear method, although it has proved to be very useful even compared with non-linear methods [23, 24]. However, correlations are obtained for the whole band. Nevertheless, the different EEG bands are rooted in different neural systems and, therefore, the synchronization of different bands can offer specific information. A very useful method to assess specific bands is coherence [35–37] defined by these equations

$$\text{coh}(\omega) = \frac{\left|\left\langle S_{ij}(\omega)\right\rangle_n\right|^2}{\left\langle S_{ii}(\omega)\right\rangle_n \left\langle S_{jj}(\omega)\right\rangle_n} \tag{6.6}$$

where S_{ij} is the cross-spectrum of channels i and j normalized by the power spectral S_{ii} and S_{jj}. The symbol $\langle \ldots \rangle_n$ means the average over n epochs. The averaging of cross-spectrum S_{ij} must be performed before the absolute value is taken. Observe that spectra and cross-spectrum are functions of the frequency.

Areas of the same band are grouped by cerebral lobe. In the case of the left hemisphere (shown as an example), we group frontal $F = \left\{ \frac{(Fp_1 - F_3) + (F_3 - C_3) + (Fp_1 - F_7)}{3} \right\}$, parieto-occipital

$$PO = \left\{ \frac{(C_3 - P_3) + (P_3 - O_1) + (T_5 - O_1)}{3} \right\}, \text{ and}$$

temporal $$T = \left\{ \frac{(F_7 - T_3) + (T_3 - T_5) + (T_5 - O_1)}{3} \right\}.$$

Channels from the right hemisphere are grouped accordingly. These areas, for both bands (j) and lobes (r), $A_j^r(t); r = F, PO, T$, are plotted as time functions and statistically compared between hemispheres. The same groups are used to compute SSE.

The total time of analysis (t_{analysis}) is obtained from this linear expression obtained from least-square fitting[1]

$$t_{\text{analysis}}(s) = 0.32Sz(MB) + 46.3 \quad (6.7)$$

For a typical 75 MB file (10 min record) we can estimate the time consumed in export + analysis as less than 2 min.

We can optionally introduce two time-markers to define different situations (e.g., pre-ictal, ictal, and post-ictal periods) in order to compare statistically the changes.

We optionally can export numerical results to an Excel® file. This last step is the most time-consuming (up to 3 min for a custom-length file). Nevertheless, there are other methods to save results faster, e.g., *.mat files.

It's interesting to observe that the information offered for Pearson's correlation coefficient and the coherence measured for the different EEG bands is different, not only in the graphs, but even for the numerical values.

For power spectra as well as for synchronization, we can represent measurements either as dynamic time-dependent variables (Fig. 6.1d), or as the mean values, averaged over the whole file (Fig. 6.1b, f). Therefore, although complementary, information obtained from both kinds of computations must be interpreted differently. In other words, average measurements are only useful if the stationarity of the record is evident (e.g., basal recordings, well-defined phases of sleep, etc.).

A very important aspect of any numerical method is its robustness, e.g., the evaluation of the method wherein the results obtained are found to be reliable, even when performed under slightly varied conditions. It is extremely important to check trials of the numerical method within the same group of EEG records under the different conditions of (i) down-sampling, (ii) windowing, or (iii) overlapping. Moreover, it is important to check whether synchronization measures are affected by the global analysis of different, non-consecutive chunks.

We have checked that the method is highly robust, at least for the limits addressed and does not vary from patient to patient.

However, although the method is robust, it must be taken into account that the numerical values of each analysis depend on its characteristics and, specifically, on the width of the window used for the analysis. The relationship between the window width, i.e., the number of points used for the computation, and the sensitivity of the x-axis in the power spectrum are inverse (see Chap. 4 for more details). However, not only the spectra are modified, but also the average values obtained for each window will be different. Therefore, it is very important that, in order to compare the numerical results of different analyses (whether consecutive in the same patient, or synchronous in different patients), the characteristics of the analysis are similar with respect to the filters used and the width of the analysis windows (Fig. 6.2).

This example shows how the numerical values for determining the same magnitude are modified with the width of the window used. As an example, we can see that the left delta coherence has a value of 0.132 ± 0.003 for a 1 s window, while with a 2 s window, the same magnitude has a value of 0.121 ± 0.003. The same occurs for all the magnitudes shown. However, obviously, the proportions between both hemispheres and between different lobes remain the same. In fact, the interhemispheric statistical differences observed when the window is 1 s are the same as for the 2 s window with similar significance levels.

[1] Obviously, this estimation depends on the features of the computer used.

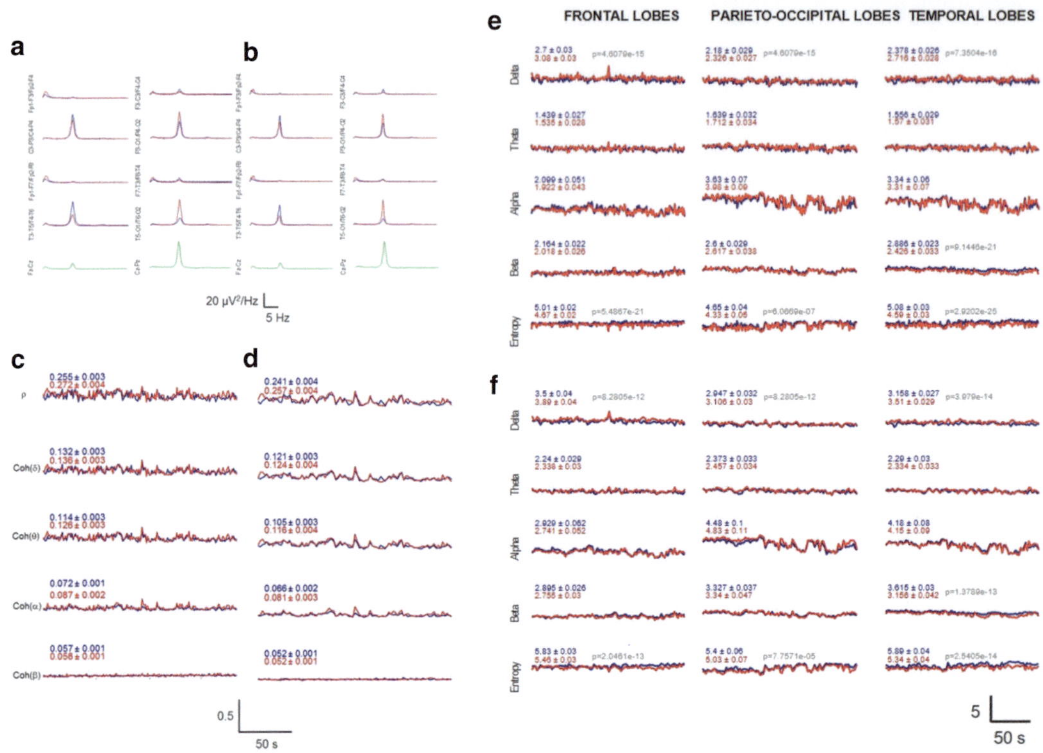

Fig. 6.2 Effect of window width on numerical values of the analysis. The analysis corresponds to 180 s EEG recording from 28-year-old woman with no pathology or medication. (**a**) Average spectra per channel calculated for a width of 1 s. (**b**) Average spectra per channel for a width of 2 s. (**c**) Dynamics of synchronization for a 1 s, and (**d**) 2 s window. From top to bottom, Pearson's correlation coefficient and the delta, theta, alpha, and beta coherences are shown. (**e**) LogPS dynamics by frontal, parieto-occipital, and temporal lobes (top labels) for the different bands (from top to bottom) delta, theta, alpha and beta and Shannon spectral entropy for 1 s, and (**f**) 2 s windows. Red = right hemisphere and blue = left hemisphere

It is therefore very important to keep in mind that different window widths can be used for the analysis and that this will modify the absolute values. This is not particularly worrying as long as all the analyses to be compared are done with the same window, but it is clear that the use of different windows will produce different values, even in the case of consecutive analyses on the same patient, and that this will be an artifact of a faulty analysis.

The main advantage of working with a programmable system is that it can be modified according to new issues. Therefore, we will describe some features that have been added to CRANE to further analyze different aspects.

6.2.1 Analysis of Different Periods

The analysis of periods supposedly involving different brain states is very important. This is common when paroxysmal or transient phenomena occur, the evolution of which is sufficiently rapid. Typical examples of such events are epileptic seizures, but other paroxysmal events can also be analyzed, such as migraine seizures, non-epileptic seizures of psychogenic origin or response to external stimuli (e.g., music).

We have introduced the possibility of using one- or two-time markers to define two or three different states in the recording. We can define these markers either by entering the time at which

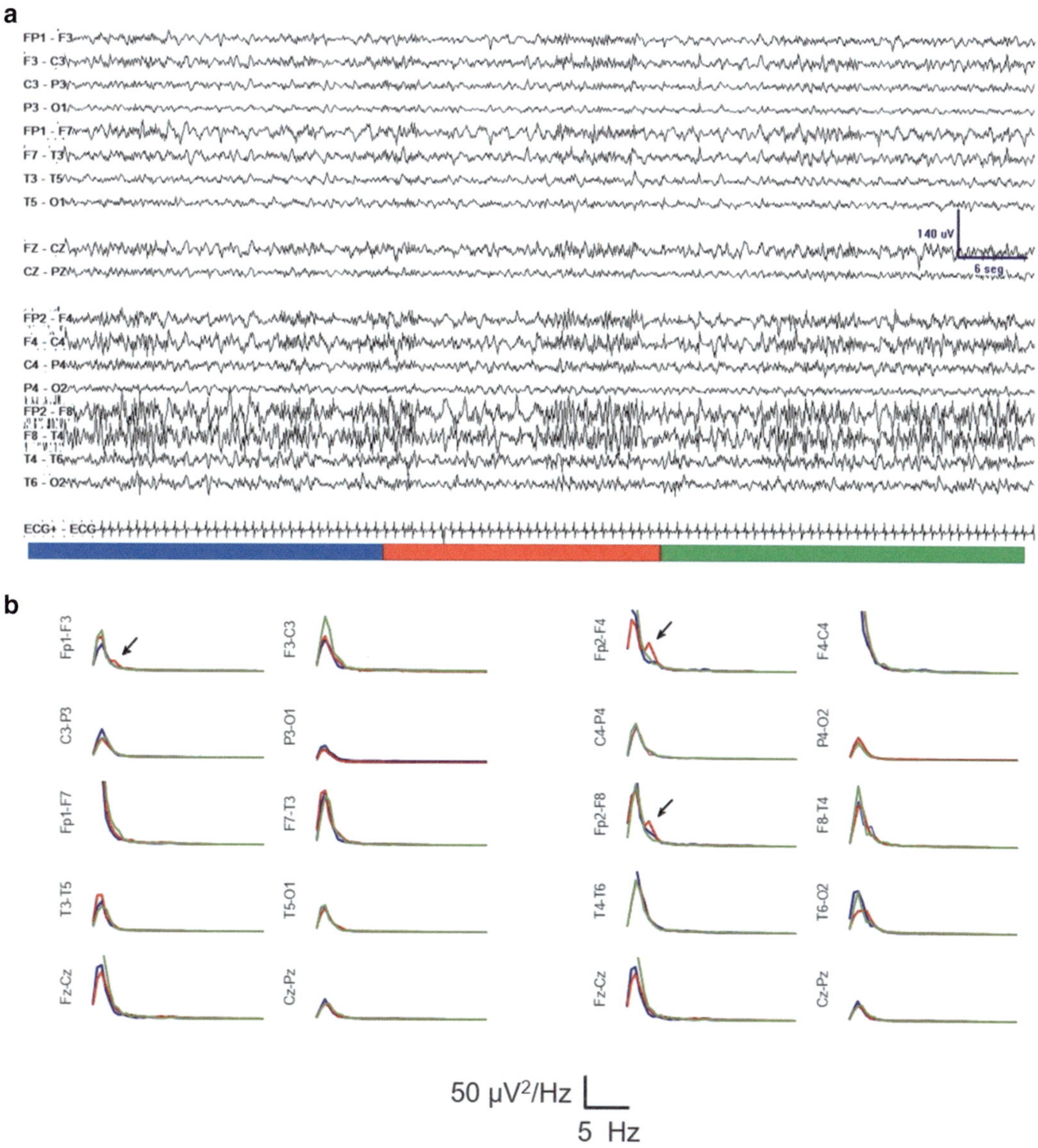

Fig. 6.3 Response to painful stimulation. (**a**) EEG recording during painful stimulation. The horizontal colored bars indicate the period before stimulation (blue), during stimulation (red,) and after the end of the stimulus (green). (**b**) Average spectra for all the channels and each of the periods indicated, following the same color code. The arrows indicate the presence of a delta component during stimulation in frontopolar regions. The patient was extubated 9 days after recording

they were observed in the EEG recording, or by clicking with the mouse on the reconstructed recording.

In Fig. 6.3, we show the utility of this approach to characterize different periods during the recording. The patient was a 28-year-old male, admitted to the intensive care unit for severe head trauma and right frontotemporal decompressive craniectomy. The recording is performed under sedation and analgesia with propofol (14 ml/h), midazolam (12 ml/h), and remifentanil (3 ml/h). Painful stimulation of the nail bed of the left hand

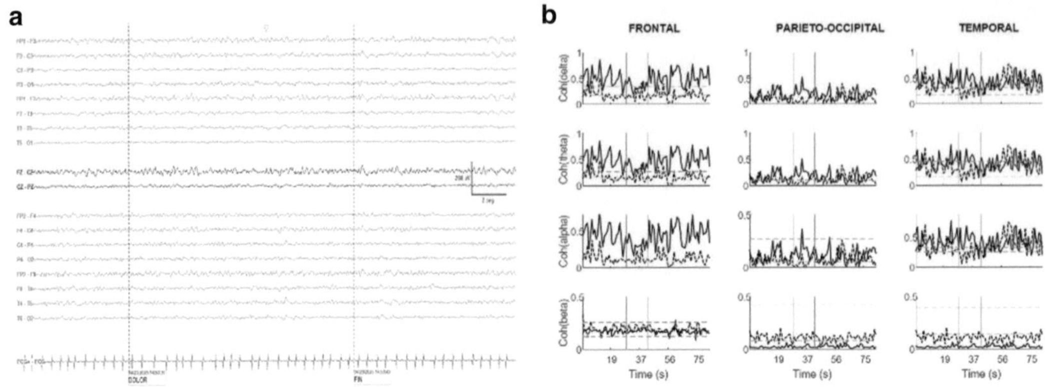

Fig. 6.4 Change in coherence during noxious stimulation in an ICU patient. (**a**) Raw recording with apparent no-response to painful stimulation (vertical lines) and (**b**) dynamics of coherences by bands (rows) for different lobes (columns). Horizontal lines are the 25–75 percentile interval for the age and vertical lines correspond to the noxius stimulation. Continuous line = left hemisphere, discontinuous line = right hemisphere

did not significantly modify the visual EEG recording. However, analysis of the average spectra, obtained before, during, and after the painful stimulation showed the existence of specific spectral changes associated with the stimulus. The presence of a 3 Hz delta component is observed in the bilateral frontopolar regions, although with greater amplitude and extension in the right anterior and lateral frontal regions. It is possible that the absolute amplitude in the right hemisphere is affected by the bone window effect, but the bilateral presence of a component that is not present before or after stimulation demonstrates a brain response to painful stimulation that is not seen in the raw recording. The existence of a brain response to pain is of great importance in establishing the patient's outcome.

Other tools, such as spectral power dynamics or synchronization, can detect modifications associated with specific states, which have significant diagnostic relevance. We will see examples of these cases when reviewing with patients in the ICU or diagnosing epilepsy.

The next example shows the use of *coh* dynamics in the evaluation of pain response in an ICU patient, treated with sedation-analgesia (propofol + remifentanil) and neuromuscular relaxation. Painful stimulation of the nail bed does not produce any apparent response in the raw recording (Fig. 6.4a). However, changes in the dynamics of delta, theta, and alpha coherence in the left frontal region, and to a lesser extent in the temporal region, are evident (Fig. 6.4b).

6.2.2 Comparison Between Consecutive Recordings

But we can not only analyze different states or periods within the same register. It is extremely important to compare records from the same patient on different days or records from different patients obtained in supposedly similar states (e.g., patients in the same sleep phase). In these cases, we have modified the program to compare the average spectra per channel and the average logPS and synchronization values.

This tool is especially useful for the assessment of small subtle changes that are well observed in the average spectrum but are very difficult to observe in *de visu* analysis. In this way, we can compare in an objective way the evolution of the same patient on consecutive days or on separate recordings several hours apart from long-term monitoring.

In many cases, especially in patients admitted to the ICU or encephalopathies, the change in the patient's brain state is very subtle between consecutive days. On the other hand, these changes have a great prognostic value on the patient's final outcome, at least in terms of brain function.

Therefore, in order to establish a prognosis on the patient, it is necessary to know if the record is the same or has changed between consecutive days and this is generally very difficult in conventional *de visu* analysis.

Fig. 6.5 [38] shows the example of a 55-year-old male admitted at ICU after acute severe respiratory syndrome (SARS-CoV-2). The patient was intubated under sedo-analgesia. Fifteen days later was released to conventional hospitalization area and developed significant cognitive and behavioral changes without focal neurological deficits. He developed insomnia and an atypical delirium with memory disturbances, temporo-spatial disorientation, and significant language skills impairment, while maintaining a good level of attention and alertness. The patient was treated with aripiprazole (10 mg/day) and quetiapine (100 mg/day). Imaging studies such as computerized tomography and magnetic resonance imaging were normal. EEG (EEG_{enc}) was nearly physiologic (Fig. 6.5a) in a first *de visu* inspection. The patient was delivered from hospital at 27 days.

Three months later, the patient was re-evaluated and a new EEG was performed (EEG_{cont}; Fig. 6.5b). He showed a complete cognitive recovery and had returned to his job as physician. Only a residual fatigability remained (after climbing stairs 1–2 floors).

De visu EEG_{enc} showed a near normal background, with presence of a low frequent presence of medium voltage sharp waves at left temporal lobe diagnosing a mild encephalopathy. The EEG_{cont} performed 3 months later, showed a quite similar background without irritative activity at left temporal lobe. Nevertheless, the numerical analysis showed more interesting and no self-evident results. Spectral analysis revealed a generalized excess of delta bands, especially at anterior regions, with a posterior dominant rhythm at 8.5 Hz and a mean amplitude of 17.1 ± 1.7 μV. On the contrary, the numerical analysis EEG_{cont} showed very different results. In fact, the posterior dominant rhythm was 9.0 Hz, with the same amplitude (21.1 ± 1.3 μV, not significant, Mann-Whitney on ranks). In both cases, amplitude of EEG was normal and not reduced, as described in *de visu* analysis. But, not only cerebral rhythms were changed during recovery, although the synchronization was modified. Interestingly, during the encephalopathy the patient presented an excessive synchronization in left hemisphere and right temporal lobe.

Therefore, even though the raw records were similar and could pass for physiological, the numerical changes were striking and explained well the patient's cognitive symptoms.

This other example shows very well the usefulness of the method for comparing records separated by a long time. A 75-year-old right-handed woman with a history of smoking. Diagnosed from small vessel leukopathy, without criteria for amyloid angiopathy, mild cognitive impairment, mood disorder, and vascular parkinsonism of probable vascular origin, under treatment with carbidopa+levodopa 75/300 mg/day. MR imaging (at 2021 and 2025) shows an extensive patchy hypodensities in the supratentorial white matter, with a tendency toward confluence, consistent with moderate-severe leukoaraiosis (Fig. 6.6a). No significant changes were observed between the two studies. An EEG was performed in 2021 due to a feeling of unsteadiness and falling. The instability persisted, and the EEG was repeated 4 years later.

The rsEEGs showed a moderately organized pattern, with faster rhythms in the anterior region and a dominant posterior rhythm around 8 or 9 Hz. No epileptiform waveforms were observed. This pattern remained very stable between 2021 and 2025 (Fig. 6.6b, c). The average spectral analysis (Fig. 6.6d) showed an anomalous pattern, with a marked increase in fast frontoparietal rhythms, especially abnormal in the midline, where both an unusual theta component at 5.5–6 Hz and an exaggerated beta component up to 24 Hz are observed. In the 2025 recording, a slight but unremarkable decrease in right frontotemporal power was observed. Most importantly, the spectral structure was preserved between the two recordings. Furthermore, there are no significant changes in synchronization between the two periods (Fig. 6.6e).

Therefore, the numerical analysis shows, beyond any subjective interpretation, that both records, separated by 4 years, represent similar brain states, a very difficult statement to make with visual analysis alone.

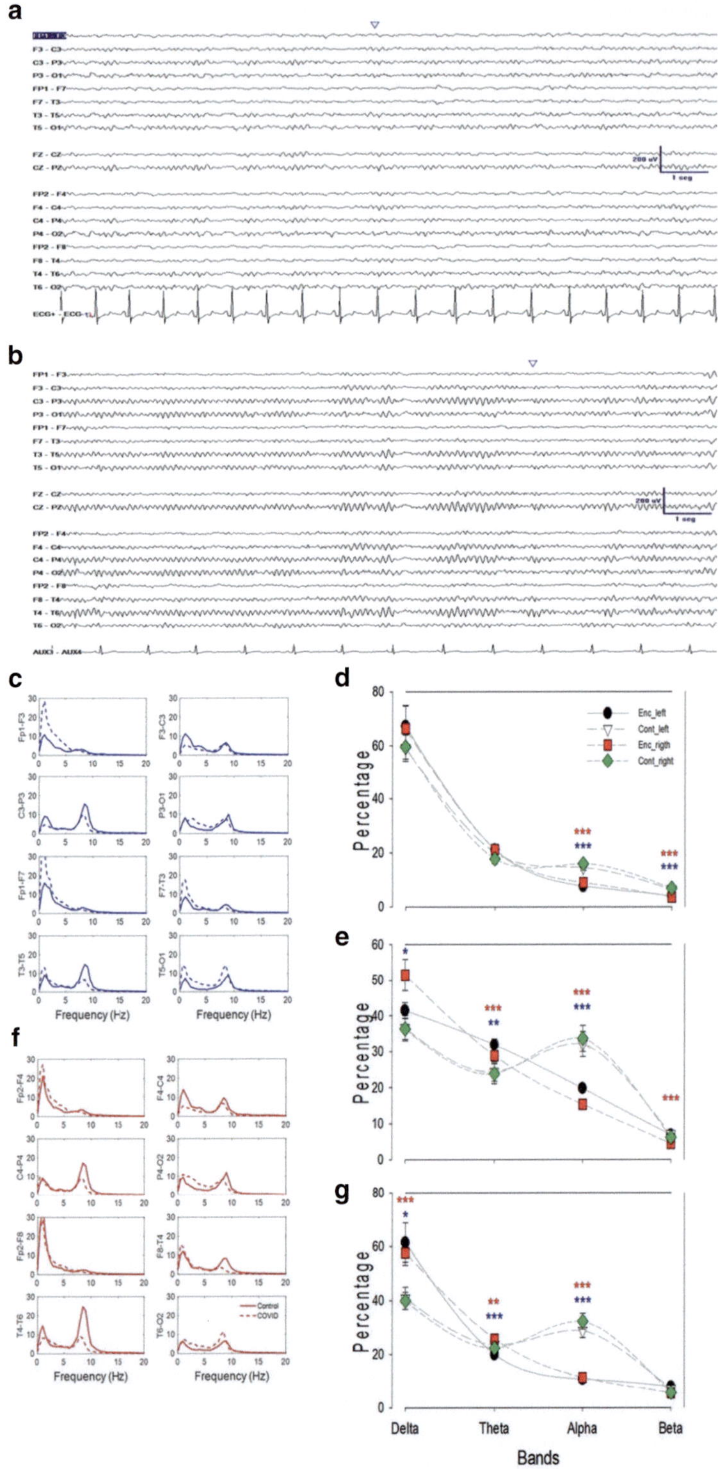

Fig. 6.5 Numerical comparison of two records of the same patient obtained 3 months apart. (**a**) EEG during the state of encephalopathy (EEGenc) and (**b**) recording after recovery 3 months later (EEGcont). (**c**) Mean spectra by channels of the left hemisphere and (**b**) right hemisphere. Dashed lines represent values from EEGenc and continuous lines EEGcont. Comparison of values for bands at (**d**) frontal, (**e**) parieto-occipital, and (**f**) temporal lobes. Filled circle = left encephalopathy; empty triangle = left control; red square = right encephalopathy and green diamond = right control. Blue asterisks mean statistically significant difference for left hemisphere and red ones for right hemisphere. * $p < 0.05$; ** $p < 0.01$; *** $p < 0.001$, Student t-test

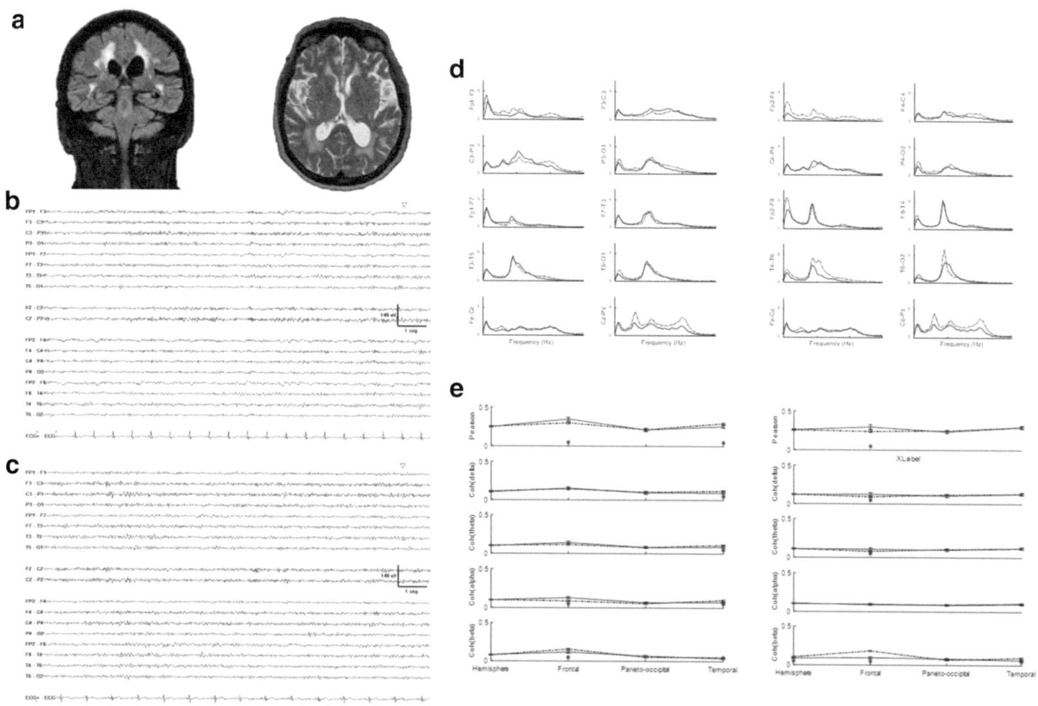

Fig. 6.6 Comparison of two records separated by 4 years. (**a**) Frontal and coronal planes of the MRI. (**b**) Raw record from 2021 and (**c**) from 2025. (**d**) Comparison of average spectra from all the channels. (**e**) Comparison of average measurements of synchronization for hemispheres and lobes. Solid lines = recording from 2021; dashed lines = recording from 2025

6.2.3 Addressing the Cross-Correlation Between Specific Channels

In general, topographic synchronization is analyzed, that is, considering an entire lobe. However, in some cases, it is very important to analyze the synchronization between two specific channels. For this reason, an application has been developed, selecting pairs of channels and performing a cross-correlation between them. This type of analysis has been chosen because, although it does not have the power of Granger causality, it allows to some extent a determination of the directionality of the correlation, so it is possible to assess not only the time difference between both, but we can also estimate which of the two signals precedes the other (see Chap. 5).

In any case, it is very important to establish whether a possible epileptic seizure is generalized or focal, since this can help decide whether to continue a preoperative evaluation study for refractory epilepsy with an invasive vEEG study or, on the contrary, to consider another type of treatment for generalized epilepsy.

In Fig. 6.7, we show the case of a male, right-handed, patient with a history of epileptic and structural encephalopathy due to Lennox-Gastaut syndrome and frontal dysplasia type I. Autism spectrum disorder with somatic mosaicism variant of the ARX gene. Carrier of vagus nerve stimulation (VNS) since 6 years ago. He reports episodes without previous aura of disconnection from the environment with manual automatisms followed by hand stiffness. Currently being treated with brivaracetam (100-0-150 mg), carbamazepine (400-0-400 mg), lamotrigine (150-0-

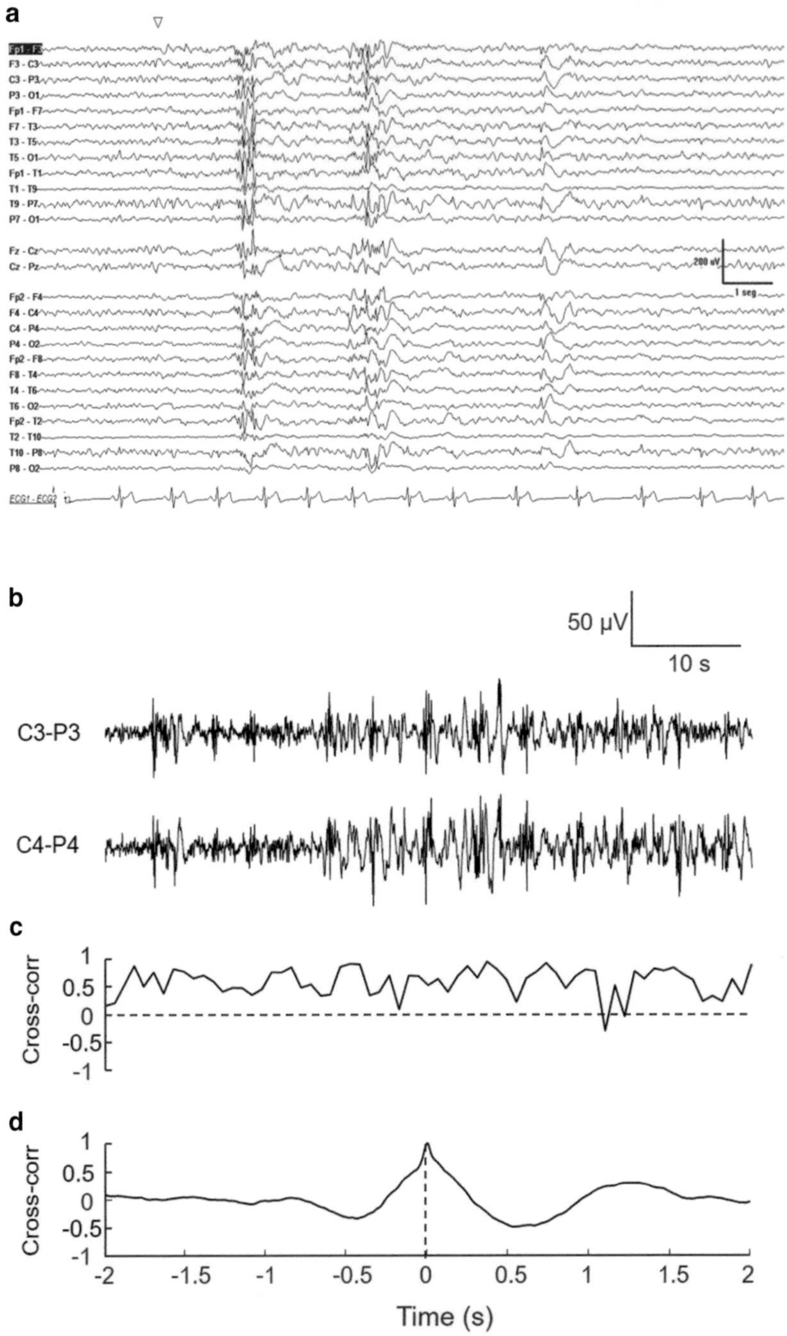

Fig. 6.7 Example of cross-correlation analysis between channel pairs. (**a**) Raw generalized discharge recording. (**b**) Longer duration recordings of other generalized discharges for the frontoparietal channels. (**c**) Dynamics of cross-correlation by channel window in (**b**). The average cross-correlation across all windows is 0.577 ± 0.031. D) Cross-correlation plot for the full recordings showing a 20 ms shift to the right and clear asymmetry

200 mg), vigabatrin (1000-0-1000 mg), and cannabidiol 100 mg/ml (2-0-2 ml).

Despite the appearance of generalized epileptogenic discharges (Fig. 6.7a), the cross-correlation analysis of the frontal channels showed the existence of a strong correlation (Fig. 6.7d) and a time difference of around 20 ms. We can observe an asymmetry in cross-correlation (Fig. 6.6d). Therefore, it is not a generalized discharge, but rather a spreading from the left hemisphere to the right.

Cross-correlation determines the synchronization structure between two signals, which provides a certain degree of directionality.

6.2.4 Long-Term Monitoring

This tool will be discussed in detail in the chapter on the monitoring of critical patients, so we will not go into details here. We will only say that we can obtain graphs showing the dynamics of the power of the bands (both absolute and logarithmic), synchronization measures, the ratio between fast bands (alpha+beta) and slow bands (delta+theta) and Shannon entropy over several days. Changes in the dynamics are useful to identify the presence of seizures or status epilepticus or focal or hemispheric vascular alterations.

The findings from this type of long-term registry will be discussed in detail in the chapter on ICU patients. An example of an mEEG lasting more than 60 h is shown in Fig. 6.8.

6.2.5 Averaging Recordings

Finally, we can also average an indeterminate number of records with a well-defined characteristic, such as a peak, the presence of an event such as a myoclonic jerk, or a sleep graph element. This allows us to obtain average waveforms, with well-defined characteristics, in which we can measure their time and voltage properties (Fig. 6.9), or use it for an event-related synchronization/desynchronization analysis.

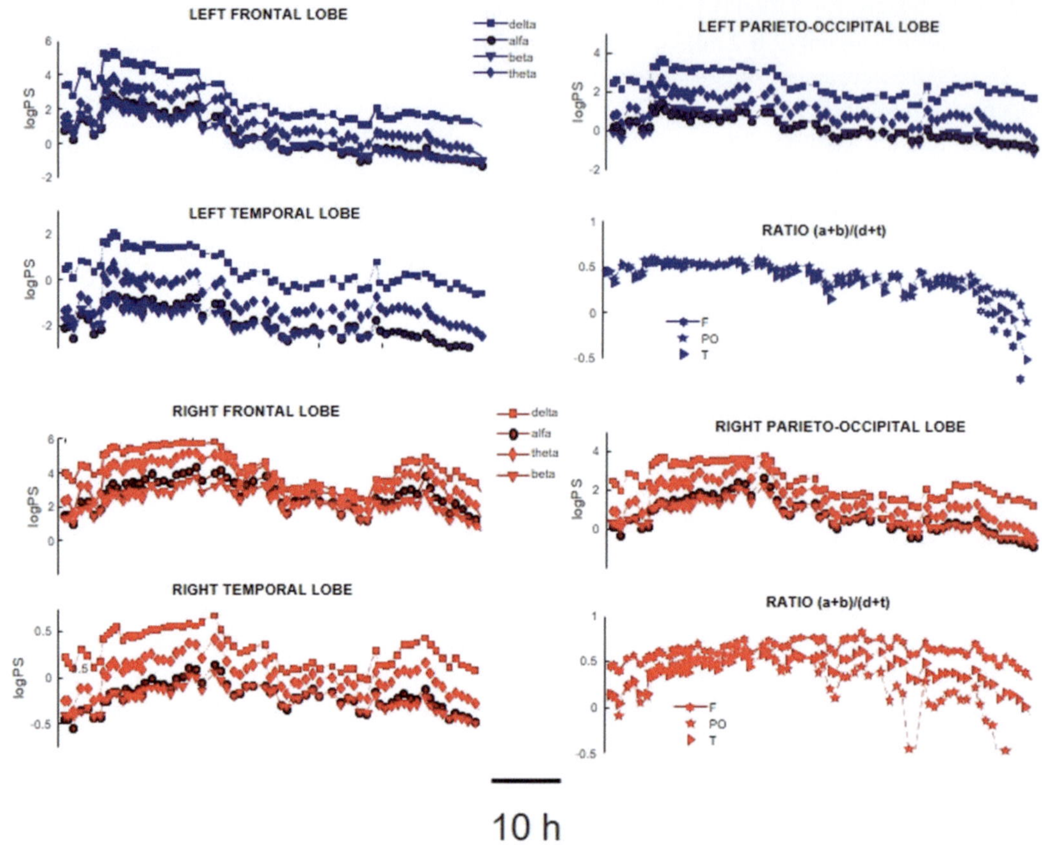

Fig. 6.8 Example of long-term monitoring (>60 h) of a patient in an ICU. The dynamics of logPS for the different EEG bands are shown, for each of the brain lobes. The dynamics of the ratios (alpha+beta)/(delta+theta) are also shown. Red = right hemisphere; blue = left hemisphere. Square = delta; diamond = theta; circle = alpha; lower triangle = beta; hexahedron = frontal lobe; five-pointed star = parieto-occipital lobe; right triangle = temporal lobe

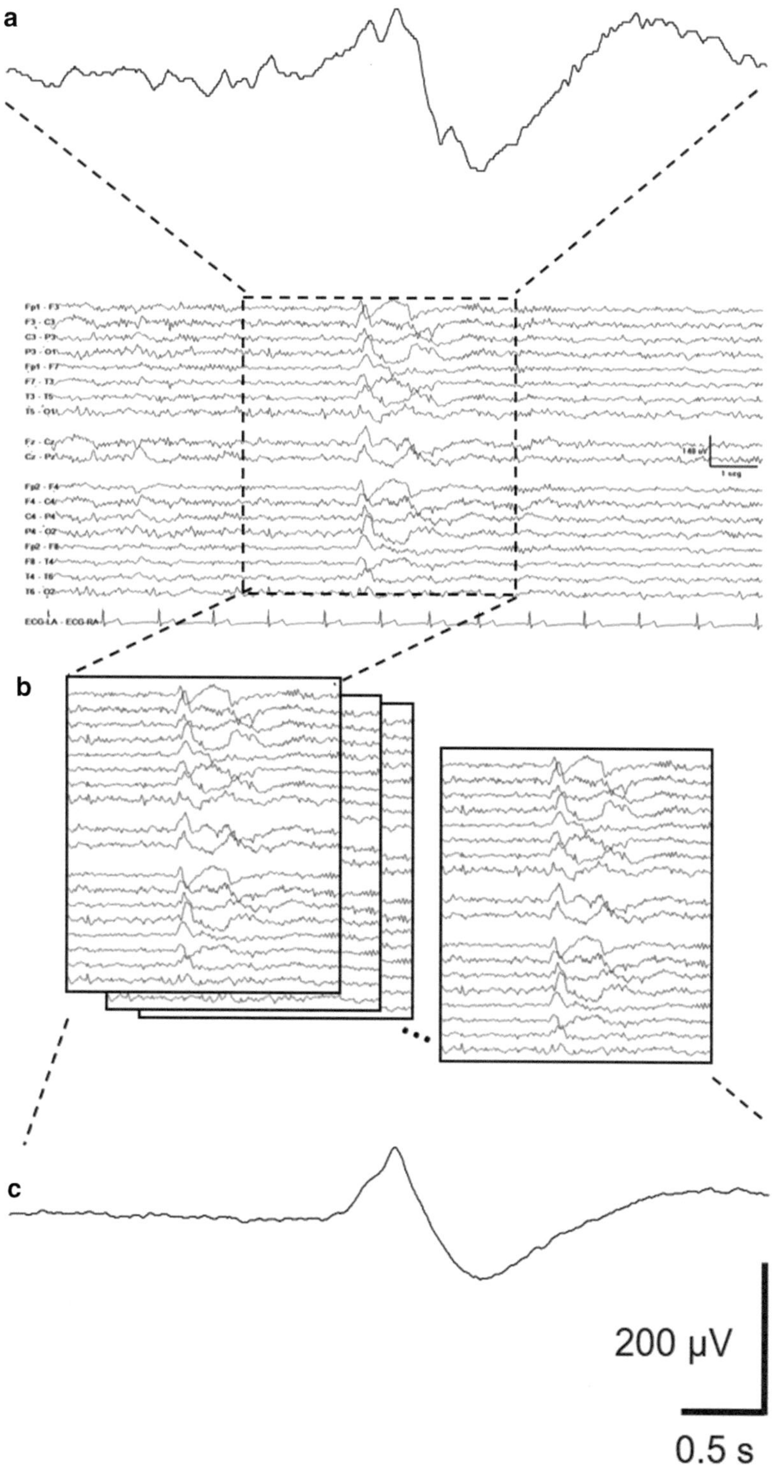

Fig. 6.9 Averaging K complexes obtained during NREM sleep. (**a**) Example of a K complex obtained from the raw recording (inset) and expanded above. (**b**) Several examples of K complexes, always obtained at the same peak and whose average results in the average K complex shown below

6.3 Concluding Remarks

It can be concluded that the use of CRANE, or a similar program, is perfectly feasible in daily clinical practice. It is a method sufficiently robust for the quality of routine recordings, which always have EOG, EMG, or patient agitation artifacts, and at the same time, sufficiently powerful to provide valuable clinical information. This information often allows us to refine the diagnosis and, in some cases, even completely modify it. For this reason, we consider its use to be irreplaceable in our hospital.

The use of homemade methods has pros and cons. Among the former, we can point out their flexibility, adapting to changes in the diagnostic needs of different types of patients. However, the main disadvantages should be noted, such as the need for programming by the clinician and the lack of standardization of the results. Since this is a home-made method, its structure and adaptation to brain states must be ensured by testing well-known brain states before using it in daily practice.

Evidently, we do not propose that this is the best method to be used in clinical practice, but we do wish to indicate that routine use is perfectly possible and desirable. The combination of magnitudes such as the spectral dynamics of EEG bands by lobe, the dynamics of synchronization metrics (correlation, cross-correlation, and coherence), and stationary or average measures, such as spectra or connectograms, offers highly useful information for making an objective and multidimensional assessment of the patient's brain status.

Finally, we believe that despite the extraordinary usefulness of numerical methods, these are nothing more than tools (very good, but tools) that help the neurophysiologist in the evaluation of the patient, which includes aspects such as clinical symptoms, etiology, pathophysiology, etc. Therefore, the clinician's deep knowledge of all these aspects, together with the accredited EEG corpus and the new numerical analysis tools will continue to be the best diagnostic method through EEG and not the exclusive use of numerical methods.

References

1. Wendt SL, Welinder P, Sorensen HB, Peppard PE, Jennum P, Perona P, Mignot E, Warby SC. Inter-expert and intra-expert reliability in sleep spindle scoring. Clin Neurophysiol. 2015;126(8):1548–56. https://doi.org/10.1016/j.clinph.2014.10.158.
2. Halford JJ, Shiau D, Desrochers JA, Kolls BJ, Dean BC, Waters CG, Azar NJ, Haas KF, Kutluay E, Martz GU, Sinha SR, Kern RT, Kelly KM, Sackellares JC, LaRoche SM. Inter-rater agreement on identification of electrographic seizures and periodic discharges in ICU EEG recordings. Clin Neurophysiol. 2015;126(9):1661–9. https://doi.org/10.1016/j.clinph.2014.11.008.
3. Halford JJ, Arain A, Kalamangalam GP, LaRoche SM, Leonardo B, Basha M, Azar NJ, Kutluay E, Martz GU, Bethany WJ, Waters CG, Dean BC. Characteristics of EEG interpreters associated with higher interrater agreement. J Clin Neurophysiol. 2017;34(2):168–73. https://doi.org/10.1097/WNP.0000000000000344.
4. Jing J, Herlopian A, Karakis I, Ng M, Halford JJ, Lam A, Maus D, Chan F, Dolatshahi M, Muniz C, Chu C, Sacca V, Pathmanathan J, Ge W, Sun H, Dauwels J, Cole AJ, Hoch DB, Cash SS, Westover MB. Interrater reliability of experts in identifying Interictal epileptiform discharges in electroencephalograms. JAMA Neurol. 2019;77(1):49–57. https://doi.org/10.1001/jamaneurol.2019.3531.
5. Herman ST, Abend NS, Bleck TP, Chapman KE, Drislane FW, Emerson RG, Gerard EE, Hahn CD, Husain AM, Kaplan PW, LaRoche SM, Nuwer MR, Quigg M, Riviello JJ, Schmitt SE, Simmons LA, Tsuchida TN, Hirsch LJ, Critical Care Continuous EEG Task Force of the American Clinical Neurophysiology Society. Consensus statement on continuous EEG in critically ill adults and children, part I: indications. J Clin Neurophysiol. 2015;32(2):87–95. https://doi.org/10.1097/WNP.0000000000000166.
6. Herman ST, Abend NS, Bleck TP, Chapman KE, Drislane FW, Emerson RG, Gerard EE, Hahn CD, Husain AM, Kaplan PW, LaRoche SM, Nuwer MR, Quigg M, Riviello JJ, Schmitt SE, Simmons LA, Tsuchida TN, Hirsch LJ, Critical Care Continuous EEG Task Force of the American Clinical Neurophysiology Society. Consensus statement on continuous EEG in critically ill adults and children, part II: personnel, technical specifications, and clinical practice. J Clin Neurophysiol. 2015;32(2):96–108. https://doi.org/10.1097/WNP.0000000000000165.
7. Admiraal MM, van Rootselaar AF, Horn J. International consensus on EEG reactivity testing after cardiac arrest: towards standardization. Resuscitation. 2018;131:36–41. https://doi.org/10.1016/j.resuscitation.2018.07.025.
8. Claassen J, Taccone FS, Horn P, Holtkamp M, Stocchetti N, Oddo M, Neurointensive Care Section of the European Society of Intensive Care Medicine.

Recommendations on the use of EEG monitoring in critically ill patients: consensus statement from the neurointensive care section of the ESICM. Intensive Care Med. 2013;39(8):1337–51. https://doi.org/10.1007/s00134-013-2938-4.
9. Hirsch LJ, Fong MWK, Leitinger M, LaRoche SM, Beniczky S, Abend NS, Lee JW, Wusthoff CJ, Hahn CD, Westover MB, Gerard EE, Herman ST, Haider HA, Osman G, Rodriguez-Ruiz A, Maciel CB, Gilmore EJ, Fernandez A, Rosenthal ES, Claassen J, Husain AM, Yoo JY, So EL, Kaplan PW, Nuwer MR, van Putten M, Sutter R, Drislane FW, Trinka E, Gaspard N. American clinical neurophysiology society's standardized critical care EEG terminology: 2021 version. J Clin Neurophysiol. 2021;38(1):1–29. https://doi.org/10.1097/WNP.0000000000000806. PMID: 33475321; PMCID: PMC8135051.
10. Vega-Zelaya L, Garnés-Camarena O, Sánz-García A, Ortega GJ, Pastor J. Chapter 3: Mathematical foundations of quantified electroencephalography. In: Clinical advances in neurophysiology. Avid Science; 2016.
11. Nunez PL, Srinivasan R. Fallacies in EEG. Electric fields of the brain. In: The Neurophysics of EEG. 2nd ed. New York: Oxford University Press; 2006a. ISBN-13 978-0-19-505038-7.
12. Vega-Zelaya L, Martín Abad E, Pastor J. Quantified EEG for the characterization of epileptic seizures versus periodic activity in critically ill patients. Brain Sci. 2020;10:158. https://doi.org/10.3390/brainsci10030158.
13. Pastor J, Vega-Zelaya L, Martin AE. Specific EEG encephalopathic pattern in SARS-CoV-2 patients. J Clin Med. 2020;9:1545. https://doi.org/10.3390/jcm9051545.
14. van der Zande JJ, Gouw AA, van Steenoven I, Scheltens P, Stam CJ, Lemstra AW. EEG characteristics of dementia with Lewy bodies, Alzheimer's Disease and Mixed Pathology. Front Aging Neurosci. 2018;10:190. https://doi.org/10.3389/fnagi.2018.00190. eCollection 2018
15. Szelenberger W, Skalski M, Franaszczuk P, Mitraszewski P, Blinowska K. Pharmaco-EEG analysis by means of FFT and FAD. Acta Physiol Pol. 1989;40(4):423–30.
16. Akin M. Comparison of wavelet transform and FFT methods in the analysis of EEG signals. J Med Syst. 2002;26(3):241–7. https://doi.org/10.1023/a:1015075101937.
17. Höller Y, Trinka E, Kalss G, Schiepek G, Michaelis R. Correlation of EEG spectra, connectivity, and information theoretical biomarkers with psychological states in the epilepsy monitoring unit – a pilot study. Epilepsy Behav. 2019;99:106485. https://doi.org/10.1016/j.yebeh.2019.106485. Epub 2019 Sep 4
18. Gast H, Schindler K, Rummel C, Herrmann US, Roth C, Hess CW, Mathis J. EEG correlation and power during maintenance of wakefulness test after sleep-deprivation. Clin Neurophysiol. 2011;122(10):2025–31. https://doi.org/10.1016/j.clinph.2011.03.003. Epub 2011
19. Lenartowicz A, Loo SK. Use of EEG to diagnose ADHD. Curr Psychiatry Rep. 2014;16(11):498. https://doi.org/10.1007/s11920-014-0498-0.
20. Jiang MJ, Zhang HJ, Li WR, Wu WQ, Huang YM, Xu DM, Qi YY, Qin KY, Zhang L, Zhang JL. Analysis of EEG Lemple-Ziv complexity and correlative aspects before and after treatment of anti-syphilis therapy for neurosyphilis. Neurol Res. 2019;41(3):199–203. https://doi.org/10.1080/01616412.2018.1520438.
21. Sanz-García A, Pérez-Romero M, Pastor J, Sola RG, Vega-Zelaya L, Monasterio F, Torrecilla C, Vega G, Pulido P, Ortega G. Potential EEG biomarkers of sedation doses in intensive care patients unveiled by using a machine learning approach. J Neural Eng. 2019;16:026031. https://doi.org/10.1088/1741-2552/ab039f.
22. Sanz-García A, Pérez-Romero M, Pastor J, Sola RG, Vega-Zelaya L, Monasterio F, Torrecilla C, Vega G, Pulido P, Ortega GJ. Identifying causal relationships between EEG activity and intracranial pressure changes in neurocritical care patients. J Neural Eng. 2018;15(6):066029. https://doi.org/10.1088/1741-2552/aadeea. Epub 2018 Sep 5. PMID: 30181428.
23. Ortega GJ, Menéndez de la Prida L, Sola RG, Pastor J. Synchronization clusters of interictal activity in the lateral temporal cortex of epileptic patients: intracranial analysis. Epilepsia. 2008;49(2):269–80. https://doi.org/10.1111/j.1528-1167.2007.01266.x.
24. Ortega GJ, Herrera-Peco I, Sola RG, Pastor J. Impaired mesial synchronization in temporal lobe epilepsy. Clin Neurophysiol. 2011;122:1106–16.
25. Palmigiano A, Pastor J, Sola RG, Ortega GJ. Stability of synchronization clusters and seizurability in temporal lobe epilepsy. PLoS One. 2012;7(7):1–12.
26. Wendling F, Congredo M, Lopes da Silva FH. In: Shomer DL, Lopes da Silva F, editors. EEG analysis: theory and practice. In: Niedermeyer's electroencephalography: basic principles, clinical applications and related fields. 7th ed. Oxford University Press; 2018. ISBN 9780190228484.
27. de Las V, Heras M, Sala-Padro J, Pedro-Perez J, García-Parra B, Hernández-Pérez G, Falip M. Utility of quantitative EEG in neurological emergencies and ICU clinical practice. Brain Sci. 2024;14(9):939. https://doi.org/10.3390/brainsci14090939. PMID: 39335433; PMCID: PMC11430096
28. John ER, Prichep LS, Easton P. Normative data banks and neurometrics: Basic concepts, methods and results of norm construction. In: Gevins AS, Remond A, editors. Handbook of electroencephalography and clinical neurophysiology, vol. I. Amsterdam: Elsevier; 1987. p. 449–95.

29. John ER. The neurophysics of consciousness. Brain Res Brain Res Rev. 2002;39(1):1–28. https://doi.org/10.1016/s0165-0173(02)00142-x.
30. Szava S, Valdes P, Biscay R, Galan L, Bosch J, Clark I, Jimenez JC. High resolution quantitative EEG analysis. Brain Topogr. 1994;6:211–9.
31. Hughes JR, John ER. Conventional and quantitative electroencephalography in psychiatry. J Neuropsychiatry Clin Neurosci. 1999;11:190–208.
32. Kondacs A, Szabo M. Long-term intra-individual variability of the background EEG in normal. Clin Neurophysiol. 1999;110:1708–16.
33. Kropotov JD. Quantitative EEG, event-related potentials and neurotherapy. Amsterdam: Academic Press; 2009.
34. Van Drongelen W. Continuous, discrete and fast Fourier transform. In: Signal processing for neuroscientists. Amsterdam: Elsevier; 2007. ISBN 13:978-0-12-370867-0.
35. Fischer MHF, Zibrandtsen IC, Høgh P, Musaeus CS. Systematic review of EEG coherence in Alzheimer's disease. J Alzheimer's Dis. 2023;91(4):1261–72. https://doi.org/10.3233/JAD-220508.
36. Wolff A, Northoff G. Temporal imprecision of phase coherence in schizophrenia and psychosis-dynamic mechanisms and diagnostic marker. Mol Psychiatry. 2024;29(2):425–38. https://doi.org/10.1038/s41380-023-02337-z. Epub 2024 Jan 16
37. Wolff A, Northoff G. Temporal imprecision of phase coherence in schizophrenia and psychosis-dynamic mechanisms and diagnostic marker. Mol Psychiatry. 2024;29(2):425–38. https://doi.org/10.1038/s41380-023-02337-z. Epub 2024 Jan 16. PMID: 38228893.
38. Pastor J, Vega-Zelaya L, Delgado Parada E, Ayuso JL. Electroencephalographic evidence of organic alteration in a patient with SARS-CoV2 induced delirium. Clin Case Study Rep. 2020; https://doi.org/10.15761/CCSR.1000152.

Part II

Practical Application of qEEG

7 Epilepsy

7.1 Introduction

Epilepsy is one of the most common neurological disorders characterized by recurrent unprovoked seizures. It has an incidence of 50–100 cases per 100,000 persons/year in developed countries, and about 50 million people around the world suffer from epilepsy [1]. In addition to its high prevalence, epilepsy is associated with high morbidity, mortality, impaired quality of life, depression, social isolation, and social stigma.

Epilepsy is clinically defined as a chronic disorder caused by a predisposition to developing epileptic seizures (ES) that persists over time, including psychological, cognitive, and social impairments caused by the disease. According to the ILAE, an ES is a transient occurrence of signs and/or symptoms due to abnormal excessive or synchronous neuronal activity in the cortex of the brain [2]. Therefore, the excessive activity of the cortex is mandatory to a positive identification [3]. Usually, antiepileptic drugs can adequately control most of the patients, although up to 30% of patients are resistant to drug treatment presenting the so-called drug-resistant epilepsy.

The Commission on Classification and Terminology from the International League Against Epilepsy (ILAE) has proposed a new approximation for the classification of seizures and epilepsy types [4]. The new recommended terminology redefines focal seizures as those originating at some points within networks spatially limited to only one hemisphere, or bilaterally distributed in the case of generalized seizures. This new perspective has been mostly inspired, on one hand, from the network analyses [5–8].

Application of qEEG to epilepsy is probably the most wide and developed field in clinical practice. There are numerical methods to identify deep sources for EM imaging, commercial software to identify ES and non-convulsive status epilepticus (NCSE) during long-term monitoring in intensive ICU patients. These patients usually have limited clinical neurological examination, either because of structural or functional altered conditions of the CNS or due to the effects of drugs used for sedo-analgesia. A very effective tool to evaluate the brain function in these conditions is EEG [9]. However, a dynamical evolution of injury is commonly observed in critically ill

J. Pastor, *Quantitative EEG Practical Guide*, https://doi.org/10.1007/978-3-032-05457-9_7

patients, due to the occurrence of ES/NCES, apoptosis, vasospasm, or other different brain insults [10]. In this sense, continuous EEG (cEEG) is a non-invasive method that allows the functional assessment of cerebral function in real time for prolonged periods of time. It has been proven to be an extraordinarily useful tool for detecting electrographic ES and NCES, modifying treatment, and assessing the functional prognosis [11]. In recent years, the qEEG has introduced elements of objectivity into the analysis of EEG records. In the ICU field, the qEEG has been applied to facilitate the interpretation of prolonged EEG recordings, as well as the identification of electrographic seizures [12–15]. The particularity of long-term EEG records of patients in ICU comprises the high frequency of artefacts that are found throughout the record and the rhythmic and periodic patterns frequently observed in these patients, which are easily confused with seizures. Both can be difficult to interpret not only on an EEG raw, but also with the qEEG tools currently used [16, 17]. The time taken for a cEEG review is one of the most commonly given reasons for the use of qEEG in ICU and other diagnostic fields. An excellent review about the use of qEEG as cEEG and also with other indications in ICU patients has been recently published and we encourage the reader to consult that work [18].

However, we have taken a different approach to use qEEG during cEEG: instead of just simplifying seizure detection, our aim is to obtain a comprehensive and efficient view of bioelectrical brain physiology in the most objective way [19]. The application of qEEG to epilepsy is wider than mere detection of ES/NCES in ICU patients. In fact, the pathophysiologic features of epilepsy can allow specific qEEG hallmarks in background activity, sleep period, or ictal activity highly useful for diagnosis and classification of epilepsy.

7.2 Background Activity in Epilepsy

In this chapter, EEG recordings with more than 19 electrodes will be commonly used. This is because we use additional electrodes for v-EEG employed for pre-surgical evaluation. The configuration of supplementary electrodes is shown in Fig. 7.1.

The presence of epilepsy is identified during *de visu* analysis in conventional rsEEG by the presence of irritative or epileptic activity, e.g., spikes, sharp-waves, poly-spikes, generalized spike-wave activity, etc. Usually, this kind of activity is clearly different from background and is easily identified by an experienced neurophysiologist (Fig. 7.2). It can be thought that diagnosis of this kind activity does not need any numerical method. Nevertheless, there is much more information of great utility beyond the presence of epileptic activity.

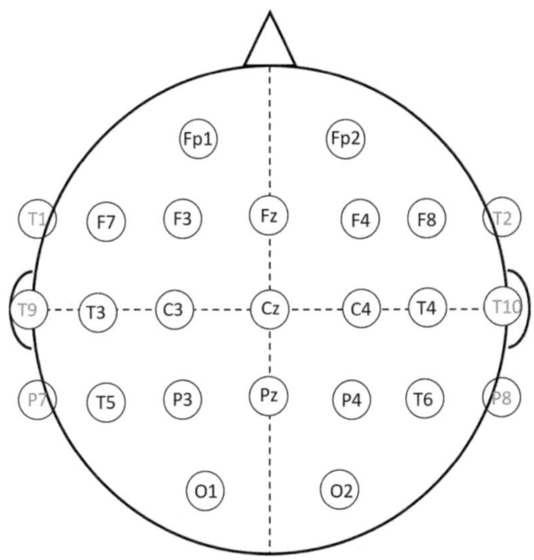

Fig. 7.1 Schematic position of 10–20 IS electrodes (black) and supplementary electrodes used for pre-surgical evaluation patients (gray)

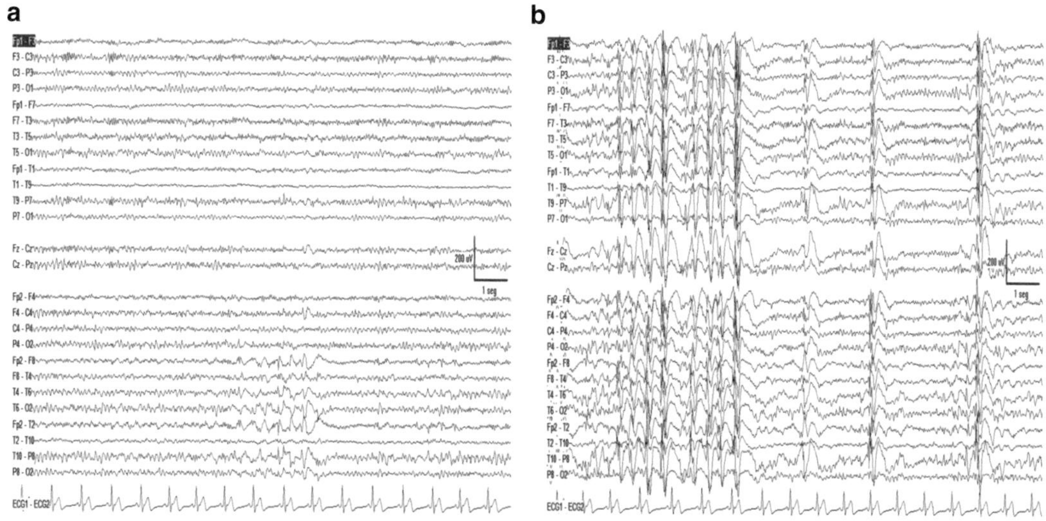

Fig. 7.2 Epileptic activity during rsEEG. (**a**) Evolutionary theta sharp-waves pattern in right temporal lobe during resting activity. (**b**) Generalized high amplitude spike-wave in a 19 year-old. Bipolar double-banana montage with accessory temporal electrodes

The presence of anomalous widespread epileptic networks can explain differences of PS obtained during rsEEG [20], even in the absence of epileptic activity in *de visu* analysis. This fact (absence of epileptic activity) is frequently observed in EEG records, which, on the other hand, exhibits a physiologic aspect. In the example shown next, it is presented the rsEEG record of a 75-year-old male patient treated with carbamacepin 800 mg/12 h and diagnosed of focal epilepsy (Fig. 7.3a). The mean power spectra for channels (Fig. 7.3b) showed anomalous components not observed in physiological recordings, located bilaterally at frontal and sagittal channels. In focal epilepsy patients, this symmetry is not observed. Moreover, the presence of unusual components in the anterior half of scalp is probably responsible for the excessive presence of synchronization in the same anterior regions, as can be observed from the anomalous connectogram (Fig. 7.3c). In patients with true focal epilepsy, the background numerical features are different and are always asymmetric. In the same way, in physiological patients, the connectogram shows a typical pattern with the most associations between channels observed at posterior and, mainly around the mid-line (see Fig. 7.1f)

The presence of findings like this is highly suggestive of generalized epilepsy instead of focal one and can help to choose the anti-epileptic drug (AED) accordingly.

But the presence of abnormal neural networks during basal state, in which no irritative activity can be recorded, is also observed in focal epileptic patients. In this example (Fig. 7.4), it is presented the rsEEG record of a 41-year-old female patient showing malformation of cortical development, dysgenesia of the corpus callosum and periventricular heterotopia, complaint of monthly episodes of with fixed gaze, change in facial expression, and disconnection, lasting a few seconds. The patient was treated with lamotrigine (250 mg/12 h) and topiramate (250 mg/12 h). The rsEEG showed a physiological *de visu* recording, but the qEEG demonstrated significant alterations.

These alterations in band composition and synchronization indicate a modification in the physiology of the right hemisphere, especially for the delta and theta bands. The average power spectra show the presence of anomalous bands in the right temporal lobe. A subsequent long-term v-EEG study revealed the presence of epileptiform activity in the right temporal region.

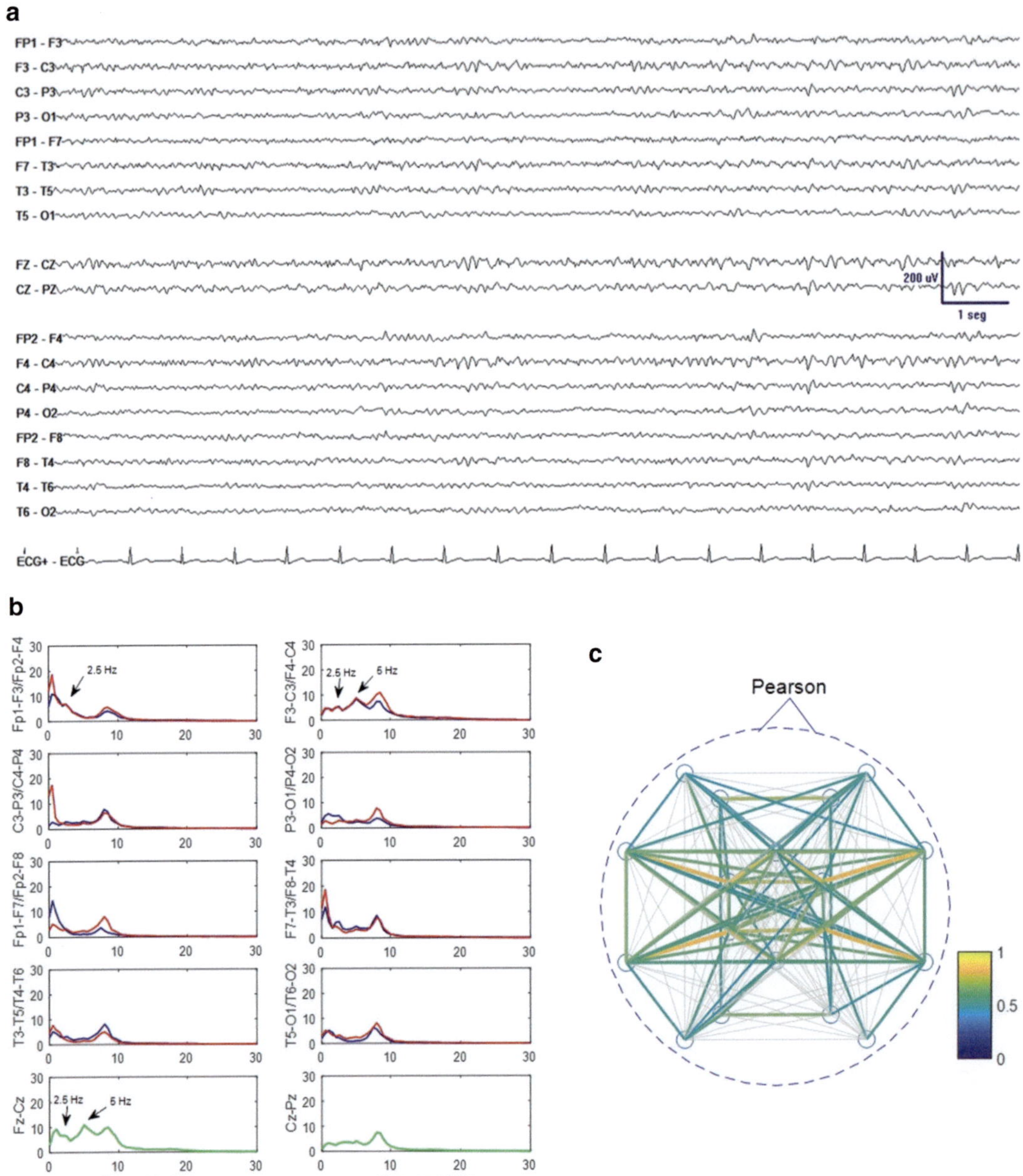

Fig. 7.3 Numerical background anomalies in a patient diagnosed supposedly from focal temporal epilepsy. (**a**) Resting state EEG with physiologic appearance. (**b**) Average power spectra for channels. Anomalous delta (2.5 Hz) and theta components (5 Hz) appearing in anterior half of scalp are indicated. (**c**) The presence of these unusual components increases the synchronization in the anterior part of scalp, showing an anomalous connectogram

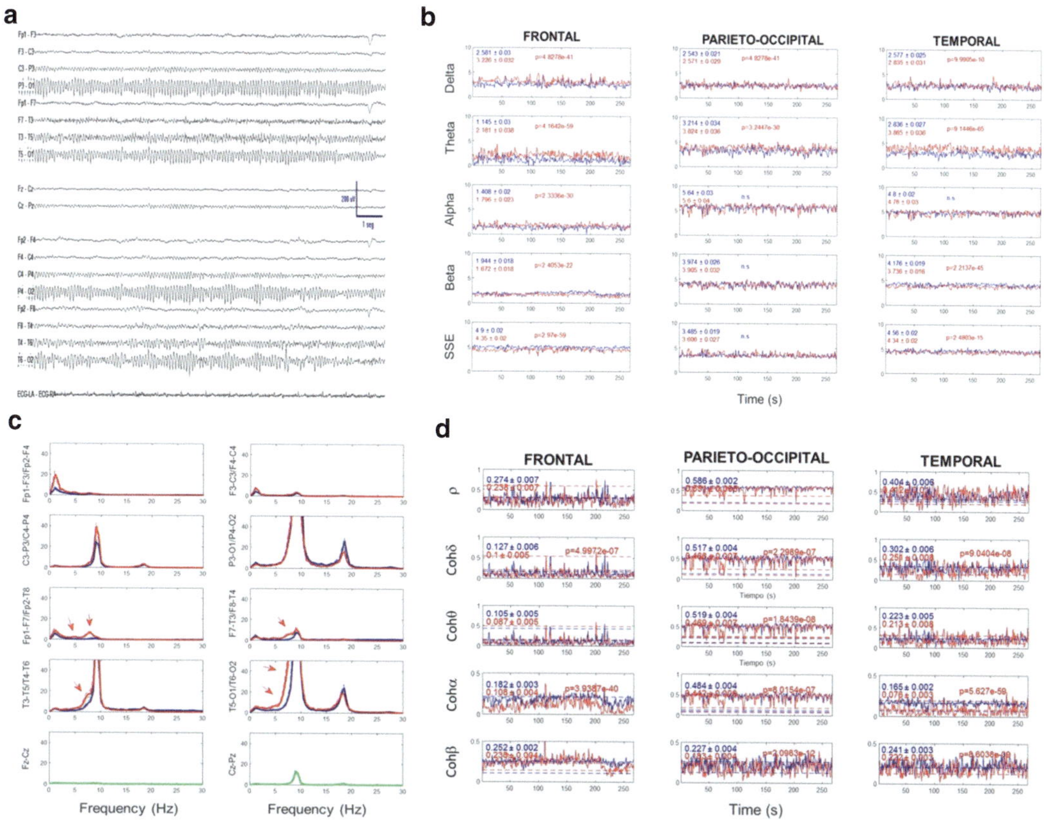

Fig. 7.4 Numerical background anomalies in a patient with cortical malformation and suspicion of epilepsy. (**a**) Resting state EEG with physiologic appearance. (**b**) Dynamics of the power spectra of bands (log-transformed) grouped by lobes, showing an excess of delta/theta right hemisphere. (**c**) Average power spectra for channels showing anomalous theta components (5 and 7 Hz) in right temporal lobe, indicated by red arrows. (**d**) Alterations in synchronization by lobes. Generalized increase in parieto-occipital lobes, although a relative decrease in synchronization can be observed in all the right hemisphere. Red = right; blue = left; green = midline. Inset numbers indicate the mean ± SEM, and a value of probability is indicated when significance between both lobes is obtained. Dashed horizontal lines indicate the confidence interval for patient age group

This other example corresponds to a male patient, right-handed, 54 years old, diagnosed with right mesial frontal epilepsy and under treatment with brivaracetam (100–0-150 mg) and eslicarbacepine (400 mg/12 h). The rsEEG (Fig. 7.5a) showed a basal trace of physiological features. However, the synchronization analysis shows a marked increase in the synchronization of all bands in the right parieto-temporo-occipital region, but especially of the parieto-occipital theta and alpha bands (Fig. 7.5b). These asynchronies are not usually observed in physiological recordings, suggesting an alteration related to epilepsy and which has an important localization value.

However, synchronization disturbances in the epileptogenic region do not always consist of hypersynchronization, but may instead consist of a decrease in synchronization.

In this other case, the example of a 65-year-old right-handed male presented a generalized tonic-clonic seizure the previous day and was brought to the emergency room. The recording revealed a right temporal intermitent rhythmic delta activity (TIRDA, Fig. 7.6a). The logPS band dynamics showed a relative increase in power in all bands in the right hemisphere (Fig. 7.6b). The average spectra showed an excess of the right middle temporal delta/theta compo-

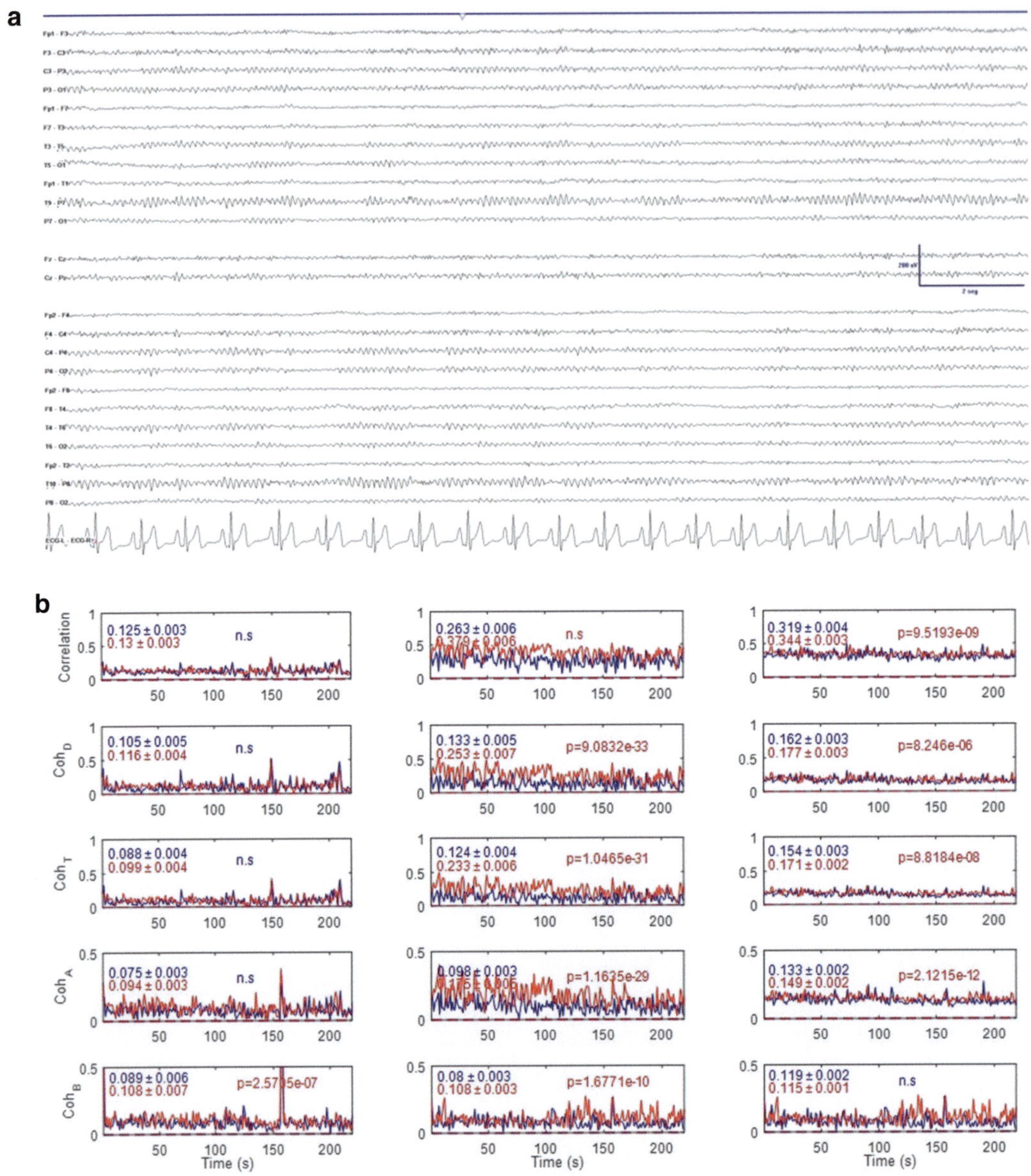

Fig. 7.5 Synchronization anomalies in a patient with right mesial frontal epilepsy. (**a**) Resting state EEG with physiologic appearance. (**b**) Dynamics of the Pearson and band coherence grouped by lobes. The relative asynchrony, mainly indicated by the excessive values in parieto-occipital right lobes. Red = right; blue = left; green = midline

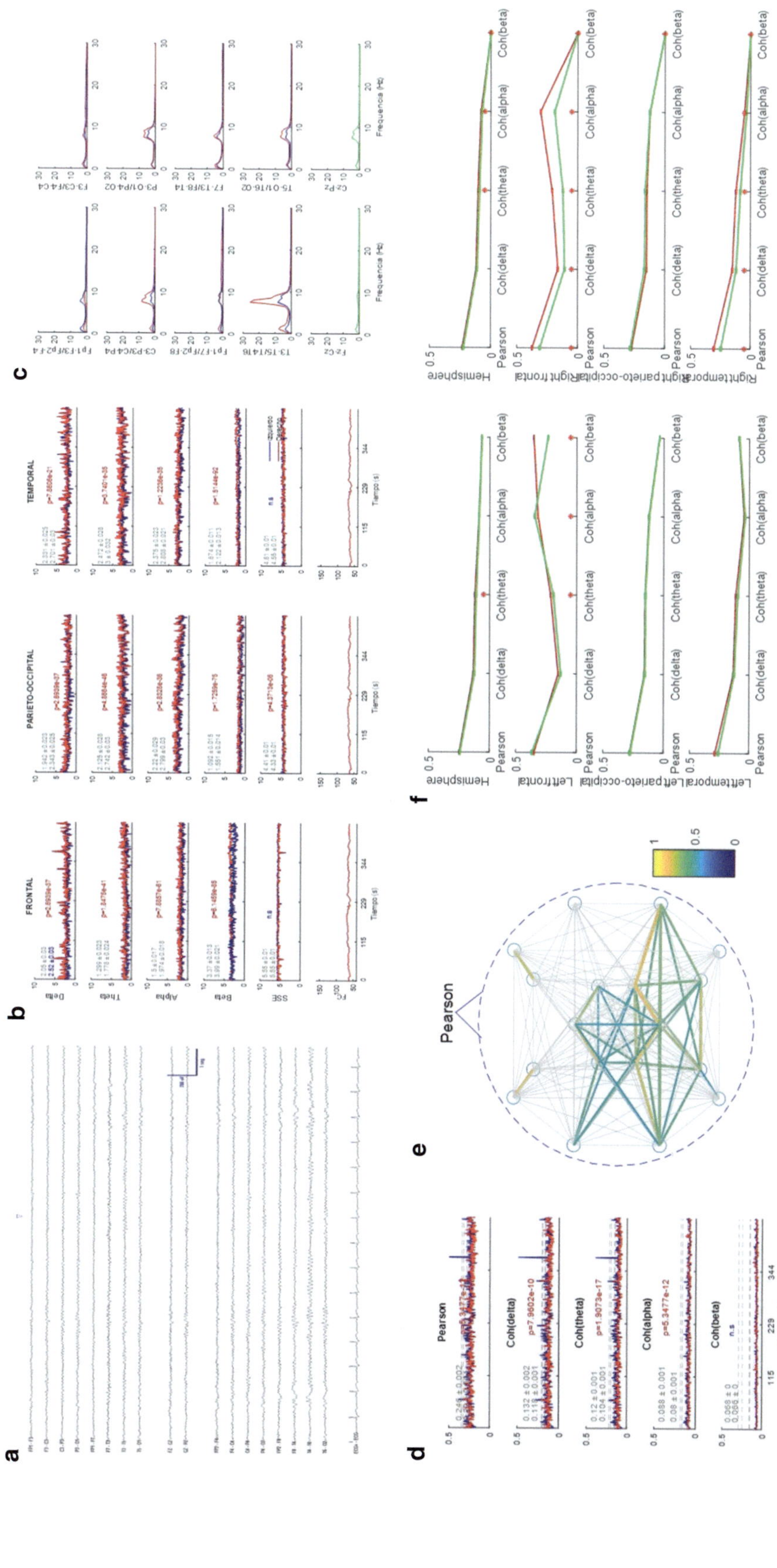

Fig. 7.6 rsEEG in a patient initially diagnosed as generalized epilepsy. (**a**) Raw recording with presence of right low-amplitude TIRDA. (**b**) Dynamics of logPS for bands and lobes. (**c**) Average power spectra for channels. (**d**) Dynamics of synchronization for hemisphere. Red = right hemisphere; blue = left hemisphere. (**e**) Connectogram for Pearson's correlation. (**f**) Measures of synchronization during resting state (red) and hyperventilation (green). Asterisk indicates statistically significance

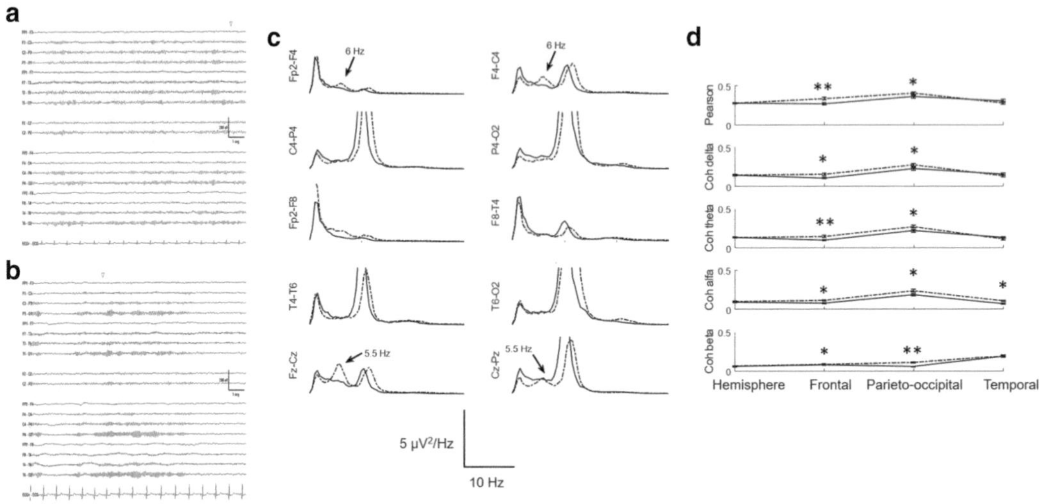

Fig. 7.7 Effect of the sleep-deprivation onto the qEEG. (**a**) rsEEG. (**b**) Sleep-deprivated EEG. (**c**) Average power spectra for the right hemisphere obtained from rsEEG (solid lines) and from sleep-deprivated (dashed lines). The alpha component is slightly displaced from 10 to 11 Hz and appears as a generalized component at 5.5 or 6 Hz of anterior predominance (arrows). (**d**) Changes in right hemisphere synchronization during sleep-deprivation (dashed lines), compared with rsEEG (solid lines). Asterisks indicated statistical significance at * $p < 0.05$ or ** $p < 0.01$, one-way ANOVA

nent (Fig. 7.6c). However, Pearson synchronization and all delta to alpha coherences were decreased in the right hemisphere (Fig. 7.6d). This alteration was greater in the frontotemporal region of that hemisphere, as seen in the connectogram (Fig. 7.6e). Furthermore, hyperventilation altered the synchronization values in the right frontal and temporal lobes, as shown in Fig. 7.6f. The probable diagnosis of focal epilepsy with second generalization was made after this recording.

EEG activation by sleep deprivation is a method to unmask the presence of epileptogenic activity that has previously gone undetected in a conventional recording [21–23]. Often, however, the sleep-deprived recording does not present clearly epileptogenic characteristics. In these cases, qEEG can be of great help by revealing the presence of activity that is barely observed in visual analysis. We show an example of a patient with these characteristics. Male, right-handed, 20-year-old patient. Referred for repeated episodes of loss of consciousness lasting 20–30 s, with subsequent exhaustion. No anti-epileptic treatment. The recording made 1 month earlier was reported as physiological, so sleep-deprived study was recommended. Sleep deprivation did not produce any pathological waveforms (Figs. 7.7a, b). However, as can be seen in Fig. 7.7c, a generalized component was observed, with a predominance in the anterior region, at 5.5 and 6 Hz, in addition to a faster dominant rhythm, which went from 10 to 11 Hz. However, not only were the spectra modified, but synchronization (Fig. 7.7d) in frontal and parietal regions also increased significantly. Just the analysis of the right hemisphere is shown in the figure for simplicity, but the results of the left hemisphere were similar.

The qEEG is a very relevant tool in those patients where the rsEEG does not show epileptiform activity. It offers valuable information about the anomalous function of brain networks, manifested as changes in power of specific bands, changes in synchronization, or presence of anomalous band in the spectra.

7.3 Background and Epileptic Activity During the Sleep

The abnormal functioning of the neural networks affected by epilepsy is observed not only during wakefulness, but also during sleep. Although during sleep, both the spectral band composition and the synchronization are modified in a striking way, the anomalies observed during wakefulness are sometimes maintained and are generally accentuated during some phases of sleep. During the different phases of sleep, the global state of the brain is modified, thus changing local and inter-regional synchronization properties [24–26]. This implies that the functional alterations of neural networks and systems observed during wakefulness are still present during sleep. An example of the maintenance of these alterations during awake and asleep can be observed in Fig. 7.8.

This is a 36-year-old male patient, treated with levetiracetam (1500 mg/12 h) and lacosamide (200 mg/12 h). Infrequent sharp waves were observed in the right anterior temporal region (not shown). Both the baseline awake (Fig. 7.8a1) and asleep (Fig. 7.8a2) recordings showed physiological and apparently symmetrical characteristics for most of the recording. However, the

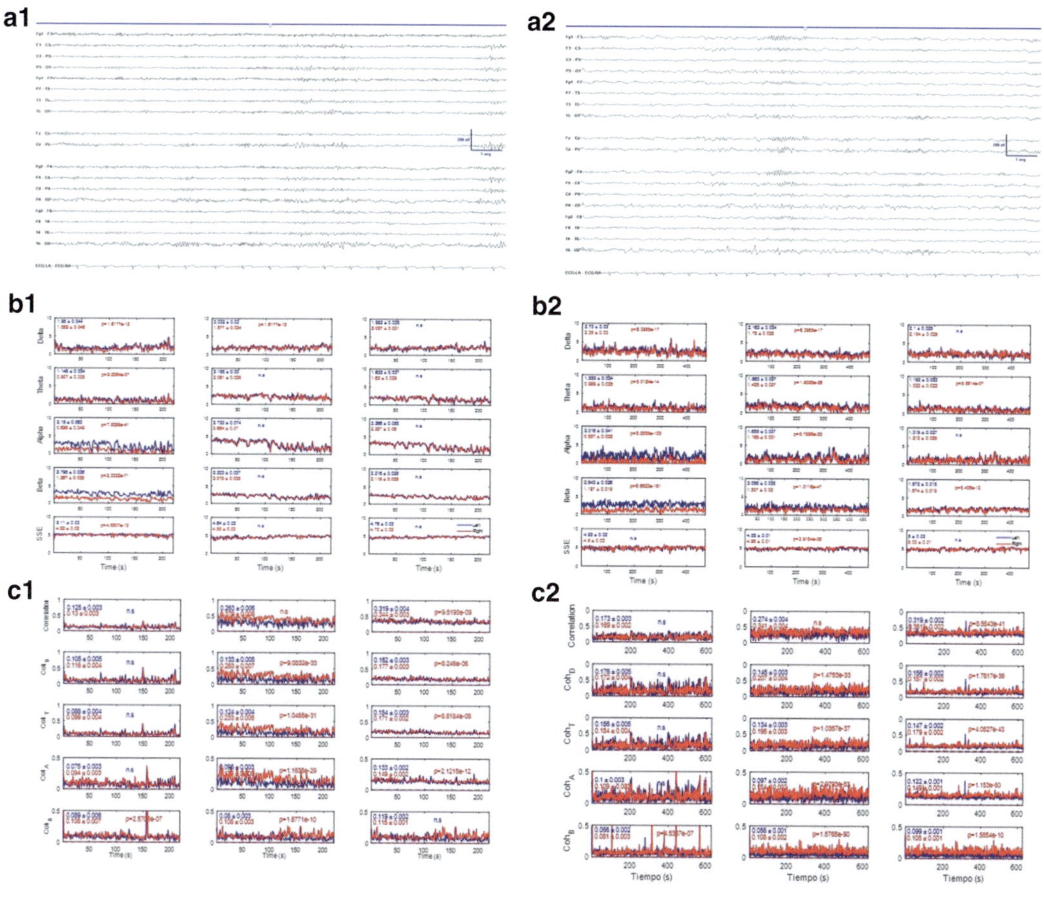

Fig. 7.8 Background abnormalities during (**a1**) wakefulness and (**a2**) NREM sleep in the absence of epileptiform activity. (**b1**) Lobar logPS dynamics during wakefulness and (**b2**) NREM and, (**c1**) dynamics of synchronization during wakefulness and (**c2**) NREM sleep. Red = right hemisphere; blue = left hemisphere. Inset numbers indicate the mean ± SEM, and a value of probability is indicated when significance between both lobes is obtained. Every column in the dynamics of logPS and synchronization corresponds to frontal (left), parieto-occipital (middle), and temporal lobes (right)

logPS of the EEG bands by lobe showed a relative decrease in right frontal alpha and beta power (Fig. 7.8b1). This asymmetry persists during sleep, but an increase in left frontal delta power was also observed in all left parieto-occipital bands (Fig. 7.8b2). During wakefulness, a relative loss of synchronization was observed in all right parieto-occipito-temporal bands (Fig. 7.8c1). This same asynchrony was maintained during NREM sleep (Fig.7.8c2).

In this case, all the differences observed for logPS during awake were increased during the NREM N2 phase, mainly in the fronto-parieto-occipital lobes. The same can be observed from the *coh* and *correlation* measures. The persistence of these asymmetries and/or asynchronies during the sleep period is important because it can help to identify the presence of pathology and help to lateralize the hemisphere and, even identify the lobe affected from the analysis of background.

The properties of the thalamo-cortico-thalamic circuits, which are of great importance in the propagation of epileptic activity, can also be modified during sleep. These changes are responsible for the increased frequency of occurrence of focal epileptic activity during NREM. A very significant increase in generalized activity in this phase of sleep is characteristic for generalized epilepsy during the NREM sleep stages. In addition to the increased frequency, the properties of these discharges are modified. Fig. 7.9 shows the example taken from a 20-year-old male, diagnosed from epilepsy after an encephalitis at 5 years old.

However, not only does the frequency of generalized bursts increase during sleep, but the very dynamics of the spike-wave was modified. Figure 7.10 shows the example of a 19-year-old male patient, diagnosed with Lennox-Gastaut syndrome and with intense epileptogenic activity in the form of generalized slow spike-wave bursts lower than 3.5 Hz. In order to perform the spectral analysis, we averaged 30 periods 10 seconds-length, containing spike-wave waveforms, synchronized with respect to the first spike, both during wakefulness and during NREM N2 sleep. Although this method degrades the end of the averaged spike-wave, the first phases are reasonably preserved. It is interesting to observe how sleep modifies the intrinsic features of the epileptogenic discharge, increasing the intrinsic period of discharge.

Therefore, numerical analysis during sleep provides very valuable information on the presence of epilepsy since the functional alterations of the neural networks and systems involved observed during wakefulness are maintained during sleep.

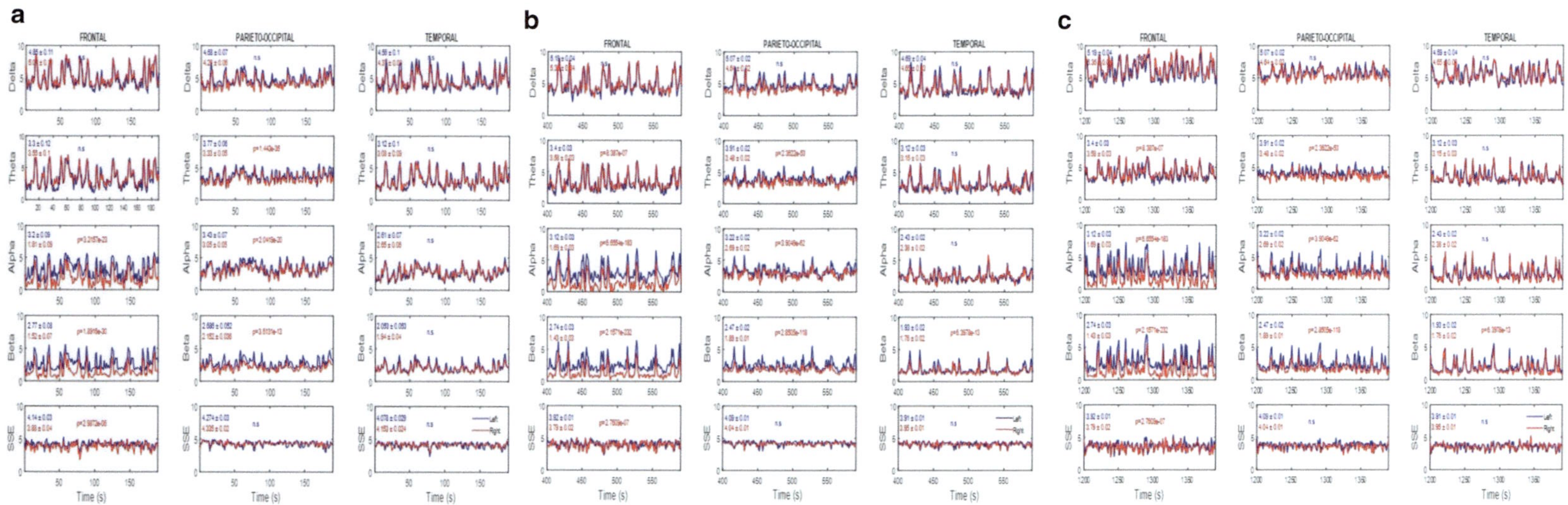

Fig. 7.9 Increased generalized epileptiform activity during sleep. (**a**) LogPS dynamics during wakefulness, (**b**) at the onset of NREM N2 sleep, and (**c**) during NREM N3 sleep. The frequency of generalized bursts (seen as spikes synchronized across all the bands) is 13, 12, and 21 in equal periods of 190 s. The generalized spike-wave bursts affect all the EEG bands. Furthermore, the asymmetries in spectral power observed during wakefulness are increased during sleep. Red = right; blue = left. Inset numbers indicate the mean ± SEM, and a value of probability is indicated when significance between both lobes is obtained. Every column in the dynamics of logPS and synchronization corresponds to frontal (left), parieto-occipital (middle), and temporal lobes (right)

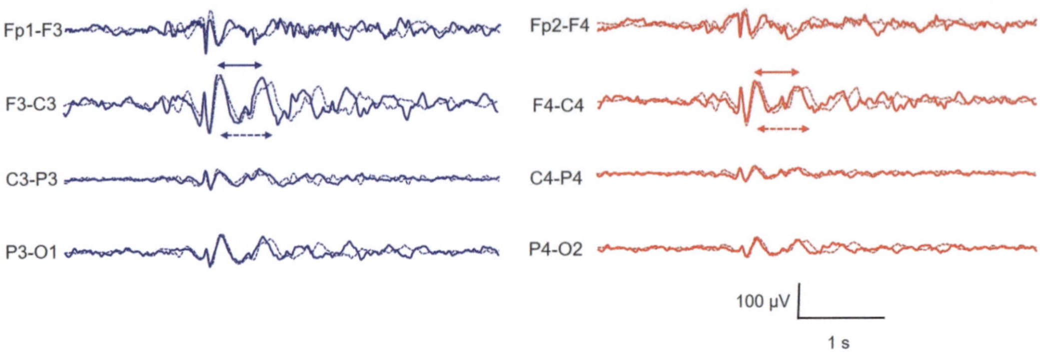

Fig. 7.10 Modification of the intrinsic properties of generalized epileptiform discharges during the wakefulness (thick line) and NREM N2 sleep stage (dashed lines). Horizontal double arrow represents the period of the first two elements for both stages

7.4 Epileptic Seizures

The presence of ES, in general terms, is easy to identify visually in EEG recordings. However, in certain patients, it is not easy to know whether it is a generalized or focal seizure and, in the last case, to identify the hemisphere of origin of the seizure. In these cases, the use of numerical analysis tools can be of great help. Although it should not be forgotten that epileptic seizures can be so artifactual (by the patient's movement, eye-ball movement, or muscle artifact) that numerical analysis is impossible.

According to the ILAE, an ES is a transient occurrence of signs and/or symptoms due to abnormal excessive or synchronous neuronal activity in the cortex of the brain [27], and the excess of neural cortical neural activity can be observed at the EEG as an increase of beta activity [28, 29]. Therefore, during an ES it should be observed an increase, either locally or generalized, in the activity of the fastest EEG band. Nevertheless, there are several facts that modify this pathophysiological hallmark of epilepsy. Briefly, we can enumerate the next:

1. More bands are implicated. Although the definition implies rapid neuronal activity in the cerebral cortex, which in scalp is recorded as a beta band, it is clear that during most ESs progressively larger sets of neural systems (and eventually the whole brain) are affected. This necessarily implies that other EEG bands are modified during the evolution of the seizure. It is therefore common to observe an increase in delta and/or theta bands during seizures too. Only seizures strictly restricted to the cortex, such as motor area seizures (Jacksonian march) or central beta seizures (see below) show an (near) exclusive increase in the beta band.
2. Biophysics of brain-scalp. The skull acts as a high resistance system [30] to the conductance of electric current compared to other structures, such as the cerebrospinal fluid, the meninges, or the scalp. Currents with a lower amplitude (such as beta or gamma bands) will not have a sufficient amplitude in the scalp to be distinguished from noise, and therefore will not generate observable potentials on the surface. See Chap. 2 Biophysics of EEG for a more detailed description.
3. EMG artifact. A large percentage of ES involve more or less intense contractions of the facial muscles, ranging from moderate or mild intensity movements, such as oromandibular automatisms during seizures with loss of consciousness, to highly intense contractions, such as myoclonic or tonic contractions of the facial muscles, or even generalized tonic-clonic seizures. This muscular activity contaminates recording, making it extremely difficult to identify truly cerebral beta activity.
4. Pathological and pharmacological state. The amplitude, morphology, and frequency fea-

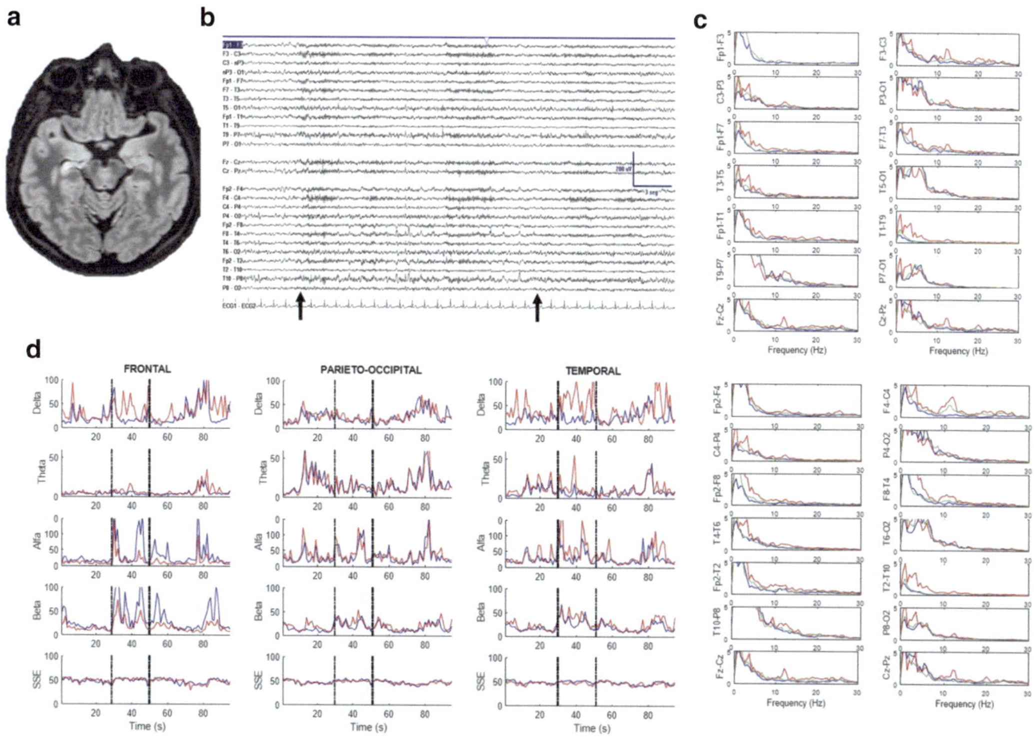

Fig. 7.11 Electrical seizure during sleep in a patient with right mesial temporal sclerosis. (**a**) MR imaging showing right mesial temporal sclerosis. (**b**) Recording with accessory channels in temporal lobes. Arrows indicate the onset and offset of the ictal pattern. (**c**) Average spectra before (blue), during (red), and after the event (green) for all channels in the recording. At the top are the left hemisphere channels, and at the bottom are the right hemisphere channels. Ordinate axis in μV2/Hz. (**d**) Band dynamics by lobe. The vertical lines indicate the seizure and correspond to the arrows in B. Red = right hemisphere; blue = left hemisphere

tures defining epileptic seizures in patients under physiological conditions, such as spontaneous seizures recorded rsEEG recordings or in a v-EEG study, are not always observable in other circumstances. The presence of another brain pathology (in addition to epilepsy) such as a previous stroke or encephalopathy, as well as the existence of sedo-analgesia in patients admitted to the intensive care unit, can give rise to ictal bioelectrical patterns with anomalous characteristics, showing waves of more blunted morphology or a lower frequency.

However, in general and provided that the appropriate biophysical and clinical conditions are met, ES result in an increase in beta band power, although this increase may be difficult to observe due to the presence of artifacts or, in other cases because the ictal onset zone is far away from the scalp, as is the case of mesial temporal or frontal lobes epilepsies. Some of the most frequently observed epileptic seizures are mesial temporal seizures. The typical ictal pattern in this type of seizure is either delta or theta [31], but some increase in beta activity is usually observable (although this is often masked by the presence of muscle artifact).

This example refers to a 55-year-old right-handed woman who has been experiencing seizures since the age of 37. She reports an aura with a sensation of coldness ascending from her feet, followed by a disconnection from her surroundings, and orofacial and hand automatisms. At the time of registration, the patient was being treated with lamotrigine (200 mg/12 h). Magnetic resonance imaging showed right mesial temporal sclerosis (Fig. 7.11a).

The recording takes place during NREM sleep. A blunted theta pattern of sharp waves was observed in the right temporal region (Fig. 7.11b). This could be thought to be interictal activity. However, numerical analysis showed a significant increase in the alpha and beta bands in right frontotemporal regions, as can be seen in the spectral analysis (Fig. 7.10c, bottom row). This increase also occurs in the right frontotemporal delta band (Fig. 7.11d).

Epileptic seizures, even those consisting primarily of a theta pattern, as are typically the seizures of mesial temporal lobe epilepsy [32], are associated with local increases in beta activity, of cortical origin.

This example shows the case of a female 56 years old, right-handed patient, with a history of left temporal pilocytic astrocytoma operated in 1996 and left hippocampal sclerosis. She reports episodes of epigastric aura, disconnection from the environment without language emission, oral automatisms and forced flexion of the right hand, since the age of 26, with a monthly frequency. The patient was treated by lacosamide (150 mg/12 h), brivaracetam (100 mg/12 h), and cenobamate (250 mg/day). The patient was referred for preoperative evaluation. Figure 7.12a shows the recording during one of the seizures suffered for the patient. A theta pattern can be observed in the left temporal region that spreads to the frontoparietal region of the same hemisphere. The dynamics of logPS by lobe is shown in Fig. 7.12b. Increased delta and theta power bilaterally is evident, but especially in the left temporal lobe. The analysis of the average spectra for the pre-ictal (blue), ictal (red), and post-ictal (green) states for each of the channels can be observed in Fig. 7.12c, d, with greater gain on the y-axis. A theta component is observed at 5.5 Hz in the temporal region and, to a lesser extent, in the left frontal and parietal regions. When the spectra are shown with greater power resolution, the presence of a beta component stands out in the left temporal region. This component is not of muscular origin (note how the muscle spectral component goes well beyond 30 Hz, for example, in the ictal spectrum of Fp1-F3).

It can, therefore, be concluded that the presence of beta activity during a seizure is a good marker of ictal activity. Conversely, its absence should raise suspicion that the recorded activity does not actually have the features of an epileptic seizure, although its presence may indicate epileptogenicity.

Occasionally, the presence of sudden GRDA during hyperventilation (or spontaneously) can be interpreted as an epileptic seizure. In this example, we present the case of a 74-year-old woman referred for an EEG due to episodes of nonspecific dizziness. The patient was treated with bromazepan (1.5 mg/12 h). During hyperventilation, a generalized 40 s lasting, non-progressive delta pattern (Fig. 7.13a), without clinical symptomatology, was observed. It resolved spontaneously without subsequent slowing or any other post-ictal manifestation. Analysis of the averaged spectra by state (13B1, left hemisphere, and 13B2, right hemisphere) shows the presence of a delta component around 3 Hz, with no increase in beta activity. The dynamics of the logPS (Fig. 7.13c) shows a marked increase in generalized delta and theta power. The beta band remains unchanged. The SSE dynamics shows a decrease (associated with the increase in delta). No changes in heart rate were observed.

In summary, the presence of lateralized rhythmic delta activity (RDA) is a possible ictal pattern, especially when this pattern is evolutionary [32]. This pattern is particularly relevant in ICU patients with brain alteration, whether due to trauma, hemorrhage, infection, or other causes. In these cases, patients, in addition to the brain pathology, may be sedated, which alters the brain's response to a seizure. Therefore, in these cases, the beta component may be absent in the seizure recording, even for numerical analysis. However, this case should be considered the exception rather than the rule. We'll see this type of crisis in Chap. 10 of ICU Patients.

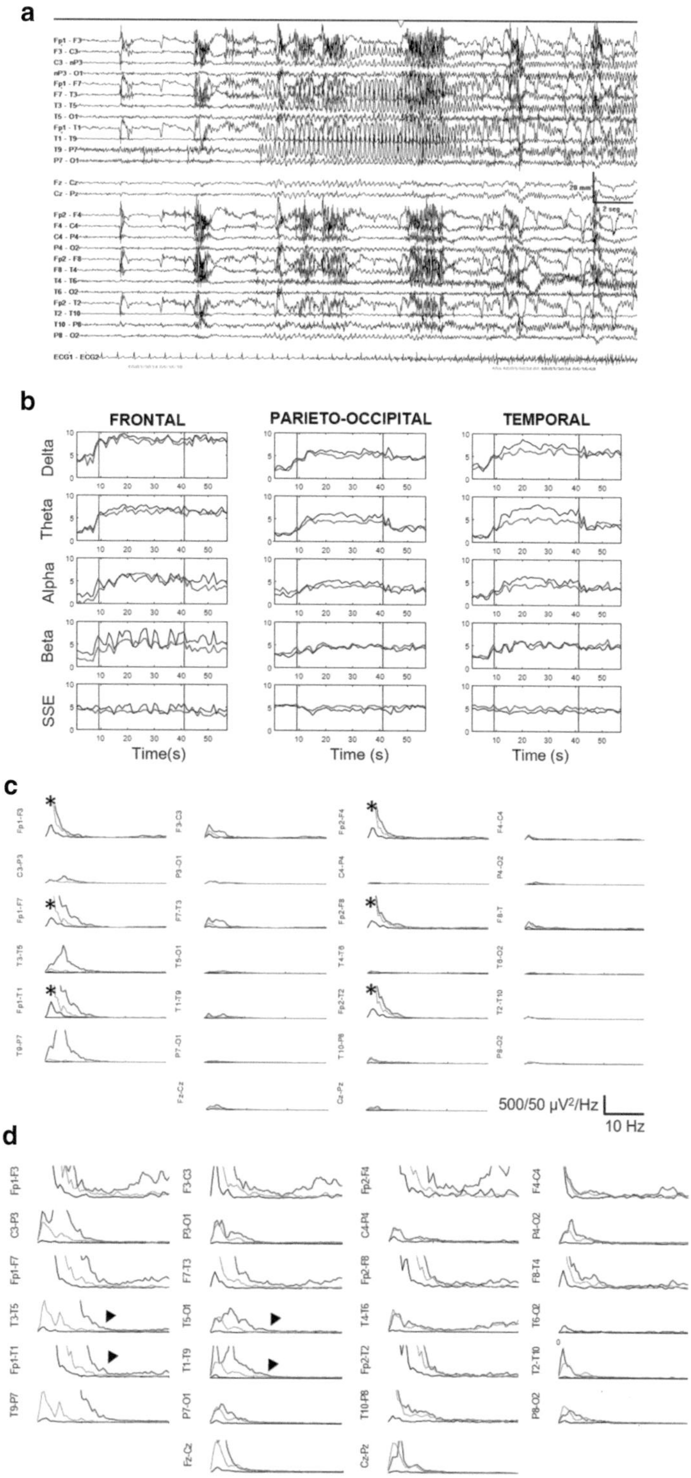

Fig. 7.12 Left temporal lobe seizure. (**a**) Raw recording during the start of a typical temporal lobe seizure in a patient with hippocampal sclerosis. A characteristic high-voltage theta ictal pattern is evident in left temporal and central channels. (**b**) LogPS dynamics for all bands, grouped by lobe. The vertical lines indicate the beginning and end of the seizure. (**c**) Average power spectra for all the channels before (blue), during (red), and after the seizure (green). Asterisk indicates EOG artefact. A prominent theta component at 5.5 Hz is observed at left temporal channels. (**d**) The same power spectra at greater gain of y-axis. Arrowheads indicate the increase of beta activity at temporal channels

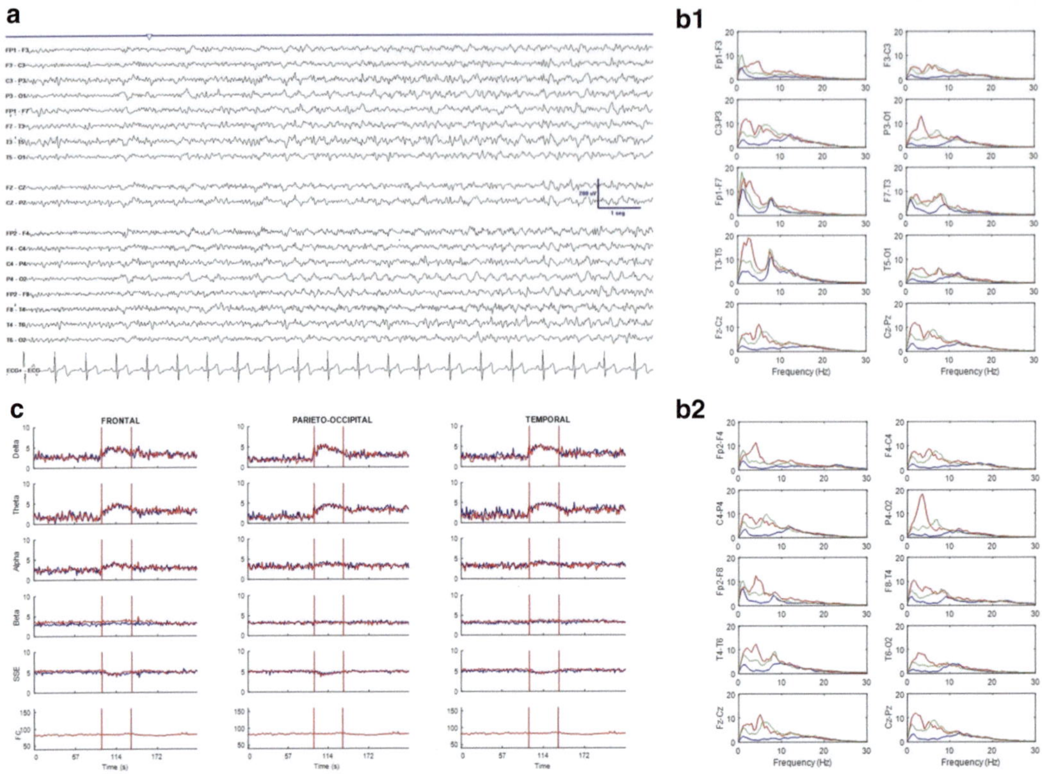

Fig. 7.13 Delta and theta generalized activity during hyperventilation. (**a**) Raw recording. (**b1**) Average spectral from left and (**b2**) right hemispheres. Blue = basal; red = burst, green = after burst. (**c**) Lobar logPS dynamics for the different bands, SSE, and heart rate. Vertical lines indicate the start and stop of the burst. Red = right hemisphere (except for HR), blue lines = left hemisphere

7.5 Central Beta Seizures

Generalized paroxysmal fast activity (GPFA) is an electroencephalographic finding characterized by the following properties [33–36]:

1. Interictal bursts of generalized fast activity, predominantly bifrontal, at 15–25 Hz and lasting longer than 1 s.
2. More common in women.
3. Observed in patients with symptomatic generalized epilepsy, especially during sleep.
4. Associated with epileptic encephalopathy in patients with drug-resistant epilepsy and intellectual disability (typically LGS), although it can be recorded in patients without epileptic encephalopathy but with drug-resistant epilepsy.

We have recently observed a type of ES with some features similar to GPFA, but appearing in a completely different clinical situation. These are patients with no history of epilepsy, with an EEG pattern of encephalopathy and alterations in the level of consciousness that are associated with the presence of an EEG pattern with ictal characteristics with a predominance of beta and alpha frequencies.

The clinical recordings, conventional and quantified, of 6 men and 3 women, aged between 64 and 88 years, admitted for meningoencephalitis (1 case), diabetic ketoacidosis (1 case), cerebrovascular accident (4 cases), or ES (3 cases) were analyzed. The seizures consisted of a low amplitude, beta pattern originating in the frontoparietal and midline regions (Fig. 7.14), remaining well-localized. On three occasions, a lateralized ictal pattern was obtained, but in the remaining cases the pattern was bilateral, as described above.

Numerically, central beta seizures showed a localized increase in beta spectral power between 15 and 25 Hz and a local increase in Pearson correlation and coherence, lasting between 15 and 55 s. On three occasions they were unilateral and on the rest bilateral. They were not associated with motor or sensory symptoms, but rather with a low level of consciousness and mild tachycardia, with an increase between 5 and 10 beats per minute (bpm). It is very interesting to observe that the background activity showed moderate encephalopathy in 8/9 cases and a normal pattern in one case, but in the last case, with very low generalized and especially frontal synchronization.

Figure 7.15 shows a real example of this type of crisis. A 78—year-old right-handed male with a right middle cerebral artery stroke and no personal or family history of epilepsy. Three days after admission, the patient experienced fluctuating episodes of decreased level of consciousness. The rsEEG showed a recording with no clear anteroposterior gradient and features that could be considered physiological (Fig. 7.15a). However, numerical analysis of this recording showed the existence of moderate encephalopathy, with a marked generalized increase in the delta component (Fig. 7.15b). Scalp connectivity was clearly altered (Fig. 7.15c), with connectivity shifted towards the anterior half, and was also asymmetrical, with decreased connectivity in the right temporo-parietal region. The recording was performed during a low level of consciousness. Periods of encephalopathy were interrupted by an amplitude-increasing beta pattern (Fig. 7.15d), associated with a slight increase in heart rate but without motor or sensory manifestations. Analysis of the ictal period reveals a well-defined alpha/beta component in some parasagittal and midline channels (Fig. 7.15e). The dynamics of the different logPS bands by lobe show a specific increase in generalized alpha and beta power.

In summary, central beta seizures have the following characteristics:

1. ES lasting >10 s and showing typical frequencies between 15 and 25 Hz.
2. Clear predominance in central regions.
3. They are generally bilateral and less frequently unilateral.
4. They are more common in men.

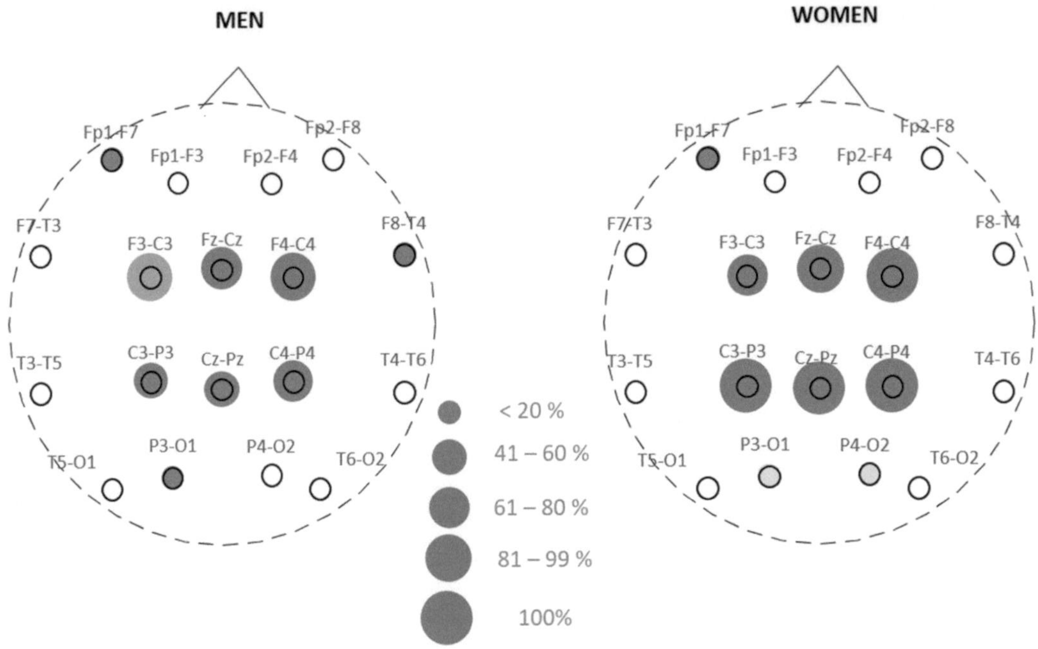

Fig. 7.14 Analysis of the topography where central beta seizures appear. The area and intensity of the circles are related to the absolute frequency of beta seizures

5. They appear during wakefulness.
6. They appear in patients with no history of epileptic encephalopathy or drug-resistant epilepsy.
7. They are generally associated with the presence of encephalopathy (89%) due to metabolic alterations or direct brain involvement.
8. They present with altered level of consciousness.

Therefore, although they share some morphological characteristics with GPFA, they appear in completely different clinical situations, in the context of an acute brain insult, and are associated with impaired level of consciousness, making them a new clinical entity that had not been previously described.

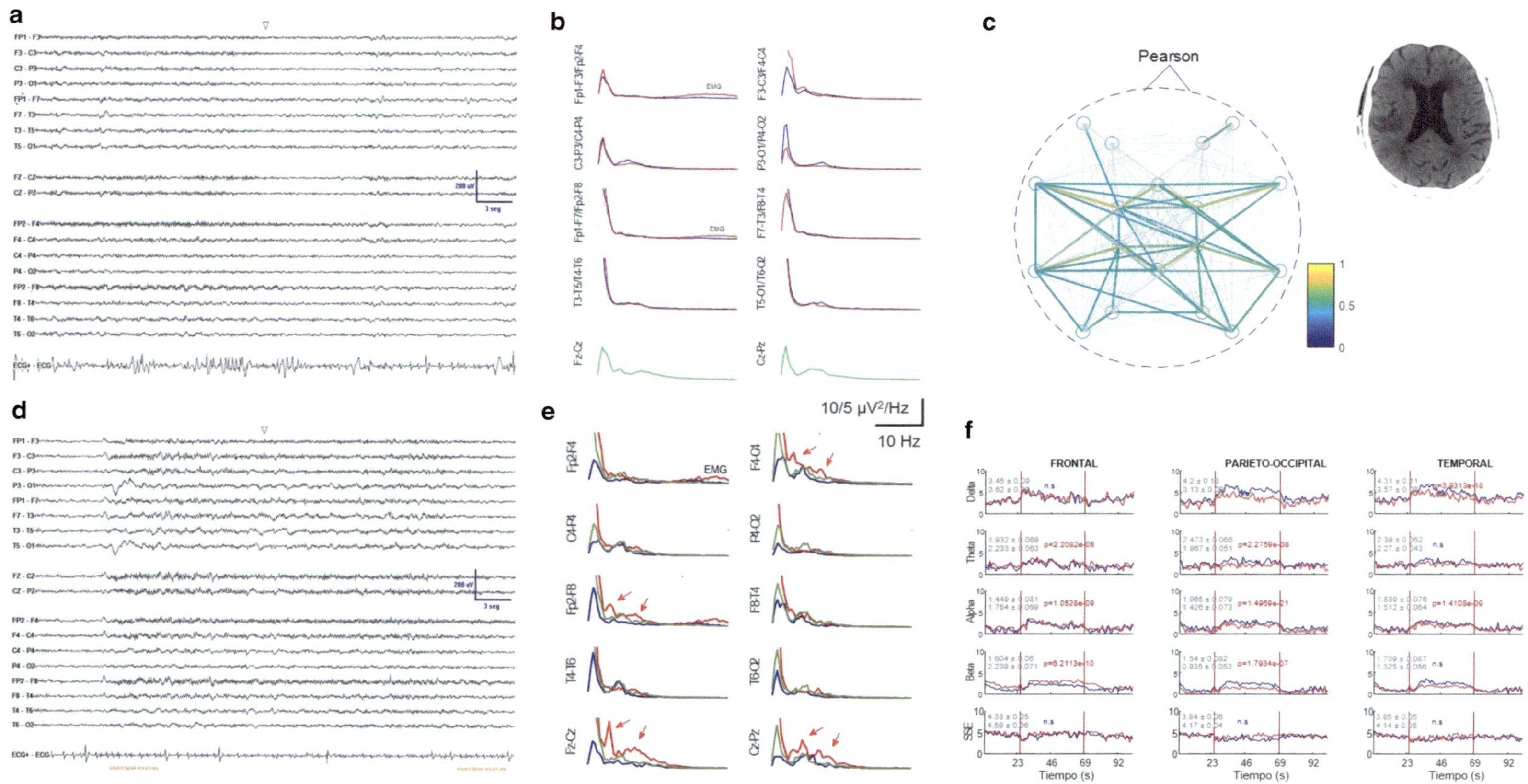

Fig. 7.15 EEG recordings in a patient with a low level of consciousness due to stroke (upper right inset). (**a**) rsEEG. (**b**) Average spectra per channel obtained from **a**. (**c**) Abnormal connectogram. (**d**) Recording showing the presence of a beta epileptic seizure. (**e**) Average spectra before (blue), during (red), and after the seizure (green). (**f**) Dynamics of the logPS of the EEG bands by lobe. Red = right; blue = left; green = midline

7.6 Status Epilepticus

EEG recording is essential for the diagnosis and monitoring of non-convulsive status epilepticus (NCSE) [37–39]. In many cases, it is the only method to identify this state in patients with altered levels of consciousness and no clinical manifestations or very subtle ones. NCSE is one of the situations where numerical analysis can be most useful. In many cases, the EEG recording is clear enough to make a positive diagnosis. However, in many others, the origin of the status epilepticus is difficult to define. In any case, the numerical analysis provides very interesting information about the characteristics of the NCSE.

The next example shows the case of an 89-year-old right-handed woman taken to the emergency department with a probable stroke and right hemispheric syndrome. During the EEG recording, the patient presented a low level of consciousness. The rsEEG showed a seemingly generalized sharp-wave pattern with a frequency of 1.8 Hz (Fig. 7.16a). The CT scan showed no evidence of ischemia, or intracranial hemorrhage, but signs of chronic small vessel ischemia and ventriculomegaly (Fig. 7.16b). The average spectral analysis (performed with a 5 s window) shows the presence of a generalized component at 1.25 Hz. However, in the right fronto-parieto-temporal region, the presence of anomalous components at 2.62 and 3.64 Hz is observed (Fig. 7.16c). The pathological brain state is shown in the connectogram (Fig. 7.16d), with anterior displacement of connectivity, which also shows a significant asymmetry, with decreased right hemispheric connectivity. This decrease in right hemisphere synchronization is shown in Fig. 7.16e. The top row shows the dynamics of the Pearson correlation for the

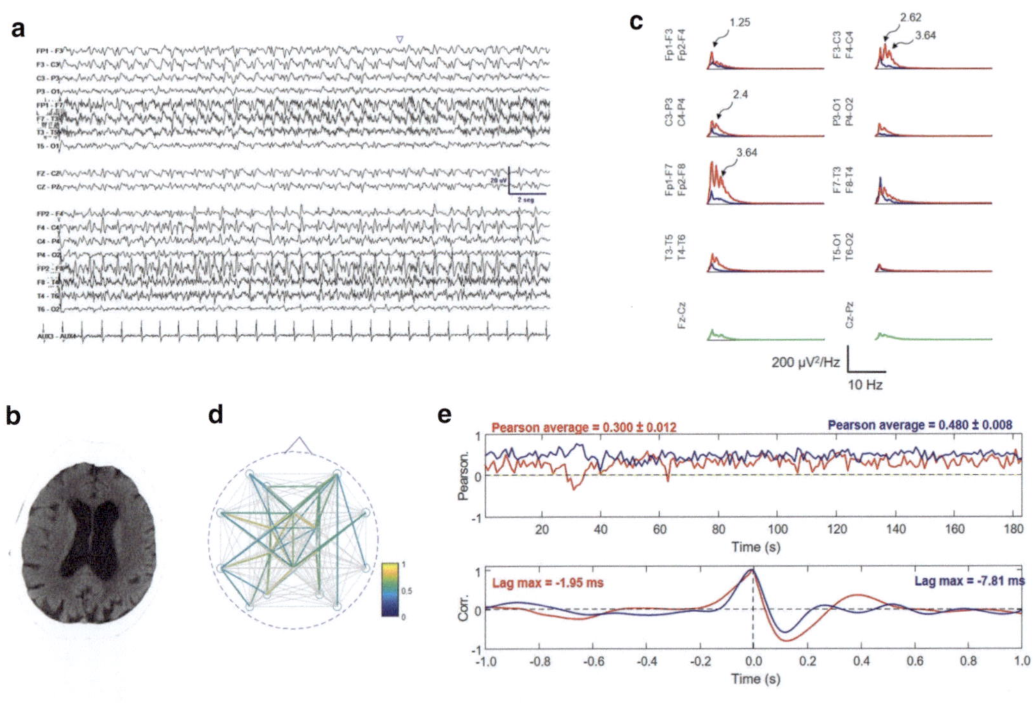

Fig. 7.16 Patient with NCSE. (**a**) Raw recording. (**b**) CT scan. (**c**) Averaged spectra per channel. Red = right hemisphere, blue = left hemisphere, green = midline. (**d**) Connectogram for Pearson correlation. (**e**) Synchronization analysis for channel pairs; the upper row shows the Pearson correlation dynamics for the F3-C3/P3-O1 pairs (blue) and for F4-C4/P4-O2 (red); the lower row shows the interhemispheric cross-correlation between the frontal (F3-C3/F4-C4, red) and parieto-occipital (P3-O1/P4-O2, blue) channels. The max lag values are indicated in the same color code

F3-C3/P3-O1 channels (blue) and F4-C4/P4-O2 (red). The bottom row shows the interhemispheric cross-correlation. Synchronization between the F3-C3/F4-C4 channels is shown in red, and between the P3-O1/P4-O2 channels in blue. In both cases, asynchrony is evident, with a right-to-left direction, although this difference is smaller in the frontal lobes (max lag = −1.95 ms) than in the parieto-occipital channels (max lag = −7.81 ms). Numerical analysis therefore shows that this is an NCSE with a right hemispheric origin, rather than a generalized one, as might be thought from the raw record.

This example is of a 54-year-old, right-handed woman with no significant medical history, who was brought to the Emergency Department after being found unconscious at home. A right frontal contusion on the scalp was observed. The CT scan showed no significant findings. Two days after the admission in ICU, she was under sedation and analgesia, with propofol 8 ml/h and remifentanil 4 mg/h. An EEG recording was performed which showed a generalized NCSE (in a patient with neuromuscular blockade, Fig. 7.17a). Spectral analysis, with a 5-s window, shows a delta component at 2.25 Hz, bilateral and predominantly anterior, with other components associated with 4.5 and 6.25 Hz. In the occipito-temporal region, only the 2.25 Hz component is observed. The connectogram shows a generalized, abnormal pattern with a marked increase in right temporal connectivity (Fig. 7.17c). Synchronization dynamics shows a marked increase in Pearson's correlation and all coherence measures in the right hemisphere, but especially in the temporal lobe ($p < 10^{-35}$). As can be seen from the cross-correlation measures (Fig. 7.17e), the activity is not truly widespread. Activity appears first in the right hemisphere, with greatest synchronization between the frontal lobes, followed by the middle temporal regions, and finally by the parieto-occipital regions (Fig. 7.17e, upper row). In all three cases, the latency indicates an apparent right hemisphere origin, although the latency is highest in the temporal regions, followed by the frontal region, and finally the occipito-parietal region. The cross-correlation shows widespread but non-homogeneous synchronization patterns, with differences in synchronization between the frontal, temporal, and occipital regions analyzed (Fig. 7.17e, bottom row).

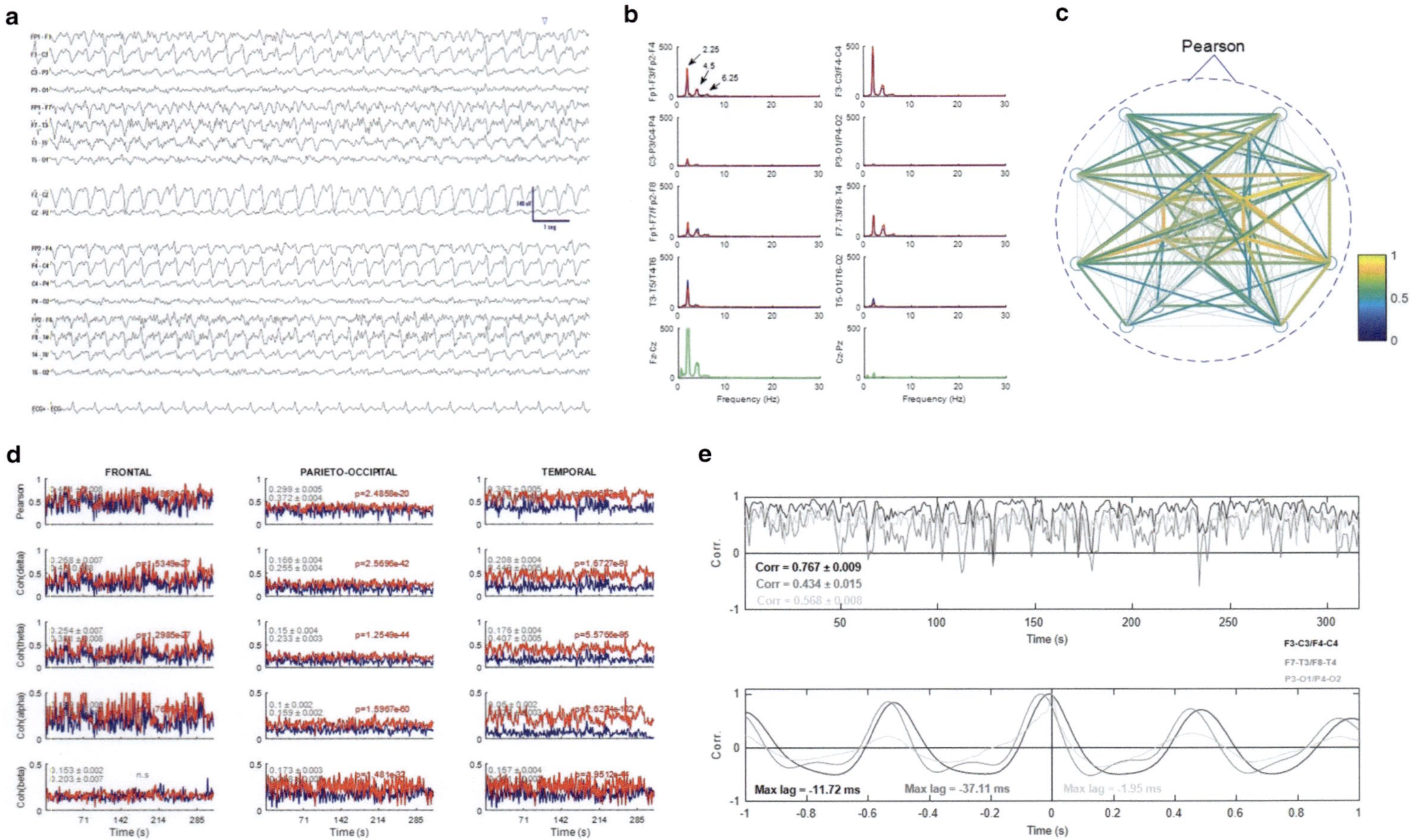

Fig. 7.17 NCSE. (**a**) Raw recording. (**b**) Averaged spectra, with 5-s windows. (**c**) Averaged connectogram for Pearson correlation. (**d**) Dynamics of synchronization measures by lobe. Red = right hemisphere; Blue = left hemisphere. (**e**) Analysis of interlobar synchronization. Black = frontal lobes, intermediate gray = temporal lobes, light gray = occipital lobes

7.7 Concluding Remarks

Epilepsy is the pathology for which EEG and its variants, such as v-EEG, cEEG in ICU monitoring, or electrocorticography (ECoG), are most commonly used. Various numerical tools are used to identify epileptic sources (e.g., the LORETA method) or to identify and quantify seizures in ICU patients (e.g., trend graphs in Persyst® software). But qEEG can do much more than this. Analysis of interictal recordings allows us to identify alterations in spectral composition or synchronization that can aid diagnosis, even in the absence of epileptogenic activity. These background alterations persist during sleep, indicating an established alteration in brain dynamics.

The diagnosis of epileptic seizures is also enhanced by qEEG, especially when subtle changes are present. However, in cases of generalized SE, an apparent hemispheric origin can also be determined. However, the artifacts that are often present during seizures (EOG, EMG, movement, etc.) can make their analysis less efficient than in other states.

In the case of central beta seizures, described in this chapter for the first time as a differential clinical entity, numerical analysis is essential, because visual identification is difficult and they can progress unnoticed.

References

1. Villanueva V, Sánchez-Álvarez JC, Peña P, Puig JS, Caballero-Martínez F, Gil-Nagel A. Treatment initiation in epilepsy: an expert consensus in Spain. Epilepsy Behav. 2010;19(3):332–42. https://doi.org/10.1016/j.yebeh.2010.07.016. Epub 2010 Sep 24. PMID: 20869920.
2. Fisher RS, van Emde BW, Blume W, Elger C, Genton P, Lee P, Engel J Jr. Epileptic seizures and epilepsy: definitions proposed by the International League Against Epilepsy (ILAE) and the International Bureau for Epilepsy (IBE). Epilepsia. 2005;46(4):470–2. https://doi.org/10.1111/j.0013-9580.2005.66104.x. PMID: 15816939.
3. Vega-Zelaya L, Martín Abad E, Pastor J. Quantified EEG for the characterization of epileptic seizures versus periodic activity in critically ill patients. Brain Sci. 2020;10:158. https://doi.org/10.3390/brainsci10030158.
4. Scheffer IE, Berkovic S, Capovilla G, Connolly MB, French J, Guilhoto L, Hirsch E, Jain S, Mathern GW, Moshé SL, Nordli DR, Perucca E, Tomson T, Wiebe S, Zhang YH, Zuberi SM. ILAE classification of the epilepsies: position paper of the ILAE Commission for Classification and Terminology. Epilepsia. 2017;58(4):512–21. https://doi.org/10.1111/epi.13709. Epub 2017 Mar 8. PMID: 28276062; PMCID: PMC5386840
5. Sanz A, Sola RG, Vega-Zelaya L, Pastor J, Ortega GJ. Network theoretical approach to describe epileptic processes. In: Hintermüller C, editor. Advanced biosignal processing and diagnostic methods. Intech; 2016. p. 85–115. ISBN 978-953-51-2519-8.
6. Bartolomei F, Wendling F, Regis J, Gavaret M, Guye M, Chauvel P. Pre-ictal synchronicity in limbic networks of mesial temporal lobe epilepsy. Epilepsy Res. 2004;61:89–104. https://doi.org/10.1016/j.eplepsyres.2004.06.006.
7. Vega-Zelaya L, Torres CV, Garnes-Camarena O, Ortega GJ, García-Navarrete E, Navas M, Sola RG, Pastor J. Electrocorticographic evidence and surgical implications of different physiopathologic subtypes of temporal epilepsy. Clin Neurophysiol. 2014;125:2349–57. https://doi.org/10.1016/j.clinph.2014.03.027.
8. Pastor J, Vega-Zelaya L. Redes neurales en epilepsia. In: Velasco AL, editor. Contribución Iberoamericana a la Epilepsia. Mexico: Editorial Alfil; 2018. p. 17–33. ISBN: 978-607-741-220-5.
9. Sutter R, Kaplan PW. Clinical and electroencephalographic correlates of acute encephalopathy. J Clin Neurophysiol. 2013;30:443–53.
10. Lorente L, Martín MM, González-Rivero AF, Pérez-Cejas A, Argueso M, Ramos L, Solé-Violán J, Cáceres JJ, Jiménez A, García-Marín V. High serum caspase-cleaved Cytokeratin-18 levels and mortality of traumatic brain injury patients. Brain Sci. 2019;9:269.
11. Pastor J, Vega-Zelaya L. Titration of pharmacological responses in ICU patients by quantified EEG. Curr Neuropharmacol. 2022; https://doi.org/10.2174/1570159X20666220411083213.
12. Stewart CP, Otsubo H, Ochi A, Sharma R, Hutchison JS, Hahn CD. Seizure identification in the ICU using quantitative EEG displays. Neurology. 2010;75:1501–8.
13. Sackellares JC, Shiau DS, Halford JJ, LaRoche SM, Kelly KM. Quantitative EEG analysis for automated detection of nonconvulsive seizures in intensive care units. Epilepsy Behav. 2011;22:S69–73.
14. Dericioglu N, Yetim E, Bas DF, Bilgen N, Caglar G, Arsava EM, Topcuoglu MA. Non-expert use of quantitative EEG displays for seizure identification in the adult neuro-intensive care unit. Epilepsy Res. 2015;109:48–56.
15. Haider HA, Esteller R, Hahn CD, Westover MB, Halford JJ, Lee JW, Shafi MM, Gaspard N, Herman ST, Gerard EE, Critical Care EEG Monitoring Research Consortium, et al. Sensitivity of quantitative EEG for seizure identification in the intensive care unit. Neurology. 2016;87:935–44.

16. Akman CI, Micic V, Thompson A, Riviello JJ Jr. Seizure detection using digital trend analysis: factors affecting utility. Epilepsy Res. 2011;93:66–72.
17. Williamson CA, Wahlster S, Shafi MM, Westover MB. Sensitivity of compressed spectral arrays for detecting seizures in acutely ill adults. Neurocrit Care. 2014;20:32–9.
18. de Las V, Heras M, Sala-Padro J, Pedro-Perez J, García-Parra B, Hernández-Pérez G, Falip M. Utility of quantitative EEG in neurological emergencies and ICU clinical practice. Brain Sci. 2024;14(9):939. https://doi.org/10.3390/brainsci14090939. PMID: 39335433; PMCID: PMC11430096
19. Pastor J, Vega-Zelaya L, Martin-Abad E. Necessity of quantitative EEG in daily clinical practice. In: Nakano H, editor. Electroencephalography. InTech; 2021. ISBN 978-1-83968-289-6.
20. Pitetzis D, Frantzidis C, Psoma E, Deretzi G, Kalogera-Fountzila A, Bamidis PD, Spilioti M. EEG network analysis in epilepsy with generalized tonic–clonic seizures alone. Brain Sci. 2022;12:1574. https://doi.org/10.3390/brainsci12111574.
21. Daley JT, DeWolfe JL. Sleep, circadian rhythms, and epilepsy. Curr Treat Options Neurol. 2018;20(11):47. https://doi.org/10.1007/s11940-018-0534-1.
22. McLeod GA, Ghassemi A, Ng MC. Can REM sleep localize the epileptogenic zone? A systematic review and analysis. Front Neurol. 2020;11:584. https://doi.org/10.3389/fneur.2020.00584. PMID: 32793089; PMCID: PMC7393443.
23. Frauscher B, Timofeev I. Chapter 14: Seizures and sleep. In: Noebels JL, Avoli M, Rogawski MA, Vezzani A, Delgado-Escueta AV, editors. Jasper's basic mechanisms of the epilepsies. 5th. ed. New York: Oxford University Press; 2024. PMID: 39637155.
24. Timofeev I, Bazhenov M, Seigneur J, Sejnowski T. Neuronal synchronization and Thalamocortical rhythms in sleep, wake and epilepsy. In: Noebels JL, Avoli M, Rogawski MA, Olsen RW, Delgado-Escueta AV, editors. Jasper's basic mechanisms of the epilepsies [Internet]. 4th ed. Bethesda: National Center for Biotechnology Information (US); 2012.
25. Souza RT, Gerhardt GJ, Schönwald SV, Rybarczyk-Filho JL, Lemke N. Synchronization and propagation of global sleep spindles. PLoS One. 2016;11(3):e0151369. https://doi.org/10.1371/journal.pone.0151369. PMID: 26963102; PMCID: PMC4786112.
26. Guo D, Thomas RJ, Liu Y, Shea SA, Lu J, Peng CK. Slow wave synchronization and sleep state transitions. Sci Rep. 2022;12(1):7467. https://doi.org/10.1038/s41598-022-11513-0. PMID: 35523989; PMCID: PMC9076647.
27. Fisher RS, Acevedo C, Arzimanoglou A, Bogacz A, Cross JH, Elger CE, Engel J Jr, Forsgren L, French JA, Glynn M, Hesdorffer DC, Lee BI, Mathern GW, Moshé SL, Perucca E, Scheffer IE, Tomson T, Watanabe M, Wiebe S. ILAE official report: a practical clinical definition of epilepsy. Epilepsia. 2014;55(4):475–82. https://doi.org/10.1111/epi.12550. Epub 2014 Apr 14. PMID: 24730690.
28. Ropohl A, Sperling W, Elstner S, Tomandl B, Reulbach U, Kaltenhäuser M, Kornhuber J, Maihöfner C. Cortical activity associated with auditory hallucinations. Neuroreport. 2004;15(3):523–6. https://doi.org/10.1097/00001756-200403010-00028. PMID: 15094516
29. Gong R, Roth RW, Chang AJ, Sinha N, Parashos A, Davis KA, Kuzniecky R, Bonilha L, Gleichgerrcht E. EEG ictal power dynamics, function-structure associations, and epilepsy surgical outcomes. Neurology. 2024;102(12):e209451. https://doi.org/10.1212/WNL.0000000000209451. Epub 2024 May 31. PMID: 38820468.
30. Nunez PL, Srinivasan R. Electric current in physical media tissue. Electric fields of the brain. In: The neurophysics of EEG. 2nd ed. Oxford University Press; 2006. ISBN-13 978-0-19-505038-7.
31. Lüders HO. Mesial temporal sclerosis. In: Lüders HO, editor. Textbook of epilepsy surgery. London: Informa Healthcare; 2008. ISBN-10: 1 84184 576 0.
32. Hirsch LJ, Fong MWK, Leitinger M, LaRoche SM, Beniczky S, Abend NS, Lee JW, Wusthoff CJ, Hahn CD, Westover MB, Gerard EE, Herman ST, Haider HA, Osman G, Rodriguez-Ruiz A, Maciel CB, Gilmore EJ, Fernandez A, Rosenthal ES, Claassen J, Husain AM, Yoo JY, So EL, Kaplan PW, Nuwer MR, van Putten M, Sutter R, Drislane FW, Trinka E, Gaspard N. American clinical neurophysiology society's standardized critical care EEG terminology: 2021 version. J Clin Neurophysiol. 2021;38(1):1–29. https://doi.org/10.1097/WNP.0000000000000806. PMID: 33475321; PMCID: PMC8135051.
33. Arzimanoglou A, French J, Blume WT, Cross JH, Ernst JP, Feucht M, Genton P, Guerrini R, Kluger G, Pellock JM, Perucca E, Wheless JW. Lennox-Gastaut syndrome: a consensus approach on diagnosis, assessment, management, and trial methodology. Lancet Neurol. 2009;8(1):82–93. https://doi.org/10.1016/S1474-4422(08)70292-8. PMID: 19081517.
34. Aydin-Özemir Z, Matur Z, Bebek N, Gürses C, Gökyiğit A, Baykan B. Long-term follow-up of adult patients with genetic generalized epilepsy with typical absence seizures and generalized paroxysmal fast activity in their EEG. Seizure. 2014;23(8):607–15. https://doi.org/10.1016/j.seizure.2014.04.017. Epub 2014 May 15. PMID: 24908140.
35. Bansal L, Vargas Collado L, Pawar K, Nagesh D, Ilyas M, Hall A, Kinnaman B, Abdelmoity A. Electroclinical features of generalized paroxysmal fast activity in typical absence seizures. J Clin Neurophysiol. 2019;36(1):36–44. https://doi.org/10.1097/WNP.0000000000000535. PMID: 30394946.
36. Das Pektezel L, Tezer FI, Saygi S. Electroclinical spectrum of generalized paroxysmal fast activity in adults without epileptic encephalopathy. Neurol Sci. 2022;43(6):3857–66. https://doi.org/10.1007/s10072-021-05808-9. Epub 2022 Jan 16. PMID: 35034232.

37. Trinka E, Leitinger M. Which EEG patterns in coma are nonconvulsive status epilepticus? Epilepsy Behav. 2015;49:203–22. https://doi.org/10.1016/j.yebeh.2015.05.005. Epub 2015 Jul 4. PMID: 26148985.
38. Leitinger M, Beniczky S, Rohracher A, Gardella E, Kalss G, Qerama E, Höfler J, Hess Lindberg-Larsen A, Kuchukhidze G, Dobesberger J, Langthaler PB, Trinka E. Salzburg Consensus Criteria for Non-Convulsive Status Epilepticus–approach to clinical application. Epilepsy Behav. 2015;49:158–63. https://doi.org/10.1016/j.yebeh.2015.05.007. Epub 2015 Jun 17. PMID: 26092326.
39. Wang X, Yang F, Chen B, Jiang W. Non-convulsive seizures and non-convulsive status epilepticus in neuro-intensive care unit. Acta Neurol Scand. 2022;146(6):752–60. https://doi.org/10.1111/ane.13718. Epub 2022 Oct 26. PMID: 36285875.

8 Intensive Care Unit Patients

8.1 Introduction

The use of qEEG has probably experienced the greater and more profound implementation in patients admitted to the ICU. Conventional EEG, video-EEG, and continuous or long-term monitoring (cEEG) have experimented an exponential increase in the number of publications and in the number of courses and meetings where special attention is devoted to them.

In fact, one of the most used commercial packages used in clinical practice (Persyst®) is primarily intended (although it can be used in other settings) for cEEG in ICU patients. An excellent description of the use of this software can be found in [1].

Critically ill patients admitted to the ICU are usually treated with sedoanalgesia and usually have an orotracheal intubation [2–5]. This situation prevents or severely limits the clinical examination of patients, which means that an undesirable or even serious clinical situation, such as cerebral ischemia or an ES/SE, can go unnoticed by the physician, leading to a severe worsening of the patient's condition. This situation can occur both in patients who have been admitted for primary brain involvement (e.g., traumatic brain injury or stroke) and in those, whose primary disease did not directly affect the CNS (e.g., a metabolic derangement). Therefore, the best way to assess the clinical situation of the brain, in these states, is by means of EEG. However, these are patients who have both conditions of primary brain involvement and sedation-analgesia. Therefore, the EEG recording is completely different from the conventional one, which can make its correct interpretation very difficult. Moreover, although it is always important to interpret the EEG properly, to reach a good diagnosis, in this type of patients, it is even more important, because the diagnosis will guide the treatment. Thus, it is as important to diagnose a status epilepticus correctly as it is not to misdiagnose a status epilepticus if, in fact, it is not.

From the point of view of diagnostic purposes, we can distinguish between recordings of short duration, i.e., conventional EEG, and cEEG, which can have a duration of several hours to several days. We will follow this methodological distinction in the development of this chapter.

8.2 Short-Term Recordings

The indications for urgent EEG are as follows [6, 7]:

J. Pastor, *Quantitative EEG Practical Guide*, https://doi.org/10.1007/978-3-032-05457-9_8

- Differential diagnosis of abnormal movements [8].
- Suspicion of seizure or non-convulsive status epilepticus [9].
- Coma of unknown origin [10, 11].
- Behavioral disturbances or psychiatric symptoms [12, 13].
- Follow-up after cardiorespiratory arrest [14, 15].
- Diagnosis of brain death [16–18].

All of these situations can occur in ICU patients.

Conventional EEG recordings in ICU inpatients are generally performed using commercially available caps, which have the 10–20 international system implemented, as is done with rsEEG recordings performed in other types of patients, such as outpatients or inpatients. Accessory channels for ECG and EMG are essential.

As previously mentioned, diagnosing an ES/SE can be very difficult in patients with severe cerebral involvement and under sedoanalgesia. In many of these patients, the background activity shows encephalopathy pattern, with abundant delta and theta rhythms, and the changes can be insidious. In cases like these, visual diagnosis can be extremely difficult.

The following example shows the case of a right-handed 64-year-old female patient presenting right hemispheric ischemic stroke of the proximal M2 segment of the right middle cerebral artery of cardio-embolic origin, treated with primary thrombectomy. Four days later, she presented loss of consciousness and was admitted to ICU. Prophylactic anticonvulsive treatment with levetiracetam 1000 mg was administered. Despite the reduction of sedation and analgesia, the patient did not awaken, so she was monitored by EEG. During the first day of monitoring, which was the fourth day in the ICU, the result shown in Fig. 8.1 was obtained.

The recordings were obtained at 9:30 a.m. (Fig. 8.1a), 10:00 a.m. (Fig. 8.1b), and 10:15 a.m. (Fig. 8.1c). Visual analysis of the recordings resulted extremely difficult, as the changes were insidious and do not fit the canonical description of an ES. However, consecutive analysis of average spectra (Fig. 8.1d) shows changes in the peak frequency of certain channels that cannot be explained as a change in the alternating pattern or as an arousal or change in state. The first spectrum (pink, corresponding to record **a**) shows a typically encephalopathic pattern, with a marked excess of the delta band at 1 Hz generalized throughout the scalp. However, 15 min later (blue) and 30 min afterward (green, corresponding to record **b**), a shift of the main delta peak to 2 Hz is observed, with another smaller peak at 4 Hz, in the left frontotemporal region. This activity is not observed in the rest of the left hemisphere (except in the parieto-occipital region) or in the contralateral hemisphere. It is very interesting to note that this frequency is not even the same throughout the left hemisphere, as can be seen in the spectrum of the anterior midline channel. The delta activity at 2 Hz, 15 min later (red, corresponding to recording **c**) returned to the initial spectra, with a maximum peak at 1 Hz and disappearance of the theta component at 4 Hz. This dynamic can only correspond to a left frontal focal epileptic seizure which, however, due to its insidious dynamics, is very difficult to diagnose in the visual analysis of the recording.

These abnormal ictal patterns, in which the presence of epileptiform waveforms (spikes, sharp waves, etc.) is not evident, are very common in ICU patients. In fact, in our experience, they are more frequent than typical ictal patterns. The presence of a primary cerebral involvement (in this case, the stroke) and significant sedation and analgesia for several days likely alter normal cerebral dynamics, causing ictal manifestations to present atypical features, as in this case. However, its diagnosis is absolutely essential for proper patient management. Therefore, in cases where seizures are highly suspected, even though they are not evident in conventional recording (not even for highly experienced neurophysiologists), numerical analysis can enable diagnosis. It is highly doubtful, however, that CSA analysis can be useful in cases like this, as the changes are truly subtle and gradual. However, consecutive average spectra allow the identification of the focal change in the spectral pattern of a single

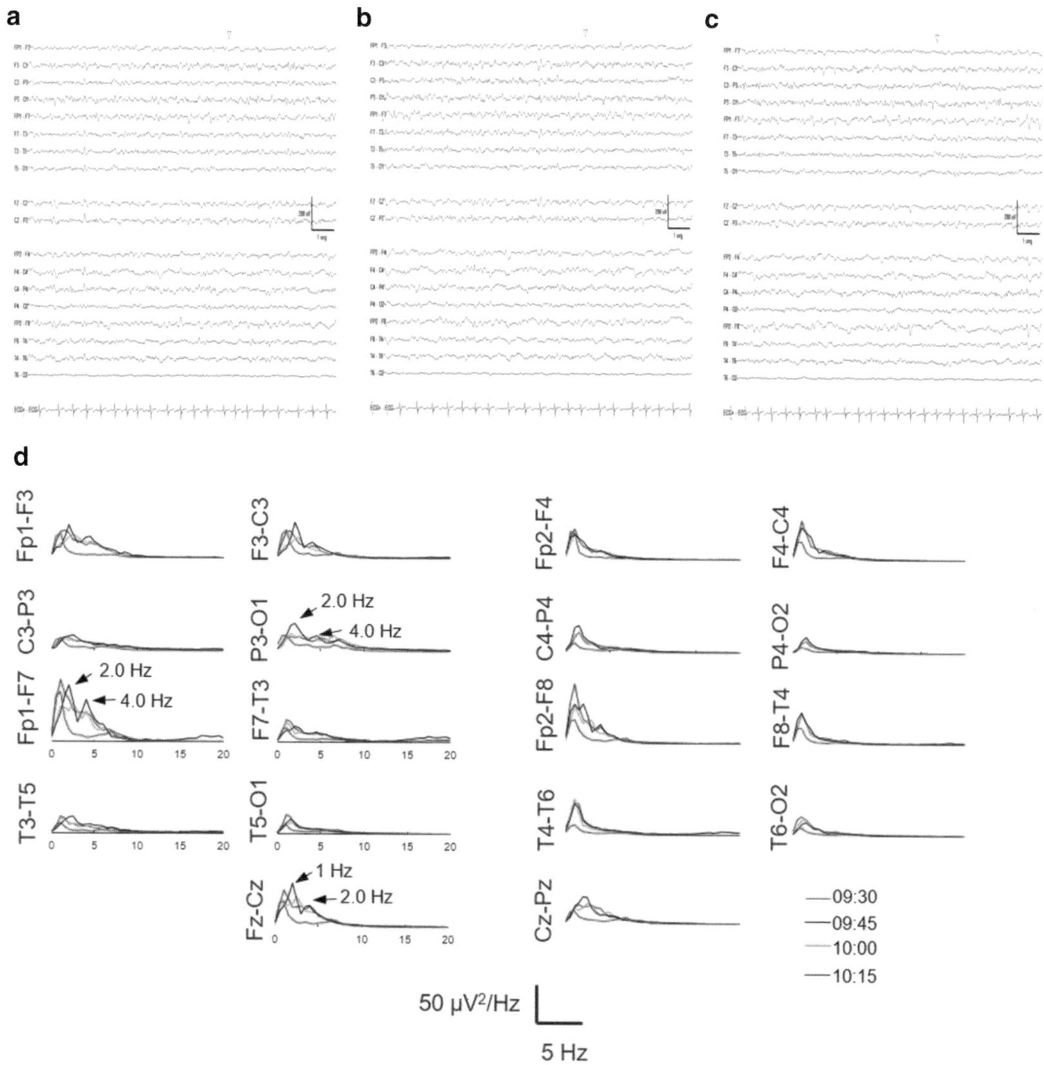

Fig. 8.1 Analysis of a slowly evolving focal seizure. (**a**) Initial recording, (**b**) at 30 min, and (**c**) at 45 min. (**d**) Average spectra for the periods shown in (**a–c**). Another period 15 min after the initial recording has also been included. The different colors indicate the periods analyzed, as shown in the legend

lobe, which, furthermore, in the case of seizures (but not in the case of status), will be reversible.

Numerical analysis is an extraordinarily helpful tool for the diagnosis of ES/SE. The increased beta activity, which indicates cortical activity, during ictal events is a good diagnostic criterion for epileptic seizures (see below). In the next example, we show the case of an 82-year-old woman, admitted at ICU for low level of consciousness and clonic movements in the left hemi-body. At the time of registration, the patient was treated with levetiracetam (1500 mg/12), and valproate in continuous infusion at 20 ml/h. Initially, with a sedation-analgesia of propofol (8 ml/h) and remifentanil (4 ml/h). A sedation window was performed. The EEG recording initially showed a disorganized pattern, with no antero-posterior gradient or posterior dominant rhythm (Fig. 8.2a). Around 10 min later, generalized high-moderate amplitude waves appeared, increasing in frequency (Fig. 8.2b). No abnormal movements were observed in the patient (due to

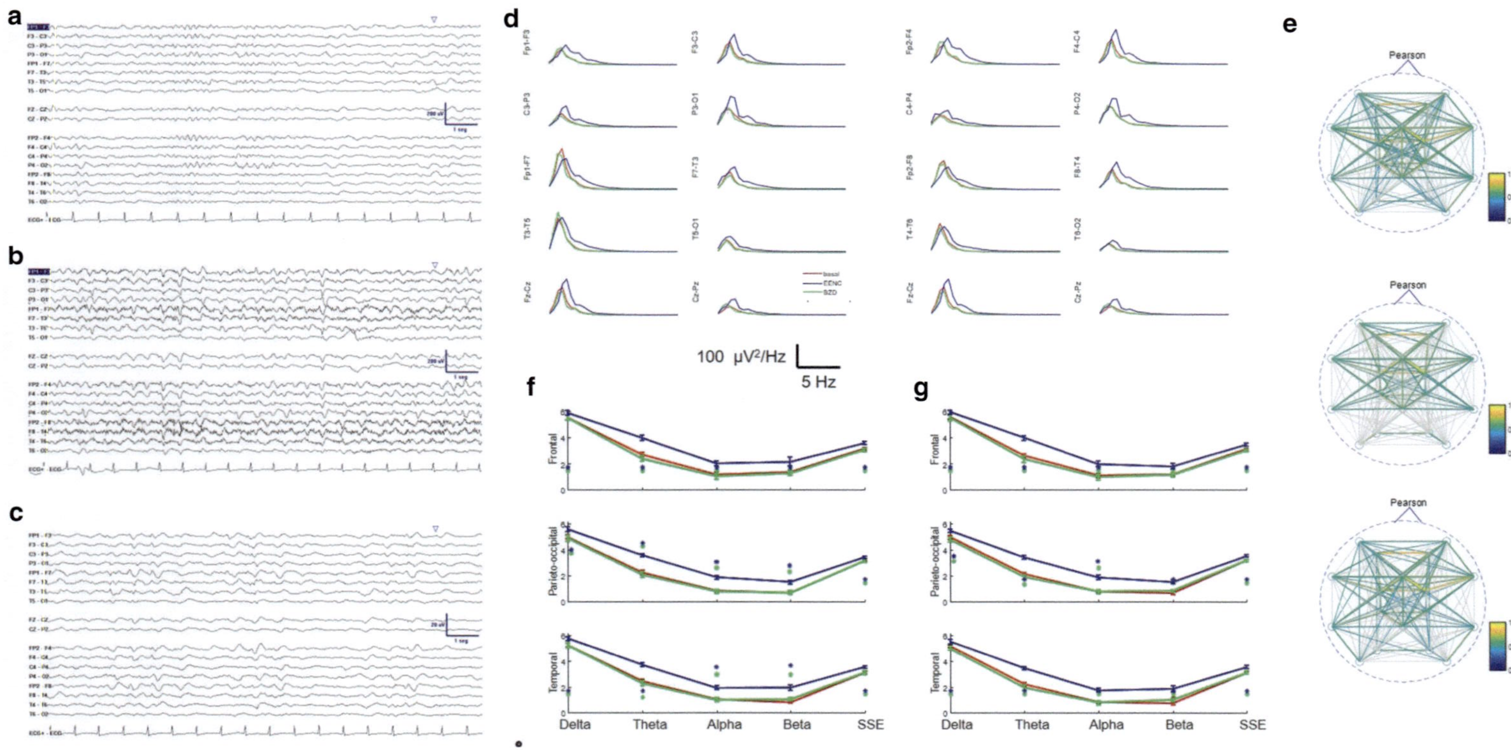

Fig. 8.2 Appearance of generalized SE during the sedation window. (**a**) Baseline recording, with sedation-analgesia. (**b**) Registration 10 min after the propofol was withdrawn. (**c**) Registration after administration of BZD. (**d**) Average spectra for the previous periods. (**e**) Connectograms of the same periods (from top to bottom) as in the left column. (**f**) Average logPS values for the left hemisphere and G) right. Red = basal; blue = DE, green = BZD

neuro-muscular blockade). When the presence of SE was verified (see below), a dose of clonazepam (3 mg i.v.) was administered. A complete remission of the SE was obtained (Fig. 8.2c) and the reintroduction of sedation-analgesia at the same doses was performed. Figure 8.2d shows the average spectra per channel for the three periods. As can be seen (blue line), the maximum peak of delta power moves to 2 Hz in an almost generalized way throughout the scalp, with a delta component also appearing at 3.5 Hz in frontal regions. This component is characteristic of the SE and does not appear in triphasic or delta waves of encephalopathy. This spectrum returns to its previous state after the application of benzodiazepine. However, it is not only the average spectra that are modified during the SE. Connectivity is also modified. The right column (Fig. 8.2e) shows the baseline connectogram (upper), during SE (middle), and after reversal (lower). The extreme connectograms (upper and lower) show the anomalous connectivity characteristic of encephalopathies (in this case due to sedation-analgesia). However, during SE this connectivity is altered and increased in the anterior half of the scalp, while it decreases at its posterior half. As mentioned above, the beta band has to be increased during an ES/SE. Of course, this increase is usually accompanied by the modification of other bands. Figures 8.2f, g show the average values of the logPS by bands, together with the SSE for the left and right hemispheres, respectively. Asterisks indicate statistical significance for the pairs of values analyzed. In this case, differences are observed between the basal/EE and BZD/EE values.

As already mentioned in Chap. 7 Epilepsy, central beta seizures can be difficult to identify and may be undertreated. These seizures are also observed in patients admitted to the ICU under sedoanalgesia. We present the case of a 43-year-old male patient admitted to the ICU with convulsive status epilepticus and right frontotemporal head trauma. He was treated with levetiracetam (1500 mg/8 h), lacosamide (50 mg/8 h), propofol (8 ml/h), and remifentanil (6 ml/h). An EEG performed 12 h after admission revealed no epileptiform waveforms or typical seizures. The background EEG (Fig. 8.3a) showed a disorganized pattern, with a generalized delta-dominant rhythm, without an anteroposterior gradient or posterior-dominant rhythm. This pattern was frequently interrupted by the presence of beta bursts, of a progressive nature and parasagittal localization, lasting between 10 and 15 s and associated with subsequent slowing and loss of background rhythm. These episodes, with ictal characteristics, were associated with a slight increase in heart rate (in the example, between 50.9 ± 1.6 bpm, up to a maximum of 58.7 bpm, at the bottom of Fig. 8.3a). Valproate perfusion (600 mg/12 h) was added and recording was repeated after 24 h (see below). The band dynamics of individual channels show a selective increase in the alpha and beta bands, with some local enhancement in the theta band (Fig. 8.3b). Furthermore, the spectral composition of the delta band changes during the seizure, although this is not evident in the raw recording, with a shift of the main delta component from 1 to 2 Hz (Fig. 8.3c). Analysis of the average spectra in more detail shows the specific increase in alpha and beta power in parasagittal regions (Fig. 8.3d).

After the introduction of valproate, the following day's recording showed the disappearance of the seizures. The tracing showed the presence of drug-induced encephalopathy due to sedation and analgesia, with some periods of voltage attenuation, but there was no evidence of central beta seizures (Fig. 8.4a). The diagnostic difficulty of these seizures has already been mentioned. Therefore, the use of quantification tools may be essential in these cases, since, in many cases, only the small alpha and beta peaks are observed in long-term recordings. Figure 8.4b shows the dynamics of the logPS bands for the frontal lobes of the previous recording (Fig. 8.3a) and the recording of the second day (Fig. 8.4a) over a 15-min period. The diagnostic difficulty of these seizures has already been mentioned. Therefore, the use of quantification tools may be essential in these cases, since, in many cases, only the small alpha and beta peaks are observed in long-duration recordings. Figure 8.4b, left column, shows the dynamics of the logPS bands for the frontal lobes of the previous recording

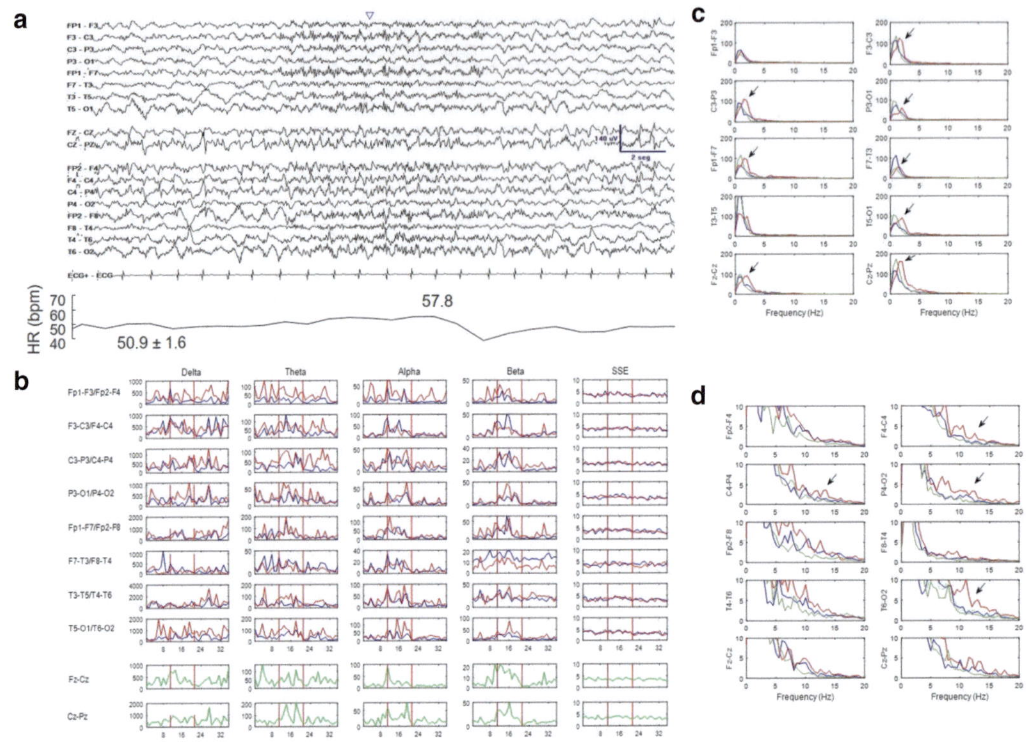

Fig. 8.3 Central beta seizures in a patient undergoing sedation and analgesia in the ICU. (**a**) Recording during one of the seizures. There is no evidence of a typical ictal pattern. The instantaneous heart rate dynamics is shown below. Numerical values are shown for clarity. (**b**) Band dynamics by channel. Vertical lines indicate the seizure period. Blue = left hemisphere; red = right; green = midline. (**c**) Average spectra before (blue), during (red), and after the seizure (green). The shift in the dominant delta frequency during the seizure is indicated by arrows. (**d**) The same spectra, shown with higher power resolution. Alpha and beta power (indicated by arrows) are observed to increase specifically during the seizure. Spectra from one hemisphere are shown for clarity

(Fig. 8.3a) and the recording (right column) of the second day (Fig. 8.4a) over a 15-min period. These peaks (attenuated in this figure due to a logarithmic transformation) indicate the presence of seizures. It is very important to note that 11 seizures were recorded in 15 min, with a total duration of 198 s, representing an ictal load of 22% (which, strictly speaking, corresponds to an electrical NCSE). Status epilepticus in the left column of the figure is associated with irregular dynamics of instantaneous heart rate and SSE. Both variables are stationary once the seizures resolve (Fig. 8.4b, right column). Comparing the average spectra of both states can be misleading (Fig. 8.4c). In both cases, the almost sole component of the spectra is the delta band, which is slightly more prominent on the second day. However, comparing both periods in greater detail (spectral inset) shows the presence of a larger alpha/beta component when seizures are present, which disappears with the control of seizures. The numerical values (in logPS) of these changes are shown in Fig. 8.4d, which demonstrates how the power in the theta to beta bands is clearly greater throughout the scalp during the presence of seizures, as well as the absolute value of the SSE. Only the delta power is greater once the seizures were controlled, but a new antiepileptic drug was introduced. After 3 days, sedation was removed, and the patient was discharged from the ICU.

The diagnosis of ES/SE epilepticus is a fundamental objective of EEG in the ICU. However, it is not the only one. The patient's functional prognosis is another very important objective. One of the most important factors in determining a

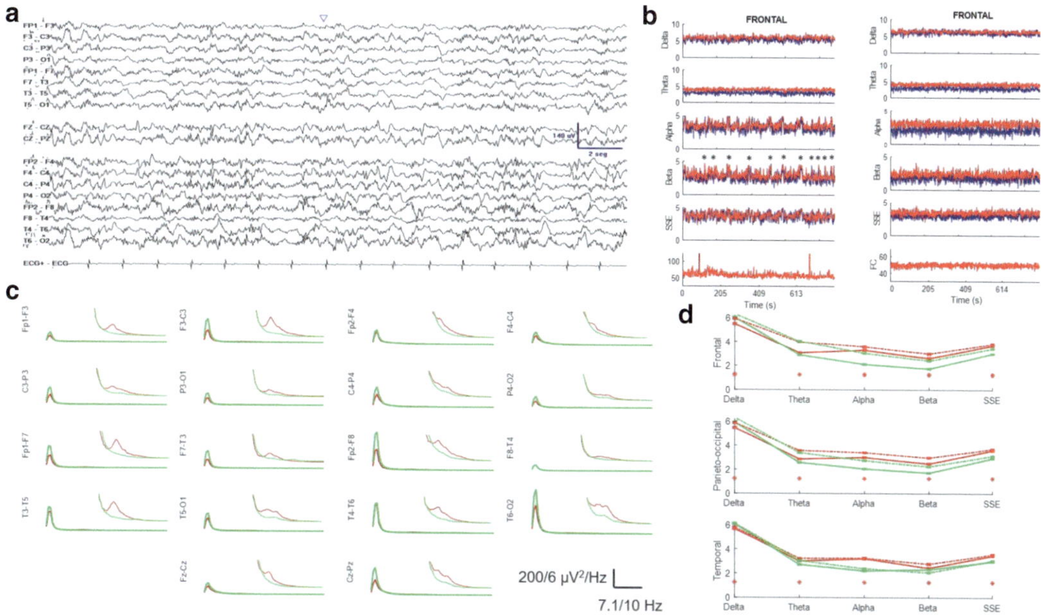

Fig. 8.4 Difficulty in diagnosing central beta seizures. (**a**) Recording the day after the introduction of valproate. No evidence of seizures. (**b**) Band-wise dynamics of logPS for the frontal lobes, SSE, and heart rate for the recordings from the first day (left column) and the second day (right column). Asterisks indicate the presence of central beta seizures. Blue = left hemisphere; red = right. (**c**) Average spectra per channel of recordings from B. Red = recording from the first day; green = recording from the second day. The insets show the spectra in detail, to observe the faster frequencies. (**d**) Comparison of the average logPS values (± SEM) for the bands in different lobes. The red lines correspond to the values from the first day, while the green ones are from the second. The solid lines correspond to the left hemisphere, and the dashed lines to the right

patient's prognosis is the patient's capability to respond to nociceptive stimulation. A lack of response implies a poor prognosis, although it should always be kept in mind that sedation and analgesia can modify cerebral response. Therefore, the lack of response in cases of deep sedation-analgesia should be viewed with caution. However, the presence of a cortical response is a good prognostic factor, so it is very important to assess its existence in detail. In a significant percentage of cases, the cortical response to painful stimulation is practically non-existent in conventional visual analysis. In these cases, numerical analysis is of extraordinary importance, as it can convert a false negative result into a true positive, thereby changing the patient's outcome.

In this patient, painful stimulation (Fig. 8.5a) to the right hand did not appreciably modify the cortical response, so one might assume an absence of response. However, numerical analysis showed completely different results. It can be observed how, during stimulation, a marked reduction in synchronization was observed in the frontal and, to a lesser extent, left temporal regions (Fig. 8.5b). However, not only cortical synchronization was modified; the average spectral component was also profoundly modified in response to pain (Fig. 8.5c). This response obviously indicates a good left hemispheric cortical response (to a lesser extent, also in the right). Within 24 h, the patient awoke without any associated neurological deficit, except for mild psychomotor agitation, which did not require specific treatment.

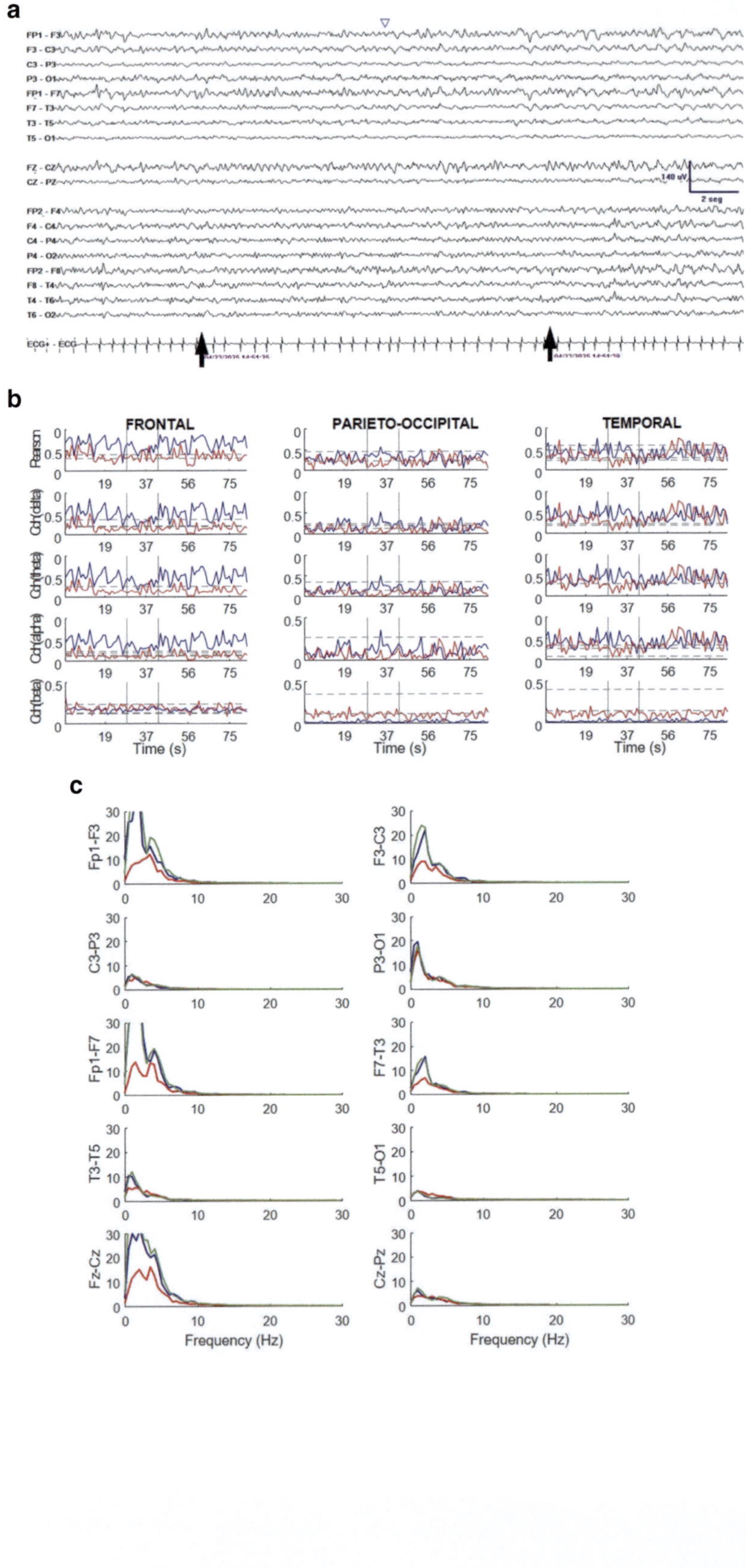

Fig. 8.5 Numerical analysis of the cortical response to painful stimulation. (**a**) Recording during painful stimulation of the nail bed (arrows). (**b**) Dynamics of the synchronization measurements. The horizontal lines show the confidence limits, while the vertical lines indicate the stimulation period. Blue = left hemisphere; red = right hemisphere. (**c**) Average spectra for the left hemisphere and midline channels before (blue), during (red), and after (green) stimulation

8.3 Long-Term Recordings: cEEG

The brain dynamics of patients admitted for long periods in the ICU are highly complex. This makes it extremely difficult to obtain an effective and rapid prognosis for these patients. This dynamic is not sufficiently well studied. However, it is essential to understand these changes in state in order to establish a diagnosis. Numerical study of these dynamics will most likely improve the diagnosis, treatment, and prognosis of these patients.

Different associations, including the American Association of Neurophysiology and the European Association of Intensive Care Medicine, and other scientific societies have made recommendations and elaborated guidelines on when to use continuous EEG monitoring. These guidelines try to establish a homogeneous language for diagnosis, treatment, and also for research purposes.

Indications for a cEEG in ICU patients are as follows [19–22]:

- Alteration of consciousness after a seizure or convulsive status epilepticus.
- Management of non-convulsive seizures or status epilepticus [23, 24].
- Alteration of the state of consciousness in patients with acute brain damage.
- Fluctuation of consciousness or low level of consciousness without known brain injury [25].
- Generalized periodic discharges (GPDs), lateralized (LPDs), or bilateral independent (BiPDs) in a short-duration EEG [26, 27].
- Procedures requiring pharmacological paralysis and at risk of developing epileptic seizures.
- Differential diagnosis of clinical paroxysmal events that simulate epileptic seizures [28, 29].
- Early detection of cerebral ischemia [30, 31].

Various numerical parameters can be monitored over consecutive days, such as the power of the spectral bands, the signal amplitude, or asymmetry [1]. We have analyzed spectral power measurements (both in absolute means and logarithmic transformation), synchronization measurements (Pearson and coherence by bands), SSE, ratio of fast and slow bands (ratio), according to this expression

$$\text{ratio} = \frac{\alpha + \beta}{\delta + \theta} \tag{8.1}$$

In addition, we use a lateralization ratio by bands and lobes ($\text{lat}_{i,\,\text{lobe}}$) that uses this formula

$$\text{lat}_{i,\text{lobe}} = \frac{\text{lPS}_i - rPS_i}{\text{lPS}_i + rPS_i}; i = \alpha, \beta, \delta, \theta, \text{SSE} \tag{8.2}$$

where lPS_i is the power spectrum for the band i at the left lobe and $r\text{PS}_i$ is the right power spectrum of the same band.

In the case of power spectra (or their logarithmic transformations), they are displayed as average values per bands, including average value of all of the lobes analyzed, with the usual color coding. In addition, the average value, along with the standard deviation, of all power values is shown for each period analyzed. This average measurement is shown in black and provides an idea of the overall dynamics.

Focal alterations in brain function can be identified early during cEEG. In this example, it is a 32-year-old man, treated by radiofrequency thermocoagulation in both amygdaloid complexes, for uncontrollable aggressiveness. The patient developed convulsive SE that required admission to the ICU, treatment with sedoanalgesia (propofol 20 ml/h and remifentanil, 4 ml/h), levetiracetam (1 g/8 h), valproate in perfusion (1.6 g/d), and diazepam (5 mg/day). At 24 h after the SE was controlled, a decrease in the ratios between left and right lobes was observed in the cEEG, in all the left hemisphere and especially in the tem-

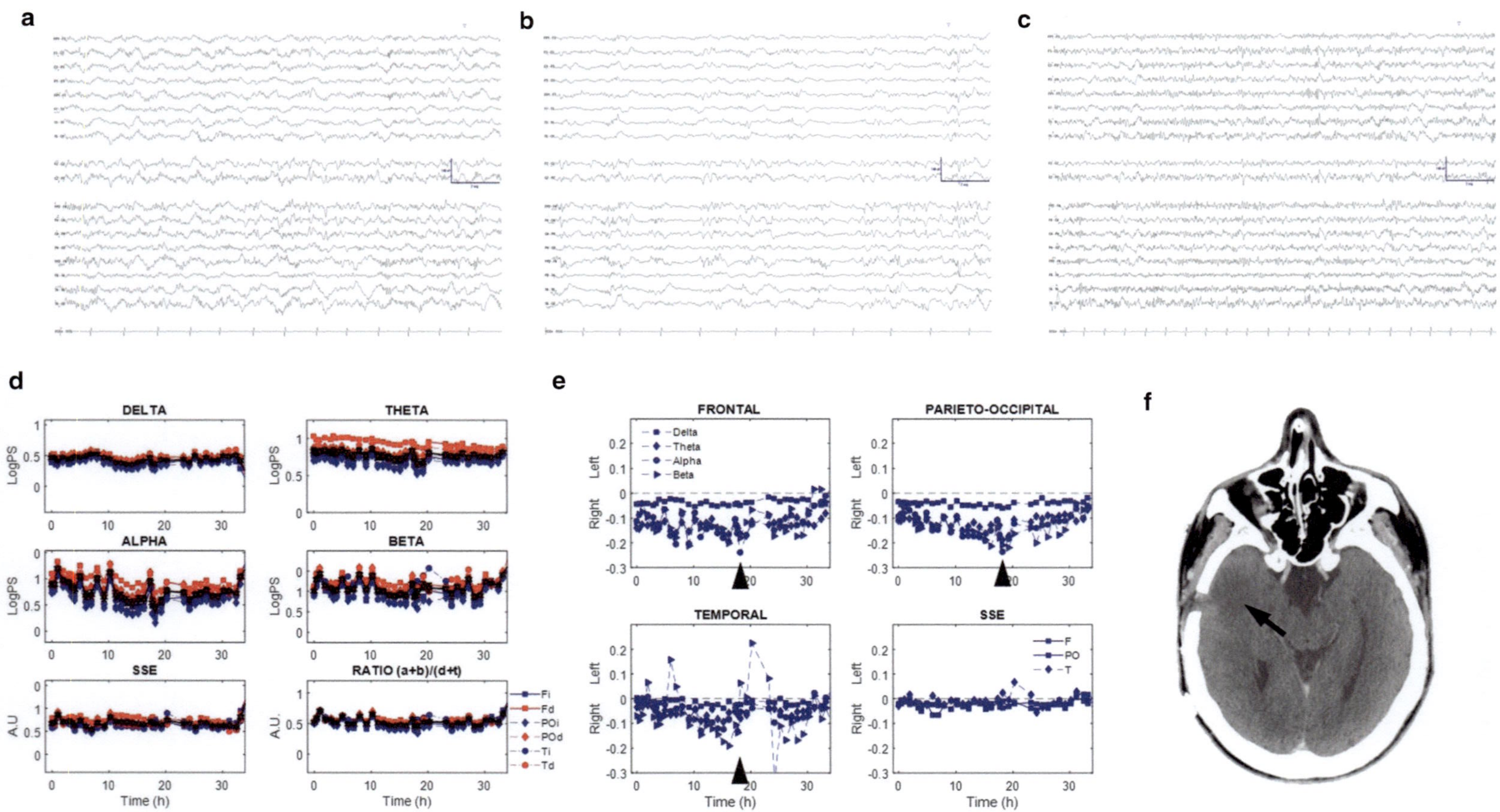

Fig. 8.6 Long-term monitoring (35 h shown) in a patient with controlled SE who developed right temporal lobe oedema. (**a**) Recording at the beginning of the monitoring (first hour). (**b**) Recording at 18 h (arrowhead in the power and lateralization dynamics). (**c**) Recording at 34 h after recording. (**d**) PS dynamics by lobe, grouped by bands. Blue = left hemisphere; red = right hemisphere; black = mean value with standard error. Arrowheads indicate the period showed at **b**. (**e**) Lateralization dynamics. **f**) Computed tomography showing the presence of oedema in the right temporal lobe (arrow)

poral lobe and, more importantly, for the theta band (Fig. 8.6a–c). An urgent CT scan was performed, revealing extracellular oedema in the right temporal lobe. This effect on brain bioelectrical activity gradually decreased until, within 15 h, activity returned to baseline levels. Twenty-four hours later, sedation and analgesia were discontinued and the patient was extubated, with no additional neurological deficits.

It is very interesting to note how the band dynamics are not very informative (Fig. 8.6d), but the lateralization ratio nonetheless shows a shift in the fronto-parieto-occipital lateralization, with a clear predominance of right delta and theta power (Fig. 8.6e). The evolution of the oedema was monitored by the reversal of lateralization. The absence of changes in the frequency ratio is also very striking.

Long-term EEG monitoring is especially useful for the detection and pharmacological monitoring of SE. The presence of changes in the dynamics of the bands (or their logarithmic transformation) is very useful in drawing attention to the possible presence of ictal changes in brain dynamics, as we can observe in the next example.

A 74-year-old right-handed female patient was admitted to the ICU with a head injury of unknown cause. Several SAH foci were present, predominantly in the right temporo-occipital, right frontal, and left parietal and paramedian occipital regions. She also had a fractured superior, inferior, and lateral left zygomatic arch and orbital floor. The patient was monitored for frequent central beta-seizures. Initial treatment consisted of levetiracetam (1500 mg/8 h) and sedation and analgesia with propofol (8 ml/h) and remifentanil (6 ml/h). The initial recording showed a drug-induced encephalopathy pattern (Fig. 8.8a), with a marked increase in power of the delta and alpha bands throughout the right hemisphere, but especially in the temporofrontal region. Two hours after the start of treatment, the propofol dose was reduced to 2 ml/h, while remifentanil was maintained. An NCSE right temporofrontal pattern appeared (Fig. 8.8b), which required a further increase in propofol, initially to 8 ml and, since it was not controlled, up to 12 ml/h (Fig. 8.8c), a dose at which the NCSE disappeared. The logPS dynamics for all bands and lobes show the increase in alpha and beta power, predominantly but not exclusively on the right, that marks the onset and maintenance of NCSE. This change is also shown in the average synchronization dynamics for the different coherence bands (Fig. 8.8e), in which the average coh(alpha) and its variation (represented by the vertical error bars) increase during NCSE and remain constant. The coh(beta) remains dispersed across lobes for much longer. Strikingly, the coh(delta) and coh(theta) decrease with the onset of the pathological state (Figs. 8.7 and 8.8).

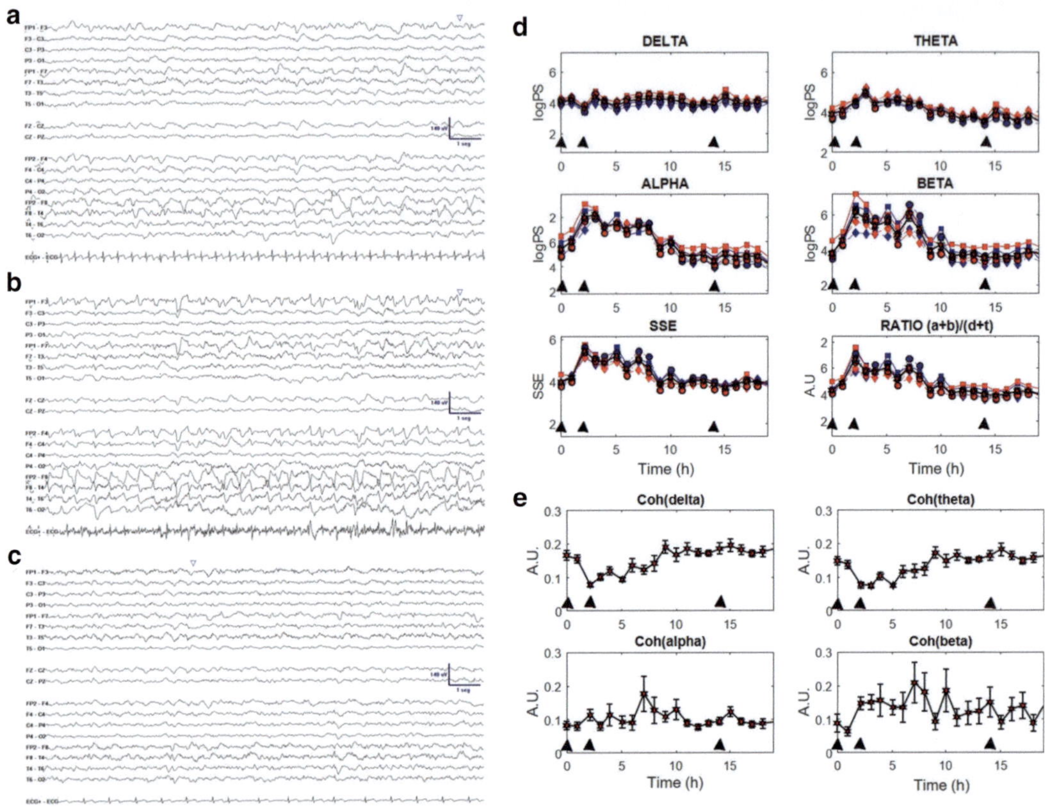

Fig. 8.7 Right hemispheric NCSE. (**a**) Recording at the start of monitoring, (**b**) at 3 h during NCSE, and (**c**) at 2 pm, with a similar pattern to the previous one. (**d**) LogPS dynamics, grouped by bands and lobes. Blue = left, red = right, black = average for the entire band. Squares = frontal, diamonds = parieto-occipital, and circles = temporal. (**e**) Dynamics of the average value and standard deviation for coherence by band. The arrowheads indicate, respectively, the recordings shown in the left column

8.4 EEG-Based Devices to Monitoring in ICU

Although this is not the focus of this work, as it is merely a reduced EEG and therefore offers only partial information, the use of numerical EEG methods in the ICU cannot be overlooked, although they were initially used in the operating room. In fact, these methods are used to monitor the depth of anesthesia through an estimation of brain activity and were initially developed for use in the operating room by anesthesiologists.

There are three main numerical methods used to determine anesthetic depth:

1. Spectral entropy determination [32–34]. This is based on a scaled measurement of a variation in SSE. Its range varies between 0 and 100. Two monopolar frontal channels (Fp1 and Fp2) are generally used for its calculation.
2. The bispectral index (BIS). This method, the details of which are not fully published, uses a transfer function between spectra obtained from brain activity and muscle activity. Its range varies between 0 and 100. Two frontopolar channels are typically used for EEG analysis and two more lateral ones for EMG analysis [35].
3. Density Spectral Array (DSA) analysis [36–39]. This method is very interesting because it provides not only a numerical value but also a colorized image of the power of each EEG band over time. This method uses four channels (Fp1, Fp2, F7, and F8).

These are very useful methods for identifying changes in brain activity, although their exclusive frontal location makes them highly nonspecific. They have the great advantage of being easy to place and interpret, so they can be used to identify the need for a detailed EEG recording.

Although several articles report a good ability to identify ES and interictal epileptiform discharges, obviously only those manifestations that are generalized or clearly predominant in the anterior region will be adequately detected. It is evident that focal seizures at posterior half of scalp or, central beta seizures, as discussed above, are not very likely to be identified.

So far, we have analyzed the clinical use of qEEG in ICU patients. Let's now look at two more examples of scientific use with clear clinical applications. These are the identification of epileptic seizures in the ICU, which we used in the previous discussion, and the identification of brain response to music, which could open a new avenue for early neurorehabilitation.

8.5 Characterization of Epileptic Seizures Versus Periodic Activity

Beta activity in the EEG is originated from cortico-cortical interactions and alpha, although generated at thalamus, has a significant participation from cortical structures (see Chap. 1, Introduction to EEG). Therefore, according to the pathophysiological definition of seizure as an abnormal increase in cortical activity, the very increase of cortical activity can discriminate between ES and non-epileptic burst of periodic activity (BPA) with, sometimes quite similar morphological properties [40, 41]. We have compared the numerical properties of ES and BPA in two groups of ICU patients [42]. ES were classified as generalized seizures (GS) in 29%, meanwhile in the rest, seizures were initially focal (FS) with later generalization. In Fig. 8.8, it's shown a typical ES (GS) and BPA visually identified.

We computed the normalized changes in mean power square of all the bands and entropy for every lobe during BPA. Only delta band significantly increased during BPA, but it did so by nearly 3 times in the frontal lobes. The rest of the bands did not significantly change during BPA. No differences between right and left lobes were observed for any band. Spectral entropy significantly decreased between 10 and 20% during BPA in the frontal lobes.

In the case of GS, the pattern induced in power spectral was completely different from BPA. There was a generalized increase in all the bands without differences between the different lobes, increasing in power from the slower (δ and θ) to the faster (α and β) bands. Delta bands

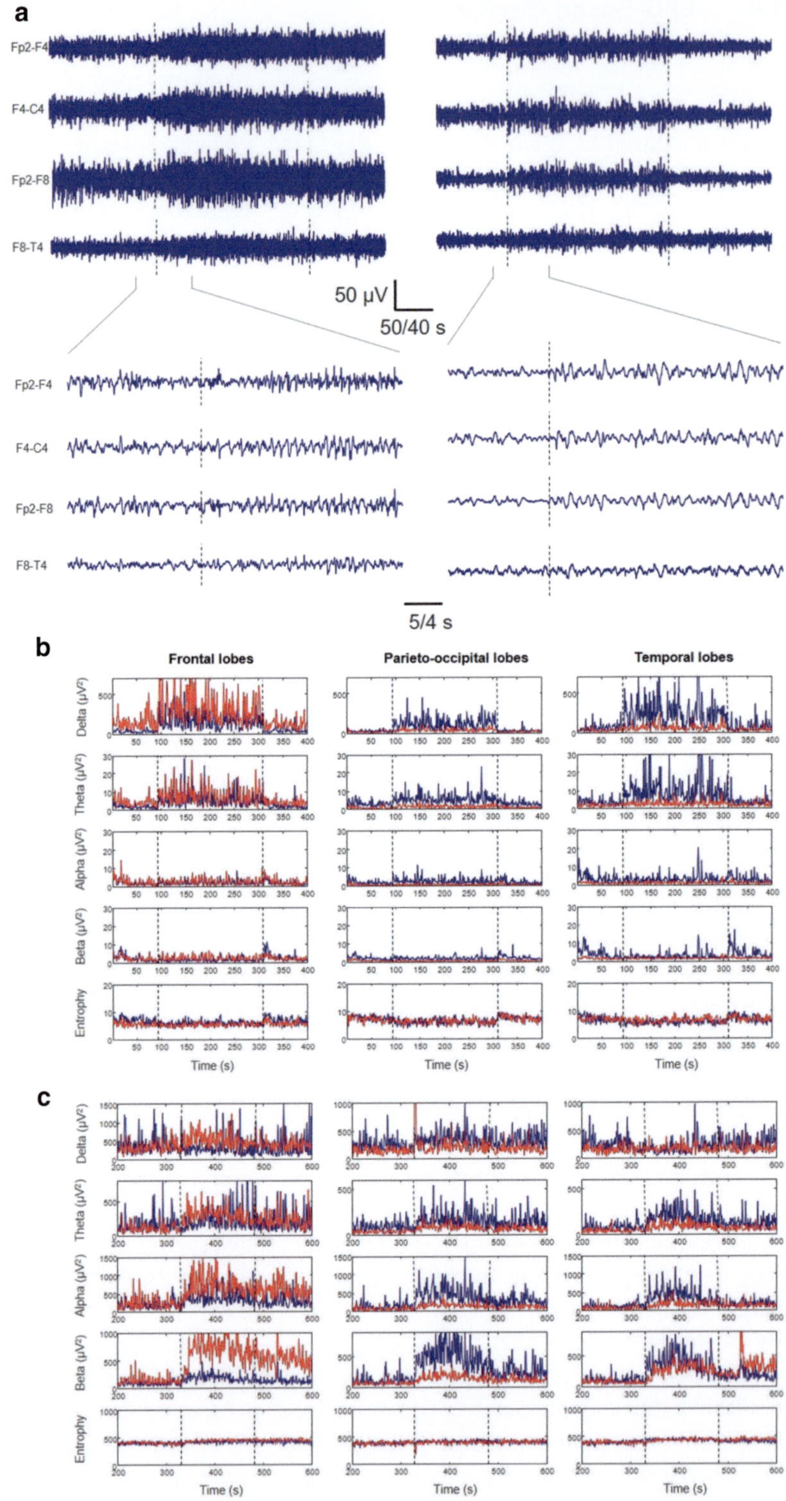

Fig. 8.8 Examples of paroxysmal activity recorded in critically ill patients. (**a**) Left column: recordings from the right frontal lobe of a patient suffering a GS. Right column: recordings from the same channels of a patient suffering a BPA. Traces are expanded below. Time-base is different for every recording. (**b**) Dynamics of bands (rows) and Se (lower row) for different lobes (columns) from the BPA above. (**c**) Dynamics of bands (rows) and Se (lower row) for different lobes (columns) from the GS above. Paroxysmal activity is indicated by vertical dotted lines. Red: right hemisphere. Blue: left hemisphere (From [42])

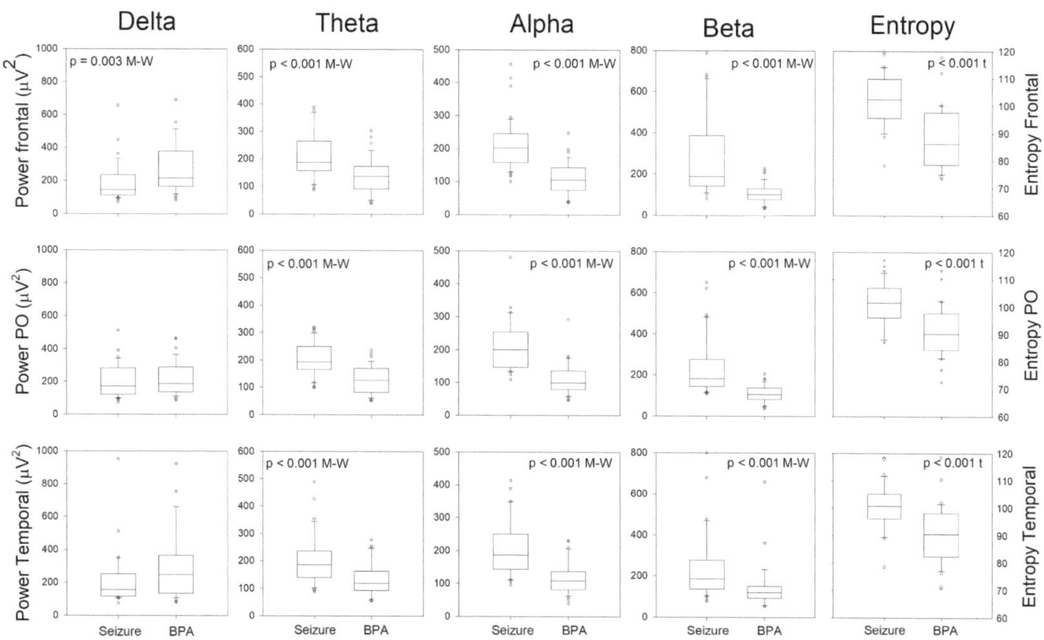

Fig. 8.9 Comparison by bands and lobes between ES and BPA. Upper row = frontal lobes; middle row = parieto-occipital lobes, and lower row = temporal lobes. Probability is indicated when the difference between ES and BPA is significant. M-W = Mann-Whitney rank sum test, t = Student-t test. From [42]

increased around 1.5 times, θ and α bands increased approximately the double mean while β bands augmented between 3 and 4 times, except at the right temporal lobe. However, SSE did not change in GS.

No differences between right and left lobes were observed for any band. Therefore, a true global and symmetrical participation for all the scalp was observed in GS. The pattern induced in the case of FS in spectral powers was also completely different from BPA and quite similar to GS. It must be realized that FS started in a well-defined region, but what we have computed was the overall activity during the previous, ictal, and post-ictal periods. We have observed a generalized increase in all the bands, increasing in power from the slower to the faster bands. Delta bands increased around 1.5 times, θ and α bands increased approximately the double and β bands augmented between 3–4 times, except for the right temporal lobe.

Taking into account that GS and FS were similar we pooled together and compared with BPA. Besides, we have used data from right and left hemispheres, considering that there are no differences between them. Delta power for frontal lobe was higher for BPA than for ES, but for the rest of the bands, the power was higher in the case of ES.

The difference between both states allowed us to define numerical criteria to separate ES from BPA (Fig. 8.9).

We have defined in a numerical way the features for BPA and ES in critically ill patients. In other words, we have identified the differences between both paroxysmal states. This finding is very important from a clinical point of view, because in a high degree of patients the morphological features of recordings are equivocal and it's extremely difficult to differentiate between seizures and non-epileptic bursts. We have used the pathophysiological definition of epilepsy and we have observed that ES always implies an increase in cortical bands (α and β), while in BPA these bands did not change.

8.6 Brain Response to Music

It has been demonstrated that neural processing of sounds, including language, continues during sleep and even under general anesthesia [43, 44]. Nevertheless, in this type of patient, the more complex cortical functions are highly impaired, so an analysis of emotional or psychological responses is meaningless. However, it is possible that the more basic neural mechanisms that give rise to the homeostatic EEG system still maintain significant functionality. Among them, the thalamic-cortico-thalamic system has an important capacity for synchronization and resonance, as manifested during NREM sleep and generalized epileptic seizures [45, 46]. Therefore, listening to music with a different rhythmic component can give rise to different brain responses through the participation of different neural circuits, especially the thalamic-cortico-thalamic circuit [47].

We selected three types of music according to their rhythmic content. From lowest to highest rhythmic content, we used (i) dodecaphonic music (Schönberg, Klavierstück Op. 33a; denoted as DodecM or sometimes Schönberg), (ii) classical music (Mozart, Sonata for two pianos in D, K. 448, denoted as ClassM or sometimes Mozart), and (iii) heavy metal music (Volbeat, The Devil's Bleeding Crown, denoted as HeavyM or sometimes Volbeat).

A 2-min duration of each of these pieces was selected and played to the patient through headphones gently placed over the ears (Fig. 8.10). A minimum 2-min period with headphones placed on-site and without music was chosen as the basal state. After the basal period, music stimulation with each of the three types of music (DodecM, ClassM, or HeavyM) was carried out, separated by periods of 2 min. The sequence was randomized and different for every patient. Therefore, we can exclude any possible bias induced by the order or appearance of the music.

Auditory intensity delivered by headphones was assessed, resulting (in dB) ClassM = 68.2 ± 4.6, DodecM = 67.6 ± 4.5, and HeavyM = 69.4 ± 1.7. This sound level corresponds to normal conversation. No differences in intensity were observed between stimuli.

We defined the response to music as a statistically significant difference (either positive or negative) in any of the musical stimulations performed (DodecM, ClassM, or HeavyM) compared with the basal period.

The EEGs recorded during music stimulation did not differ, during de visu inspection, from basal bioelectrical activity (Fig. 8.11a, b). However, numerical analysis revealed significant changes induced by music (compare with Fig. 8.11c, d). In Fig. 8.11e, f, as an example, we have superimposed mean spectra for every channel obtained at basal conditions (blue lines) during ClassM (red lines) and HeavyM (green lines) from the same patient (#2).

It is very relevant that ClassM decreased the delta component over the whole scalp. However, the more significant results were probably the changes induced in spectra by HeavyM in most channels, especially in the fronto-temporal right regions, with increases in frequencies higher than delta. However, in this example, neither SSE (basal = 3.20/3.14, ClassM = 3.25/3.22, DodecM = 3.26/3.19, and HeavyM = 3.21/3.19,

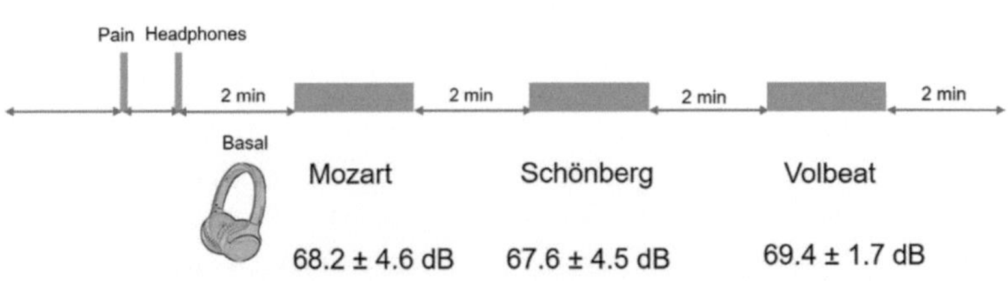

Fig. 8.10 Study protocol. Timeline and key milestones are shown. Painful stimulation and headphone placement were performed before the start of musical stimulation. Musical stimulation is indicated by the rectangle periods and the musical type is indicated

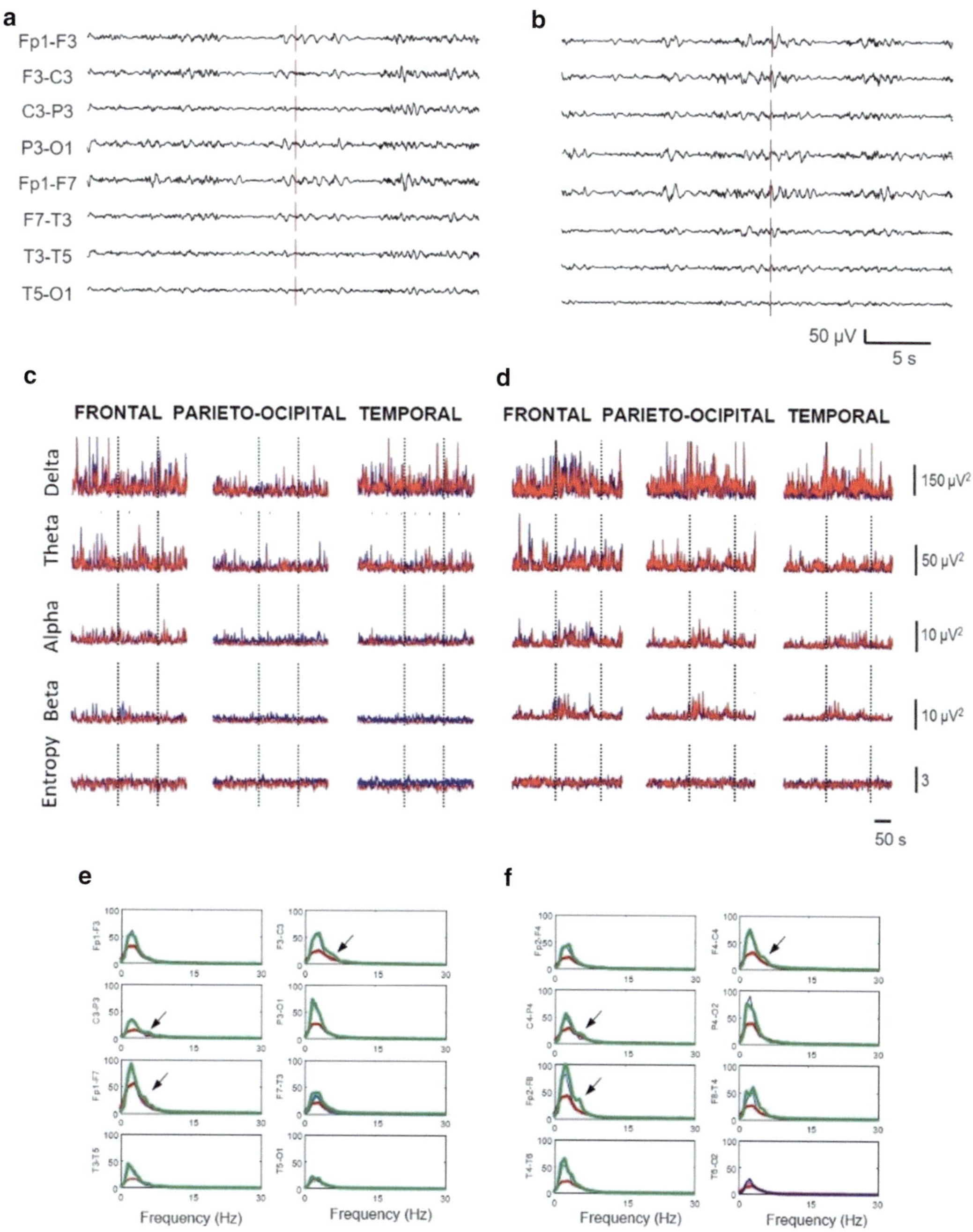

Fig. 8.11 Relevance of qEEG in obtaining the results compared with de visu recordings. All the data are from the same patient (#2). (**a**) EEG recording (left hemisphere) during the start of stimulation with ClassM (vertical line) and (**b**) recording of the same patient at the beginning of stimulation with HeavyM (vertical line). (**c**) Dynamics of all the bands at the different lobes (indicated at the upper row) during the ClassM and (**d**) HeavyM. Red lines = right hemisphere; blue lines = left hemisphere. Dotted vertical lines indicate the period of stimulation. (**e**) Left scalp and F) right scalp showing mean spectra obtained at basal condition (blue lines), during the stimulation with ClassM (red lines) and with HeavyM (green lines). Arrows indicate the appearance of a theta component at 4.5 Hz with HeavyM at both central regions

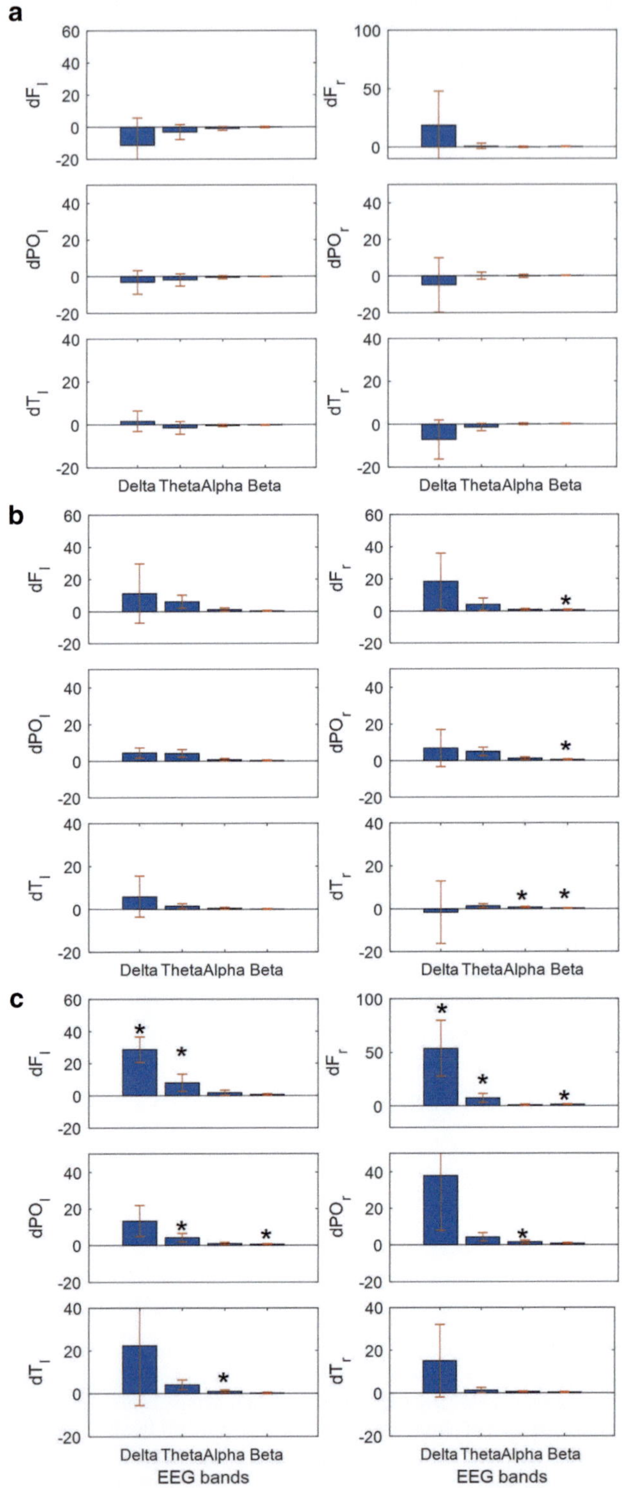

Fig. 8.12 Bar graphs showing the change in mean power spectra for the different bands at different lobes. (**a**) ClassM, (**b**) DodecM, and (**c**) HeavyM. Asterisks indicate statistical significance ($p < 0.05$) with respect to basal activity (Wilcoxon rank test)

respectively, for left/right hemispheres) nor correlation were modified by stimulation (basal = 0.613/0.591, ClassM = 0.640/0.588, DodecM = 0.608/0.585, and HeavyM = 0.614/0.586, respectively, for left/right hemispheres).

Although not fully analyzed, we observed that HeavyM stimulation induced the appearance of well-defined components at specific frequencies (see Fig. 8.11e, f) and rather than only modifying the power of preexisting bands.

The patterns of response were completely different for the left frontal, right parieto-occipital, and both temporal lobes. It is interesting to observe that only 2/18 possibilities were not significant and only for ClassM vs. DodecM in the right frontal lobe and DodecM vs. HeavyM in the left parieto-occipital lobe. All the comparisons between ClassM and HeavyM were highly significant in all the lobes (Fig. 8.12).

Although the data were obtained from a small set of patients, we have observed that different types of music induce a specific response in the brain activity from unconscious patients suffering from primary brain pathology and under sedoanalgesia. Unexpectedly, heavy metal was the most stimulating music, whereas classical music showed a tendency to reduce brain activity. Beyond the anecdote that one type of music activates more or less brain activity, what is truly relevant is that this work demonstrated, for the first time, the possibility that music therapy can activate the brain in patients with severe brain involvement and sedation-analgesia, so that it can be used as an early rehabilitation method in these cases.

8.7 Concluding Remarks

The brain dynamics of patients admitted to the ICU are severely altered compared to conscious patients, such as those with epilepsy, stroke, encephalopathy, etc. ICU patients are subjected to sedation and analgesia and often have primary brain involvement. In other cases, they may have severe ionic or metabolic abnormalities.

This impairment of brain physiology means bioelectrical manifestations are not as those usually observed, so the changes in the pattern present a slow dynamic and both the waveforms and their frequency can be more variegated and slower. All of this makes diagnosis difficult using visual EEG. Numerical analysis can determine the existence of these changes, thus modifying the initial diagnosis and the patient's functional prognosis.

Furthermore, ICU patients often have neuromuscular blockade, so the recording quality is very good, which facilitates analysis. Thus, even epileptic seizures can be analyzed with confidence.

Finally, it is important to point out that qEEG in cEEG monitoring not only provides valuable information in the case of epileptic seizures, but is also very important for the analysis of other pathologies, as evolution of encephalopathies, focal ischemia, or titration of treatment.

References

1. Veciana de las Heras M, Sala-Padro J, Pedro-Perez J, García-Parra B, Hernández-Pérez G, Falip M. Utility of quantitative EEG in neurological emergencies and ICU clinical practice. Brain Sci. 2024;14:939. https://doi.org/10.3390/brainsci14090939.
2. Pastor J, Vega-Zelaya L. Titration of pharmacological responses in ICU patients by quantified EEG. Curr Neuropharmacol. 2022; https://doi.org/10.2174/1570159X20666220411083213.
3. Alkhachroum A, Appavu B, Egawa S, Foreman B, Gaspard N, Gilmore EJ, Hirsch LJ, Kurtz P, Lambrecq V, Kromm J, Vespa P, Zafar SF, Rohaut B, Claassen J. Electroencephalogram in the intensive care unit: a focused look at acute brain injury. Intensive Care Med. 2022;48(10):1443–62. https://doi.org/10.1007/s00134-022-06854-3. Epub 2022 Aug 23. PMID: 35997792; PMCID: PMC10008537.
4. Thy M, Gaudemer A, Vellieux G, Sonneville R. Critical care management of meningitis and encephalitis: an update. Curr Opin Crit Care. 2022;28(5):486–94. https://doi.org/10.1097/MCC.0000000000000980. Epub 2022 Aug 17. PMID: 35975963.
5. Nuwer MR, Emerson RG, Hahn CD. Principles and Techniques for lont-term EEG recording (Epilepsy monitoring unit, Intensive care Unit, ambulatory). In: Shomer DL, Lopes da Silva F, editors. Niedermeyer's Electroencephalography: basic principles, clini-

cal applications and related fields. 7th ed. Oxford University Press; 2018. ISBN 9780190228484.
6. Guérit JM, Amantini A, Amodio P, Andersen KV, Butler S, de Weerd A, Facco E, Fischer C, Hantson P, Jäntti V, Lamblin MD, Litscher G, Péréon Y. Consensus on the use of neurophysiological tests in the intensive care unit (ICU): electroencephalogram (EEG), evoked potentials (EP), and electroneuromyography (ENMG). Neurophysiol Clin. 2009;39(2):71–83. https://doi.org/10.1016/j.neucli.2009.03.002. Epub 2009 Apr 11. PMID: 19467437.
7. Shellhaas RA. Seizure classification, etiology, and management. Handb Clin Neurol. 2019;162:347–61. https://doi.org/10.1016/B978-0-444-64029-1.00017-5. PMID: 31324320.
8. van Rooij LG, van den Broek MP, Rademaker CM, de Vries LS. Clinical management of seizures in newborns: diagnosis and treatment. Paediatr Drugs. 2013;15(1):9–18. https://doi.org/10.1007/s40272-012-0005-1. PMID: 23334995.
9. Trinka E, Leitinger M. Which EEG patterns in coma are nonconvulsive status epilepticus? Epilepsy Behav. 2015;49:203–22. https://doi.org/10.1016/j.yebeh.2015.05.005. Epub 2015 Jul 4. PMID: 26148985.
10. Kondziella D, Bender A, Diserens K, van Erp W, Estraneo A, Formisano R, Laureys S, Naccache L, Ozturk S, Rohaut B, Sitt JD, Stender J, Tiainen M, Rossetti AO, Gosseries O, Chatelle C, EAN Panel on Coma, Disorders of Consciousness. European Academy of Neurology guideline on the diagnosis of coma and other disorders of consciousness. Eur J Neurol. 2020;27(5):741–56. https://doi.org/10.1111/ene.14151. Epub 2020 Feb 23. PMID: 32090418.
11. Ardeshna NI. EEG and coma. Neurodiagn J. 2016;56(1):1–16. https://doi.org/10.1080/21646821.2015.1114879. PMID: 27180503.
12. Kelly LJ, Sexton VL. Anti-NMDAR encephalitis. Neurodiagn J. 2016;56(2):101–8. https://doi.org/10.1080/21646821.2015.1137729. PMID: 27373057.
13. Dal Canto G, Pellacani S, Valvo G, Masi G, Ferrari AR, Sicca F. Internalizing and externalizing symptoms in preschool and school-aged children with epilepsy: focus on clinical and EEG features. Epilepsy Behav. 2018;79:68–74. https://doi.org/10.1016/j.yebeh.2017.10.004. Epub 2017 Dec 15. PMID: 29253677.
14. Kaplan PW. Electrophysiological prognostication and brain injury from cardiac arrest. Semin Neurol. 2006;26(4):403–12. https://doi.org/10.1055/s--2006-948321. PMID: 16969741.
15. Selioutski O, Roberts D, Hamilton R, Ghosh H, Nickels J, Konig Toro F, Kruppenbacher M, Auinger P, Kaplan PW, Birbeck GL. Continuous EEG monitoring predicts a clinically meaningful recovery among adult inpatients. J Clin Neurophysiol. 2019;36(5):358–64. https://doi.org/10.1097/WNP.0000000000000594. PMID: 31491786.
16. Szurhaj W, Lamblin MD, Kaminska A, Sediri H, Société de Neurophysiologie Clinique de Langue Française. EEG guidelines in the diagnosis of brain death. Neurophysiol Clin. 2015;45(1):97–104. https://doi.org/10.1016/j.neucli.2014.11.005. Epub 2015 Jan 14. PMID: 25687591.
17. Walter U, Fernández-Torre JL, Kirschstein T, Laureys S. When is "brainstem death" brain death? The case for ancillary testing in primary infratentorial brain lesion. Clin Neurophysiol. 2018;129(11):2451–65. https://doi.org/10.1016/j.clinph.2018.08.009. Epub 2018 Sep 1. PMID: 30209020.
18. Koenig MA, Kaplan PW. Brain death. Handb Clin Neurol. 2019;161:89–102. https://doi.org/10.1016/B978-0-444-64142-7.00042-4. PMID: 31307622.
19. Herman ST, Abend NS, Bleck TP, Chapman KE, Drislane FW, Emerson RG, Gerard EE, Hahn CD, Husain AM, Kaplan PW, LaRoche SM, Nuwer MR, Quigg M, Riviello JJ, Schmitt SE, Simmons LA, Tsuchida TN, Hirsch LJ, Critical Care Continuous EEG Task Force of the American Clinical Neurophysiology Society. Consensus statement on continuous EEG in critically ill adults and children, part I: indications. J Clin Neurophysiol. 2015;32(2):87–95. https://doi.org/10.1097/WNP.0000000000000166. PMID: 25626778; PMCID: PMC4435533
20. Bauerschmidt A, Martin A, Claassen J. Advancements in the critical care management of status epilepticus. Curr Opin Crit Care. 2017;23(2):122–7. https://doi.org/10.1097/MCC.0000000000000392. PMID: 28207600.
21. Kubota Y, Nakamoto H, Egawa S, Kawamata T. Continuous EEG monitoring in ICU. J Intensive Care. 2018;6:39. https://doi.org/10.1186/s40560-018-0310-z. PMID: 30026951; PMCID: PMC6050674.
22. Banoczi W. ICU-cEEG monitoring. Neurodiagn J. 2020;60(4):231–71. https://doi.org/10.1080/21646821.2020.1824982. Epub 2020 Nov 18. PMID: 33207129.
23. Struck AF, Tabaeizadeh M, Schmitt SE, Ruiz AR, Swisher CB, Subramaniam T, Hernandez C, Kaleem S, Haider HA, Cissé AF, Dhakar MB, Hirsch LJ, Rosenthal ES, Zafar SF, Gaspard N, Westover MB. Assessment of the validity of the 2HELPS2B score for inpatient seizure risk prediction. JAMA Neurol. 2020;77(4):500–7. https://doi.org/10.1001/jamaneurol.2019.4656. Erratum in: JAMA Neurol. 2020 Apr 1;77(4):527. doi: 10.1001/jamaneurol.2020.0140. PMID: 31930362; PMCID: PMC6990873.
24. Aslan-Kara K, Demir T, Satılmış Ü, Peköz T, Bıçakcı Ş, Bozdemir H. Prognostic indicators of non-convulsive status epilepticus in intensive care unit. Acta Neurol Belg. 2023;123(4):1257–66. https://doi.org/10.1007/s13760-022-01981-6. Epub 2022 Jun 3. PMID: 35657480.
25. Zawar I, Ghosal S, Hantus S, Punia V. Indications for continuous electroencephalographic (cEEG) monitoring: what do they tell us? Epilepsy Res.

2023;190:107088. https://doi.org/10.1016/j.eplepsyres.2023.107088. Epub 2023 Jan 20. PMID: 36731271.
26. Sen-Gupta I, Schuele SU, Macken MP, Kwasny MJ, Gerard EE. "Ictal" lateralized periodic discharges. Epilepsy Behav. 2014;36:165–70. https://doi.org/10.1016/j.yebeh.2014.05.014. Epub 2014 Jun 13. PMID: 24935085.
27. Bitar R, Khan UM, Rosenthal ES. Utility and rationale for continuous EEG monitoring: a primer for the general intensivist. Crit Care. 2024;28(1):244. https://doi.org/10.1186/s13054-024-04986-0. PMID: 39014421; PMCID: PMC11251356.
28. Gallego I, Ezquiaga E, Betancor D, Sola RG, Pastor J. Crisis no epilépticas de origen psicógeno en una unidad de cirugía de la epilepsia [Psychogenic non-epileptic seizures in an epilepsy surgery unit]. Rev Neurol. 2011;52(8):449–56. Spanish. PMID: 21425097.
29. Vega-Zelaya L, Alvarez M, Ezquiaga E, Nogeiras J, Toledo M, Sola RG, Pastor J. Chapter 10: Psychogenic non-epileptic seizures in a surgical epilepsy unit: experience and a comprehensive review. In: Holmes MD, editor. Epilepsy topics. InTech; 2014. p. 193–224. https://doi.org/10.5772/57439. ISBN 978-953-51-1630-1.
30. Muniz CF, Shenoy AV, O'Connor KL, Bechek SC, Boyle EJ, Guanci MM, Tehan TM, Zafar SF, Cole AJ, Patel AB, Westover MB, Rosenthal ES. Clinical development and implementation of an institutional guideline for prospective EEG monitoring and reporting of delayed cerebral ischemia. J Clin Neurophysiol. 2016;33(3):217–26. https://doi.org/10.1097/WNP.0000000000000281. PMID: 27258445; PMCID: PMC4907266
31. Koren J, Herta J, Baumgartner C. Critical care EEG. Fortschr Neurol Psychiatr. 2020;88(1):52–63. https://doi.org/10.1055/a-1008-7783. German.. Epub 2020 Jan 27. PMID: 31986552.
32. Rautee R, Sampson T, Sarkela M, Melto S, Hovilehto S, van Gils M. Application of spectral entropy to EEG and facial EMG frequency bands for the assessment of level of sedation in ICU. Conf Proc IEEE Eng Med Biol Soc. 2004;2004:3481–4. https://doi.org/10.1109/IEMBS.2004.1403977.
33. Maksimow A, Snapir A, Särkelä M, Kentala E, Koskenvuo J, Posti J, Jääskeläinen SK, Hinkka-Yli-Salomäki S, Scheinin M, Scheinin H. Assessing the depth of dexmedetomidine-induced sedation with electroencephalogram (EEG)-based spectral entropy. Acta Anaesthesiol Scand. 2007;51(1):22–30. https://doi.org/10.1111/j.1399-6576.2006.01174.x. Epub 2006 Nov 1. PMID: 17073855.
34. Walsh TS, Ramsay P, Lapinlampi TP, Särkelä MO, Viertiö-Oja HE, Meriläinen PT. An assessment of the validity of spectral entropy as a measure of sedation state in mechanically ventilated critically ill patients. Intensive Care Med. 2008;34(2):308–15. https://doi.org/10.1007/s00134-007-0858-x. Epub 2007 Sep 27. PMID: 17898996.
35. Bilgili B, Montoya JC, Layon AJ, Berger AL, Kirchner HL, Gupta LK, Gloss DS. Utilizing bi-spectral index (BIS) for the monitoring of sedated adult ICU patients: a systematic review. Minerva Anestesiol. 2017;83(3):288–301. https://doi.org/10.23736/S0375-9393.16.10886-7. Epub 2016 Jun 17. PMID: 27314595.
36. Swisher CB, White CR, Mace BE, Dombrowski KE, Husain AM, Kolls BJ, Radtke RR, Tran TT, Sinha SR. Diagnostic accuracy of electrographic seizure detection by neurophysiologists and non-neurophysiologists in the adult ICU using a panel of quantitative EEG trends. J Clin Neurophysiol. 2015;32(4):324–30. https://doi.org/10.1097/WNP.0000000000000144. PMID: 26241242.
37. Lalgudi Ganesan S, Stewart CP, Atenafu EG, Sharma R, Guerguerian AM, Hutchison JS, Hahn CD. Seizure identification by critical Care providers using quantitative electroencephalography. Crit Care Med. 2018;46(12):e1105–11. https://doi.org/10.1097/CCM.0000000000003385. PMID: 30188384
38. Hernández-Hernández MA, Cherchi MS, Torres-Díez E, Orizaola P, Martín-Láez R, Fernández-Torre JL. Bispectral index monitoring to detect delayed cerebral ischemia after aneurysmal subarachnoid hemorrhage. J Crit Care. 2022;72:154154. https://doi.org/10.1016/j.jcrc.2022.154154. Epub 2022 Sep 22. PMID: 36152563.
39. Shiraki A, Yamamoto H, Ohno A, Kumai S, Suzui R, Sawamura F, Kawaguchi M, Suzuki T, Maki Y, Ito Y, Nakata T, Kidokoro H, Numaguchi A, Natsume J. Color density spectral array findings on continuous EEG during therapeutic hypothermia in children with acute encephalopathy. Brain Dev. 2024;46(10):313–9. https://doi.org/10.1016/j.braindev.2024.08.002. Epub 2024 Sep 6. PMID: 39242349.
40. Fisher RS, van Emde BW, Blume W, Elger C, Genton P, Lee P, Engel J Jr. Epileptic seizures and epilepsy: definitions proposed by the International League Against Epilepsy (ILAE) and the International Bureau for Epilepsy (IBE). Epilepsia. 2005;46(4):470–2. https://doi.org/10.1111/j.0013-9580.2005.66104.x. PMID: 15816939.
41. Hirsch LJ, Fong MWK, Leitinger M, LaRoche SM, Beniczky S, Abend NS, Lee JW, Wusthoff CJ, Hahn CD, Westover MB, Gerard EE, Herman ST, Haider HA, Osman G, Rodriguez-Ruiz A, Maciel CB, Gilmore EJ, Fernandez A, Rosenthal ES, Claassen J, Husain AM, Yoo JY, So EL, Kaplan PW, Nuwer MR, van Putten M, Sutter R, Drislane FW, Trinka E, Gaspard N. American clinical neurophysiology society's standardized critical Care EEG terminology: 2021 version. J Clin Neurophysiol. 2021;38(1):1–29. https://doi.org/10.1097/WNP.0000000000000806. PMID: 33475321; PMCID: PMC8135051.
42. Vega-Zelaya L, Martín Abad E, Pastor J. Quantified EEG for the characterization of epileptic seizures versus periodic activity in critically ill patients. Brain Sci. 2020;10:158. https://doi.org/10.3390/brainsci10030158.

43. Iselin-Chaves IA, Willems SJ, Jermann FC, Forster A, Adam SR, Van der Linden M. Investigation of implicit memory during isoflurane anesthesia for elective surgery using the process dissociation procedure. Anesthesiology. 2005;103(5):925–33. https://doi.org/10.1097/00000542-200511000-00005. PMID: 16249665.
44. Kouider S, Andrillon T, Barbosa LS, Goupil L, Bekinschtein TA. Inducing task-relevant responses to speech in the sleeping brain. Curr Biol. 2014;24(18):2208–14. https://doi.org/10.1016/j.cub.2014.08.016. Epub 2014 Sep 11. PMID: 25220055; PMCID: PMC4175175.
45. Jain SV, Kothare SV. Sleep and epilepsy. Semin Pediatr Neurol. 2015;22(2):86–92. https://doi.org/10.1016/j.spen.2015.03.005. Epub 2015 Mar 21. PMID: 26072338.
46. Roliz AH, Kothare S. The interaction between sleep and epilepsy. Curr Neurol Neurosci Rep. 2022;22(9):551–63. https://doi.org/10.1007/s11910-022-01219-1. Epub 2022 Jul 8. PMID: 35802300.
47. Pastor J, Vega-Zelaya L, Canabal A. Pilot Study: The differential response to classical and heavy metal music in intensive Care unit patients under Sedo-analgesia. J Integr Neurosci. 2023;22(2):30. https://doi.org/10.31083/j.jin2202030. PMID: 36992596.

9 Encephalopathies

9.1 Introduction

Encephalopathy, also called altered mental status or confusional state, refers clinically, descriptively, or pathologically to a state of impaired cognition, generally acute or subacute [1]. Frequently, focal cognitive deficits such as aphasia or amnesia can mimic global encephalopathy syndromes. Delirium, a sometimes severe status, is a subtype of encephalopathy in which the patient develops acutely altered mental status characterized by waxing and waning attention [2]. Acute or infectious toxic encephalopathy refers to a type of reversible brain syndrome caused by several factors, such as a metabolic disorder, systemic toxemia, or hypoxia during an acute infection [3–6]. The pathological changes include cerebral oedema, with no evidence of inflammation in cerebrospinal fluid. Symptoms are complex and diverse. EEG abnormalities are associated to encephalopathies associated with seizures, mental impairments, and movement disorders.

There are numerous causes responsible for, or associated with, encephalopathies. In general, an etiological diagnosis is not possible by EEG, since the response is usually a slowing of the pattern, which may be generalized or focal and may be associated with epileptiform activity. However, in some cases, the pattern is relatively specific enough to allow a diagnostic orientation. In fact, several studies have shown that a specific pattern can distinguish among several causes of encephalopathy [1, 7, 8] and between dementia and encephalopathy assessed by means of qEEG [9]. Despite these promising results, specific diagnosis of the various causes of encephalopathy remains a distant goal. It is in this field that systematic and extensive investigation of EEG numerical characteristics could be of great help.

J. Pastor, *Quantitative EEG Practical Guide*, https://doi.org/10.1007/978-3-032-05457-9_9

The most common types of encephalopathy assessed by EEG are briefly discussed next.

9.2 Metabolic Encephalopathies

This is a heterogeneous group of pathologies that are highly prevalent in everyday clinical practice [10]. The main types are as follows:

9.2.1 Hepatic Encephalopathy (HE)

Liver damage results in a porto-cava shunt, with significant alterations in protein metabolism and increase of nitrous substrate, although the pathophysiological mechanism by which these alterations affect brain function is not fully understood.

The degree of slowing of the EEG tracing is related to the level of ammonia in the blood [11] and correlated with the appearance of other EEG patterns, as triphasic waves (TW) or intermittent generalized rhythmic delta activity (GRDA). The progression of the disease is associated with greater disorganization of the background, with an increase in the asynchronous bihemispheric theta and delta components [12]. TW are not pathognomonic of HE, occurring in 25% of cases [13]. There has been debate about whether this is an ictal pattern, although this is currently considered not to be the case [14, 15]. However, the presence of sharp waves and spikes with epileptic characteristics is common in HE.

It has been shown that spectral analysis-based EEG classifications in HE can be related to the severity of liver failure and cognitive findings. It has also been shown that the numerical-based classification was more repeatable than the qualitative visual one [16].

9.2.2 Uremic Encephalopathy (UE)

Despite extensive research, there is no globally accepted description of the pathophysiology, although there is evidence of the involvement of different uremic toxins in the alteration of brain function [17]. Acute UE is characterized by the presence of confusion, agitation, and coma, while chronic state presents with lethargy, cognitive alterations, agitation, and sleep disturbances, making its staging very difficult [10, 18]. Epileptic seizures are not uncommon [19, 20].

As we saw for HE, the EEG pattern is also nonspecific. It is usually associated with an irregular low-voltage background, with progressive slowing of the posterior dominant rhythm, the presence of TW and spikes that may or may not be associated with myoclonus, ES, or SE [21]. Occasionally, dialysis disequilibrium syndrome, probably caused by cerebral oedema, may be observed during the initiation of hemodialysis treatment and characterized by irritability, nausea, headache, hallucinations, blurred vision, asterixis, and generalized seizures. During chronic dialysis treatment, dialysis-related encephalopathy syndrome, characterized by language and memory disturbances, psychiatric symptoms, and epileptic seizures, may also appear [18]. In these cases, a disorganized and slow background is frequently observed in the EEG, with frequent generalized rhythmic delta bursts of frontal predominance, which may be associated with fronto-central spike-wave [21].

9.2.3 Electrolyte Disturbance-Related Encephalopathy

Neuronal excitability is closely regulated by ionic concentrations and permeabilities [22], so changes in these magnitudes will have an obvious impact on brain physiology. Ionic disturbances are one of the most common causes of acute encephalopathy, especially observed in patients admitted to the ICU.

EEG in hypernatremic encephalopathy shows a diffuse nonspecific slowing [23], being deeper (generalized delta between 0.5 and 2 Hz) in the case of hyponatremia (< 116 mg/dl), in which periodic delta patterns [24], TW or periodic lateralized epileptic discharges (PLD) and seizures or status epilepticus have also been observed [25].

Calcium regulation is also essential for the control of neuronal excitability. Hypercalcemia (especially >12 mg/dl) can result in a slow EEG with the presence of high-voltage delta bursts lasting between 1 and 4 s, and TW may also be observed [26]. Epileptiform discharges are generally not observed. In patients with hypocalcemia (<6 mg/dl), on the contrary, ES are very frequent and may be the first clinical manifestation [27]. The EEG recording usually shows a high-amplitude delta pattern and scattered focal spikes [28].

Synchronization impairment is therefore a key feature of encephalopathies. In these patients, there is increased synchronization in anterior regions and a marked reduction in synchronization in the posterior half. The following example shows the EEG analysis of an 84-year-old patient with toxic-metabolic encephalopathy due to ionic dysregulation, dehydration, and urinary infection. The patient was admitted for bradypsychia and fluctuating levels of consciousness. As can be seen in the logPS dynamics by band there are no interhemispheric differences (Fig. 9.1a), nor are they observed in the average spectra per channel (Fig. 9.1b). The average spectra show a main delta component at 1 Hz, with a weak theta component maximum at 5 Hz. The connectogram (Fig. 9.1c) shows an abnormal pattern, with decreased synchronization in the posterior half of the scalp and a marked shift in synchronization toward the anterior region. This pattern is highly characteristic of encephalopathies. Figure 9.1d shows a complementary numerical result. Synchronization dynamics graphs for the Pearson correlation and band coherence are shown for both hemispheres. These graphs show an increase in synchronization for the delta and theta bands, as well as for the Pearson correlation, while coh(alpha) and coh(beta) values (cortical bands) are clearly decreased. The dashed horizontal lines show the confidence intervals. This excess of anterior synchronization is observed in Fig. 9.1e. The thick lines show the cross-correlation for the F3-C3/F4-C4 channel pairs, while the thin lines show the cross-correlation for the P3-O1/P4-O2 channels. The upper row shows the dynamics of the zero-lag cross-correlation windows for the analyzed 200 s period. The average value is statistically significant for the frontal channel pair (0.631 ± 0.022), while it is not for the occipital channel pair (0.062 ± 0.036). The bottom row shows the overall cross-correlation for the entire record, not by window. Synchronization is observed to be very good for the frontal channels, with a maximum at 1.95 ms and minimums at −0.40 and 0.42 s. However, cross-correlation between the parieto-occipital lobes is severely altered.

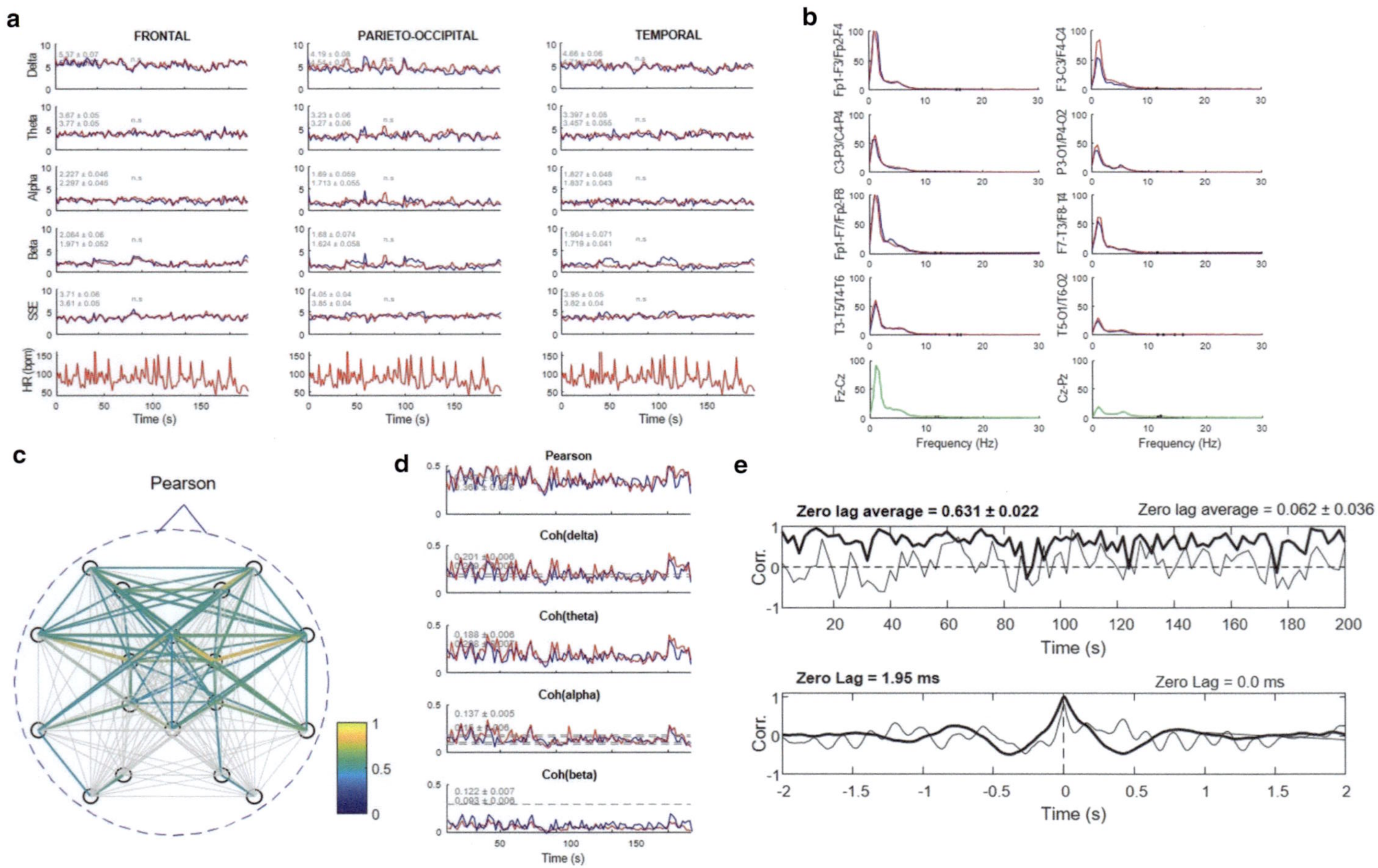

Fig. 9.1 Numerical properties of toxic-metabolic encephalopathy. (**a**) Dynamics of the logPS bands by lobe. The bottom row shows the dynamics of the instantaneous heart rate, indicating the presence of atrial fibrillation. (**b**) Average spectra by channel. (**c**) Average connectogram. (**d**) Dynamics of the synchronization variables. Confidence intervals for patient age are indicated as dashed horizontal lines. (**e**) Dynamical analysis of Pearson correlation coefficient (upper row) and for the global cross-correlation (bottom row) for pairs of frontal (thick line) and occipital (thin line) channels. Red (except in A, bottom row) = right hemisphere; blue = left hemisphere

9.3 Vitamin-Deficit-Related Encephalopathies

EEG in encephalopathies due to niacin (vitamin B3), pyridoxine (vitamin B6), or cobalamin (vitamin B12) deficiency is poorly defined. Usually, there is a pattern of nonspecific slowing [10], with the only exception of a pattern of generalized spike activity and epileptic seizures in the case of vitamin B12 deficiency, which are rapidly reversible by treatment with pyridoxine [29].

For a neurophysiologist, there is a type of encephalopathy caused by thiamine (vitamin B1) deficiency especially relevant, because EEG is quite useful in the diagnosis.

9.3.1 Wernicke's Encephalopathy (WE)

Thiamine deficiency is common in chronic alcoholism, nutritional deficiencies, metabolic disorders, malabsorption, or elimination during hemodialysis. Vitamin B1 is essential for the synthesis of gamma-aminobutyric acid (GABA) and acetylcholine [30, 31].

WE is characterized by cerebellar and oculomotor symptoms. The EEG recording shows a progressive slowing of the background, reaching a generalized delta [32]. In more advanced stages, spike-wave and sharp waves also appear which may be associated with ES [33].

9.4 Endocrine Disorders-Related Encephalopathies

9.4.1 Encephalopathies by Glycaemia Alterations

Diabetes mellitus can cause a marked increase in blood glucose, with severe osmotic disturbances affecting the basal ganglia regions [34], which is why movement disorders, seizures, and ES are frequently observed [35]. The EEG recording shows a globally slowed pattern, interspersed with spikes. In cases of non-ketogenic hyperglycemia, rolandic epileptogenic spikes can be seen, which can be induced by contralateral movements or in occipital regions, which are associated with visual seizures [36].

Alterations in the level of consciousness are more frequent in hypoglycemia, which is associated with a marked slowing of the recording [37] and, occasionally, the presence of spikes and SE [38]. Given the association between the level of hypoglycemia and EEG changes, EEG monitoring has been suggested in patients at high risk of hypoglycemia [10].

9.4.2 Encephalopathies by Adrenocortical Alterations

Decreased glucocorticoid production causes Addison's disease, which presents with fatigue, muscle weakness, loss of appetite, weight loss, and abdominal pain [39]. The opposite condition, caused by increased glucocorticoid production (Cushing's syndrome), presents with weight gain in the trunk and face, but not in the extremities, fat accumulation in the back of the neck, hyperhidrosis, thickening of the skin, and muscle weakness [40]. Pheochromocytoma is a tumor of the adrenal medulla that causes high blood pressure and sympathetic crisis due to overproduction of catecholamines [41].

EEG recordings during the most severe cases of encephalopathy show nonspecific slowing and occasionally loss of cortical reactivity [42]. The changes are generally less striking in Cushing's cases, although loss of continuity and other changes may be observed during sleep [43]. In cases of pheochromocytoma, the recording may be normal, unless hypertension-related posterior reversible encephalopathy (PRES) appears, associated with posterior slowing [44].

9.4.3 Sheehan's Syndrome

It's a severe hypopituitarism due to necrosis of the anterior pituitary gland in a mother after birth. There are few studies on this type of encephalopathy.

9.4.4 Encephalopathies by Thyroid Alterations

Hypothyroidism can range from subtle cognitive disturbances, insomnia, and anxiety, to severe encephalopathy, seizures, and coma [45]. The EEG in hyperthyroidism shows an increase in the frequency of the occipital alpha rhythm and the mu rhythm, decreasing the epileptogenic threshold in known epileptics, although it is not usually associated with the presence of new seizures [46].

Hypothyroidism is associated with tiredness, weight gain, and myxedema coma. The EEG may be completely normal, although in severe cases it presents posterior slowing, with the presence of a low-amplitude pattern, especially delta and theta, which become more striking during myxedema coma, in which TW and periodic sharp waves similar to prion encephalopathy may also be observed [47].

9.5 Intoxication-Related Encephalopathies

9.5.1 Drug-Related Encephalopathies

Except for some types, such as anesthetics, the characteristics of these conditions are not well defined. The administration of anesthetics results in an EEG tracing similar to the evolution of a significant encephalopathy and coma. Initially, abundant rapid frontal rhythms are observed, which are replaced by alpha, theta, and delta rhythms, which subsequently give rise to a burst-suppression pattern and, in extreme cases, may reach an isoelectric tracing. The presence of epileptiform manifestations is frequent.

Lithium intoxication is associated with fatigue, tremor, or muscle weakness, although at serum concentrations >2 mEq/L, renal failure, hyperthermia, and encephalopathy with cortical and subcortical symptoms may occur [48]. EEG changes include slowing, increased spectral frequency, disorganization, and bilaterally synchronous paroxysmal delta bursts, spikes, and TW. Cases of lithium-induced Creutzfeldt-Jakob syndrome and NCSE have been described [49]. In lithium encephalopathy, an increase in the relative power of delta and theta (especially in the right hemisphere) has been described, with a decrease in occipital alpha power [50].

The association between blood clozapine levels and the presence of EEG abnormalities has been described [51] with the presence of a generalized delta and theta component, especially in anterior and parietal regions and decrease of alpha and mean beta/total spectrum frequency [52] although with presence of beta components >21 Hz. Doses higher than 600 mg/day are associated with an increased risk of seizures [53]. The effect of other antipsychotics on the EEG, such as quetiapine, olanzapine, or risperidone, is much less studied.

Neuroleptic malignant syndrome is a potentially life-threatening condition caused by an adverse reaction to neuroleptics and usually presents with autonomic dysregulation, tremor, rigidity, fever, altered level of consciousness, and leukocytosis. There are very few studies on the EEG pattern, although the presence of TW, bilateral spike-wave discharges, and NCSE has been described [54].

Antidepressants (fluoxetine, paroxetine, citalopram, venlafaxine, mirtazapine, reboxetine, etc.) may produce seizures in known epileptics [55]. These drugs increase both fast and slow activities. Tricyclic antidepressants can alter the spectral structure and promote the appearance of polyspike bursts [56] and even NCSE [57, 58].

9.5.2 Illicit Drugs-Related Encephalopathies

The characteristics of the different types of encephalopathy that can be observed in response to excess or chronic use of these drugs are not well defined. Episodes of encephalopathy have been reported with lysergic acid diethylamine (LSD) and cannabis, although EEG recordings have been insufficiently studied [10].

Alcohol withdrawal in dependent patients is associated with hallucinosis and delirium tremens [59]. The association between alcoholism and the first epileptic seizure is debated [60]. Subacute encephalopathy with seizures in alcoholics (SESA) has been described as a subtype of localization-related NCSE [61]. Changes in blood alcohol levels are associated with increased amplitude of theta, alpha, and beta in frontal regions and increased alpha frequency in parieto-occipital regions [62].

Cocaine is an alkaloid able to block the dopamine reuptake transporter, increasing dopamine levels. Its use is associated with stroke, orbital ischemia, oculomotor paralysis, intracranial hemorrhage, vasculitis, and severe dyskinesia [63, 64]. An association has been observed between the ingested dose and the beta power [65], although in chronic consumers an absolute reduction of the theta and beta bands was observed in parieto-occipital regions. The incidence of epileptic seizures is very high in these patients [66].

Quantitative EEG studies in chronic cocaine encephalopathy have described a significant reduction of theta and increases of beta bands in occipital regions [65], a finding that does not fit well with the classic EEG pattern of encephalopathy.

9.5.3 Toxins-Related Encephalopathies

EEG changes during encephalopathies caused by mercury, lead, aluminum, etc., are nonspecific.

Carbon monoxide poisoning can cause death. Symptoms are nonspecific until devastating cerebral involvement develops, which, if not fatal, produces very severe brain damage. The EEG recording is nonspecific, with generalized slowing and the presence of sharp waves [67].

9.6 Epileptic Encephalopathies

Epileptic encephalopathies are a group of childhood epileptic disorders in which epileptic anomalies are thought to contribute to the progression of the brain disorder [68]. These include early myoclonic encephalopathy, early infantile epileptic encephalopathy, severe myoclonic epilepsy of infancy (Dravet syndrome), West syndrome (infantile spasm), Landau-Kleffner syndrome, and Lennox-Gastaut syndrome (LGS), among others [69–71].

This is a heterogeneous group of diseases with different clinical features, etiology, and natural history. Therefore, their study is beyond the scope of this chapter. The EEG characteristics of LGS will only be briefly described, as they have very well-defined characteristics. See for a detailed description [72, 73].

The hallmark of the EEG is the presence of generalized, frontally predominant, spike-and-wave bursts at 1–2.5 Hz. This activity increases during NREM sleep. Another important feature is the presence of rapid-rhythm runs between 10 and 20 Hz.

The background may appear almost physiological (Fig. 9.2a). However, numerical analysis reveals the presence of encephalopathy (Fig. 9.2b), with a generalized increase in delta power at 1–1.5 Hz, slowing of the dominant component posterior to 7 Hz, a significant presence of a generalized beta band, especially parasagittal, and the appearance of anomalous spectral components in the midline at 3 and 9 Hz. The basal connectogram (Fig. 9.2c) shows an anomalous pattern, spread throughout the scalp, and a significant increase in central links.

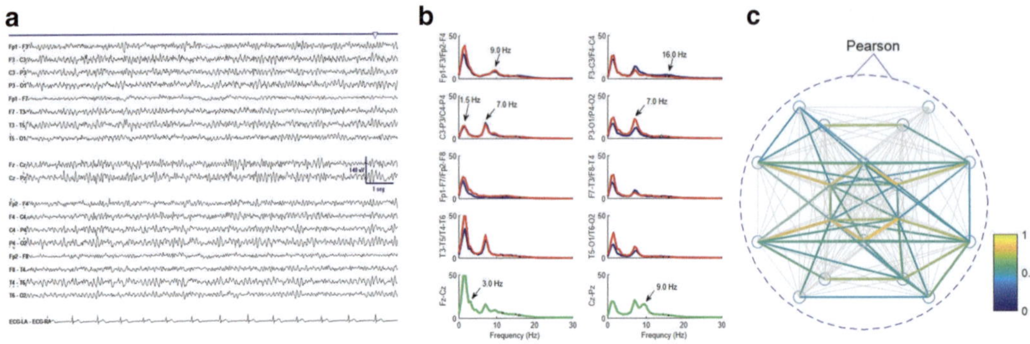

Fig. 9.2 A 19-year-old male patient with Lennox-Gastaut syndrome. (**a**) rsEEG. (**b**) Averaged spectra by channel. (**c**) Connectogram for Pearson's correlation coefficient

9.7 Encephalopathies by Cerebral Inflammatory Processes

Inflammatory encephalopathies can be divided into infectious and immune-mediated. The range of EEG manifestations can vary from an almost normal recording to the presence of highly pathological waveforms [74].

9.7.1 Infectious Encephalitis

Encephalitis is caused by brain inflammation produced by direct infection with an infectious agent. It is often associated with altered levels of consciousness, variable neurological deficits, and seizures. There are three infectious agents that cause specific EEG patterns.

9.7.2 Herpes Simplex Type 1 (HSV-1) Encephalitis

HSV-1 has a predilection for the temporal lobe and fronto-orbital area. Initially, the EEG shows lateralized polymorphic delta slowing with hypervoltage-like sharp complexes, which rapidly evolve into a PLD at 0.3–1 Hz [75]. In fatal cases, a burst-suppression pattern with polyspikes will appear. Generalized or atypical TWs may be observed, intermixed with multifocal sharp waves.

9.7.3 Creutzfeldt-Jakob Disease (CJD)

It is an uncommon, degenerative brain disorder caused by abnormal prion protein. Symptoms include rapidly progressive dementia, psychiatric abnormalities, and movement disorders. The recording is initially nonspecific, with moderate slowing and disorganization. It later usually evolves into a pattern of sharp-wave complexes with a frequency of around 1 Hz. These complexes last around 200–400 ms. These complexes are typically reactive and disappear during sleep. Myoclonus is commonly associated with these complexes [76, 77].

The Heidenhain variant begins with progressive visual disturbances, which are associated with significant abnormalities in the posterior scalp region [78, 79].

An example of CJD encephalopathy will be shown next. A 78-year-old woman comes to the hospital due to rapidly progressive cognitive deterioration, highlighting bradypsychia and significant loss of language, although she responds with monosyllables and maintains articulation.

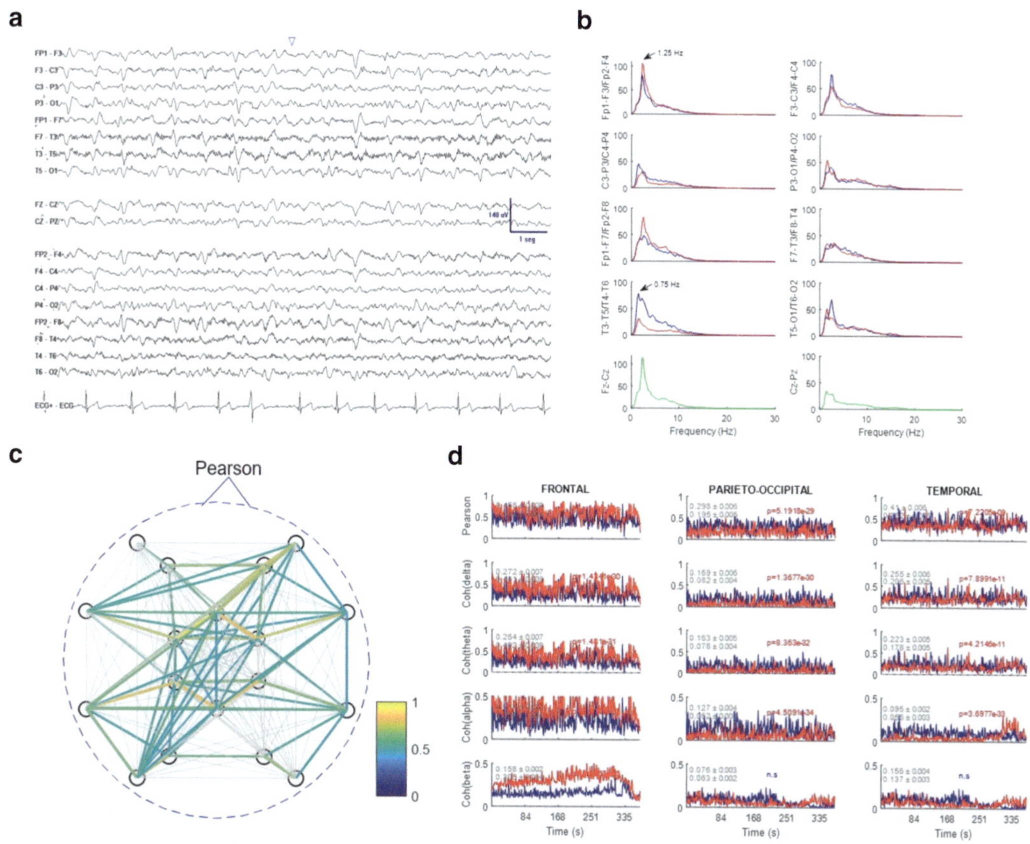

Fig. 9.3 EEG and analysis in a patient with CJD. (**a**) Raw recording. (**b**) Mean average spectra. (**c**) Connectogram severely anomalous, with significant interhemispheric asymmetry. (**d**) Dynamics of synchronization measures grouped by lobe. Red = right hemisphere; blue = left hemisphere

The rsEEG showed an asymmetric pattern of periodic sharp-wave complexes, with a lesser presence of waveforms in the right fronto-parietal region (Fig. 9.3a). No myoclonus was observed during the recording. The periodic pattern disappeared with eyelid opening and closing. The average spectra show a severely anomalous pattern, with the presence of a delta component at 1.25 Hz in the anterior half of the scalp, an absence of a dominant posterior rhythm, and an excess of beta rhythm up to 18 Hz in the posterior half (Fig. 9.3b). Brain synchronization is also significantly altered, as observed in the connectogram (Fig. 9.3c) and confirmed by the correlation and coherence dynamics grouped by lobe, which shows a relative increase in right frontal synchronization and a relative decrease in right parieto-temporal synchronization.

9.7.4 Other Types of Encephalitis

Subacute sclerosing panencephalitis (SSPE) is caused by measles encephalitis. Its EEG pattern is pathognomonic, with the presence of Radermecker-Posner complexes. However, it is very rare in the Western world [79].

Other encephalitides are caused by HIV infection, adenovirus, mycoplasma, rubella, Borrelia burgdorferi, etc. In general, the EEG pattern shows a nonspecific pattern of slowing and alteration of the basal pattern [80–84].

9.8 Infectious Meningitis

Inflammation of the meninges can have a variety of causes, including bacterial, fungal, autoimmune, or tumors. In most cases, the presence of slow waves of moderate to high voltage is observed. EEG findings indicate brain involvement, as the meninges do not generate bioelectrical activity [74].

9.8.1 Bacterial Meningitis

Most cases are caused by meningococci, pneumococci, or Haemophilus influenzae. The recording usually shows a generalized, nonspecific slowing, predominantly frontal. Electrographic seizures may be observed [89]. In cases of subdural abscess or empyema, focal activity in the form of LPD is common.

Basal meningitis is usually caused by tuberculosis, Listeria, syphilis, or borreliosis. The finding usually consists of generalized slowing, which may be asymmetric, with the presence of acute paroxysmal activity and TW [85].

9.8.2 Other Infectious Meningitis

Other agents responsible for meningitis include cryptococcus, cysticercosis, and candidiasis. Viruses also cause meningitis, which sometimes presents with no changes in the EEG ("aseptic" meningitis).

9.9 Immune-Mediated Inflammatory Diseases

9.9.1 Limbic Encephalitis

It presents with subacute memory impairment and behavioral changes, hallucinations, sleep disorders, and temporal seizures. It is often associated with anti-NMDA receptor antibodies and may be associated with ovarian tumors. The EEG pattern usually shows diffuse slowing. In one-third of cases, the "extreme delta brush" pattern is usually seen, characterized by delta activity at 1–3 Hz with the presence of a superimposed beta rhythm. This pattern does not change during sleep and is nonreactive [86, 87].

Autoimmune encephalitis due to voltage-gated potassium channel—complex/Lig1 antibodies often presents with faciobrachial dystonic seizures before the onset of encephalitis [88]. The identification of status epilepticus by the presence of rhythmic delta activity is still debated [74, 89].

9.9.2 Other Immune-Related Encephalitis

Other autoimmune diseases are associated with electroencephalographic changes, although they are generally nonspecific, consisting of slowing and disorganization. Among them, we have systemic lupus erythematosus [90, 91], which can present with epileptogenic activity and even seizures. Antiphospholipid syndrome is characterized by recurrent thrombosis and miscarriage in pregnant [92] and Behcet's disease, characterized by genital and oral aphthous ulcers, uveitis, and CNS involvement [93].

Rasmussen's encephalitis is a progressive epileptic encephalopathy that typically affects one hemisphere. The EEG recording shows unilateral slowing with epileptic spikes associated with contralateral myoclonus, which is described as epilepsia partialis continua [94].

9.10 Numerical Features of Encephalopathy

Alterations in the level of consciousness may have different causes, such as ES, focal brain involvement (of vascular or tumoral origin), psychiatric or encephalopathic pathologies. Rapid and efficient differential diagnosis of this state is essential to allow for an etiological diagnosis, establish the appropriate treatment, and avoid possible undesirable iatrogenic effects (such as, for example, admitting a patient to the ICU for a presumed SE).

One of the most reliable and fastest tools for reaching this diagnosis is the EEG. However, in many cases, the raw recording presents features that are not well defined, which make the diagnosis difficult. In these cases, the qEEG is mandatory. On the other hand, particularly clear tracings do not require the help of numerical tools for diagnosis. However, the use of these tools always provides complementary information about characteristics of the patient's brain state that are not obvious and can be very helpful.

Although much work remains to be done to define the numerical characteristics of the different types of encephalopathies, we can provide some data that are very important for explaining the pathophysiological and clinical manifestations of the condition.

This example shows the case of a 56-year-old woman with a history of obesity, low back pain, diabetes, chronic obstructive pulmonary disease, and attempted suicide by taking drugs. She was brought to the emergency department with a decreased level of consciousness and severe disorientation. The patient was in the process of separating and reported severe alcohol intake in the previous days, with several episodes of loss of consciousness and vomiting. Laboratory tests showed mild leukocytosis (10,400/mm^3). A brain CT scan and lumbar puncture were normal.

A loading dose of levetiracetam (2 g) was administered, followed by 1 g every 8 h thereafter due to suspected epileptic origin. EEG recording was performed 48 h later. An antero-posterior gradient was absent, with no posterior dominant rhythm. No epileptiform activity was observed (Fig. 9.4a). Analysis of the averaged spectra (Fig. 9.4b) shows a widespread delta component at 1 Hz, although the presence of faster theta and even beta components is also evident. However, no dominant rhythm component is observed. In addition to the generalized delta component throughout the scalp, the other defining feature of the encephalopathy is the synchronization disruption. In this case (Fig. 9.4c), an abnormal connectogram is observed, with reduced synchronization in the posterior half and a marked increase in synchronization in the anterior half of the scalp, in addition to a significant decrease in coherence in the alpha and beta bands.

This case concerns an 89-year-old right-handed woman with a history of manic-depressive disorder, undergoing treatment with lithium carbonate, and a right temporo-parietal ischemic stroke 4 years prior. She was brought to the emergency department with a 3-day history of renal failure, decreased level of consciousness, and memory problems, with dystonic paroxysmal movements of the upper limbs. The rsEEG shows a disorganized pattern, with no anteroposterior gradient or dominant posterior rhythm. The background shows a pattern of moderate-amplitude delta and theta slowing (Fig. 9.5a). The average spectral analysis by channel shows a highly altered pattern, with a large frontal delta component, while no such component is observed in the posterior half of the scalp. No posterior dominant rhythm component is present either (Fig. 9.5b). Connectivity is severely impaired, with anterior shifting of connectivity and severe reduction in posterior connectivity. Decreased right parieto-temporal connectivity is also observed as a result of the previous stroke (Fig. 9.5c).

Although the raw records are similar, the numerical analysis is markedly different, both in the spectral pattern and in the connectogram. The spectral composition, in both cases, shows an increase in the delta component and a marked decrease in cortical rhythms >7 Hz. However, the extent of the excess delta is not the same, being greater in Fig. 9.4, where, in addition, the predominance is located peripherally in the temporo-parietal area. Synchronization, typically, presents a displacement towards anterior regions, with a significant decrease in synchronization in the posterior 1/3 of the scalp.

The combined analysis of spectral patterns across the scalp and synchronization characteristics constitutes a promising method for the typological classification of encephalopathies.

Finally, an extremely useful application is the ability to compare two consecutive recordings or those separated by extended periods of time. EEG recordings in patients with encephalopathy present nonspecific characteristics, and their temporal changes are very difficult to appreciate, as

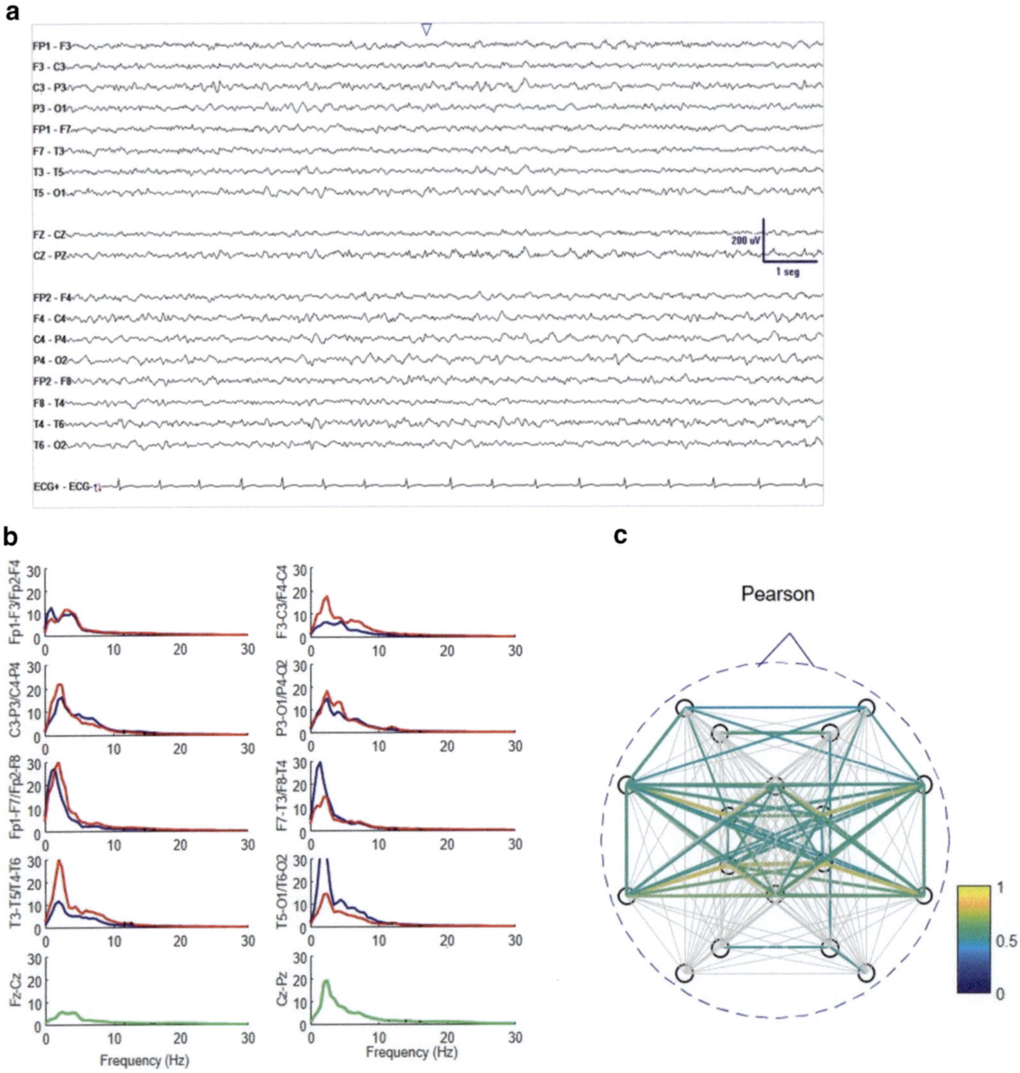

Fig. 9.4 Mild encephalopathy in a 56-year-old woman. (**a**) EEG raw trace. (**b**) Average spectra for all channels. Red = right hemisphere, blue = left hemisphere, green = midline. (**c**) Average connectrogram of the whole recording

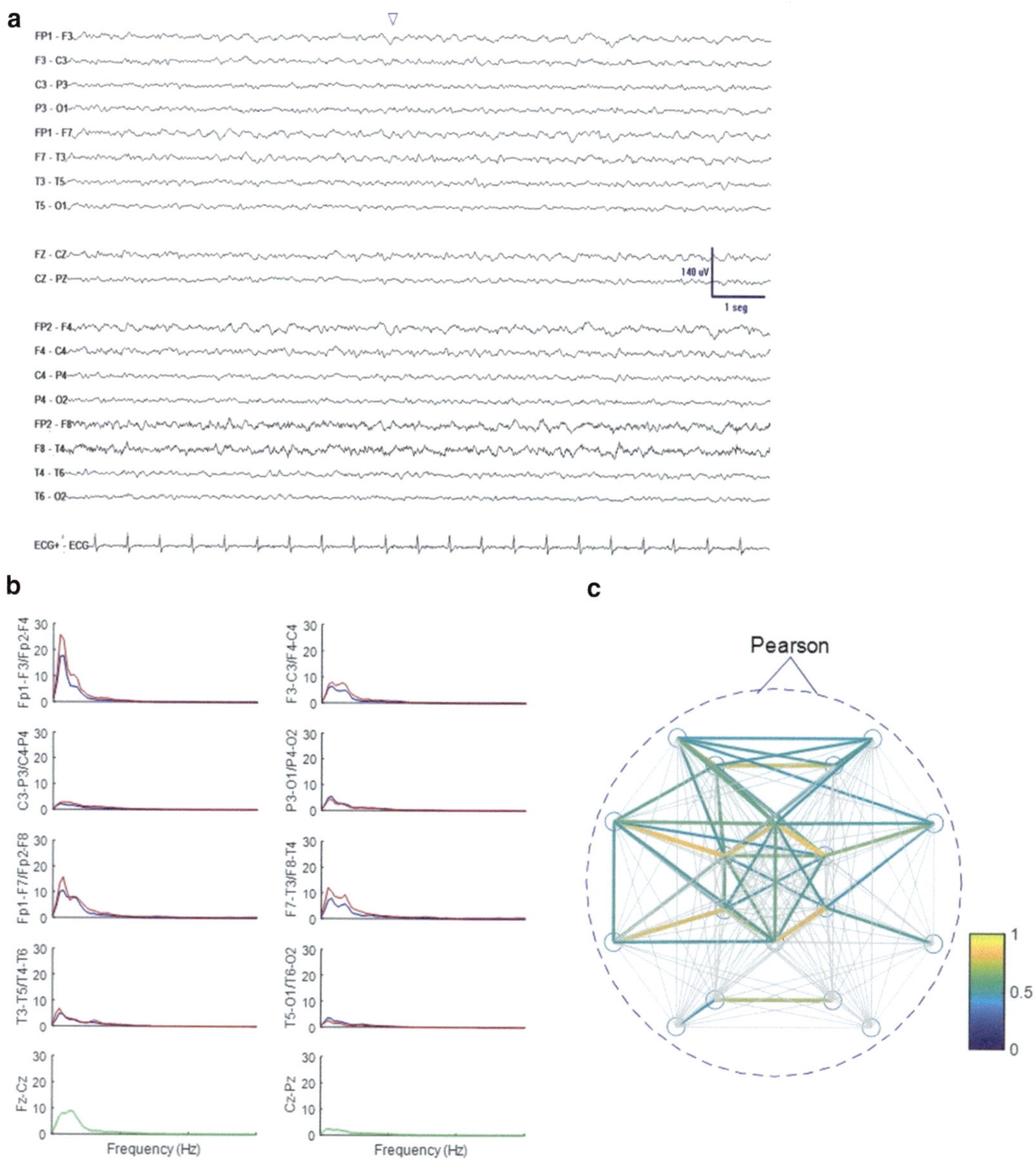

Fig. 9.5 Mild encephalopathy in a 89-year-old woman. (**a**) EEG raw trace. (**b**) Average spectra for all channels. Red = right hemisphere, blue = left hemisphere, green = midline. (**c**) Average connectrogram of the whole recording

they involve slow frequencies. However, it is very important to be able to assess whether, in a new recording, the bioelectrical characteristics of brain activity are the same, better, or worse than in the previous recording. Numerical analysis allows us to make a repeatable and objective assessment of changes in the patient's brain state.

This next example shows the case of a 74-year-old right-handed man admitted for evaluation of a three-day history of decreased level of alertness, under treatment with mirtazapine and antihypertensives. There are no signs of infection. General and liver biochemistry were normal, although he showed hypokalemia and metabolic alkalosis. A CT study revealed possible normal pressure hydrocephalus. The initial recording, made on the second day of admission (Fig. 9.6a, left) showed a disorganized pattern, without an anteroposterior

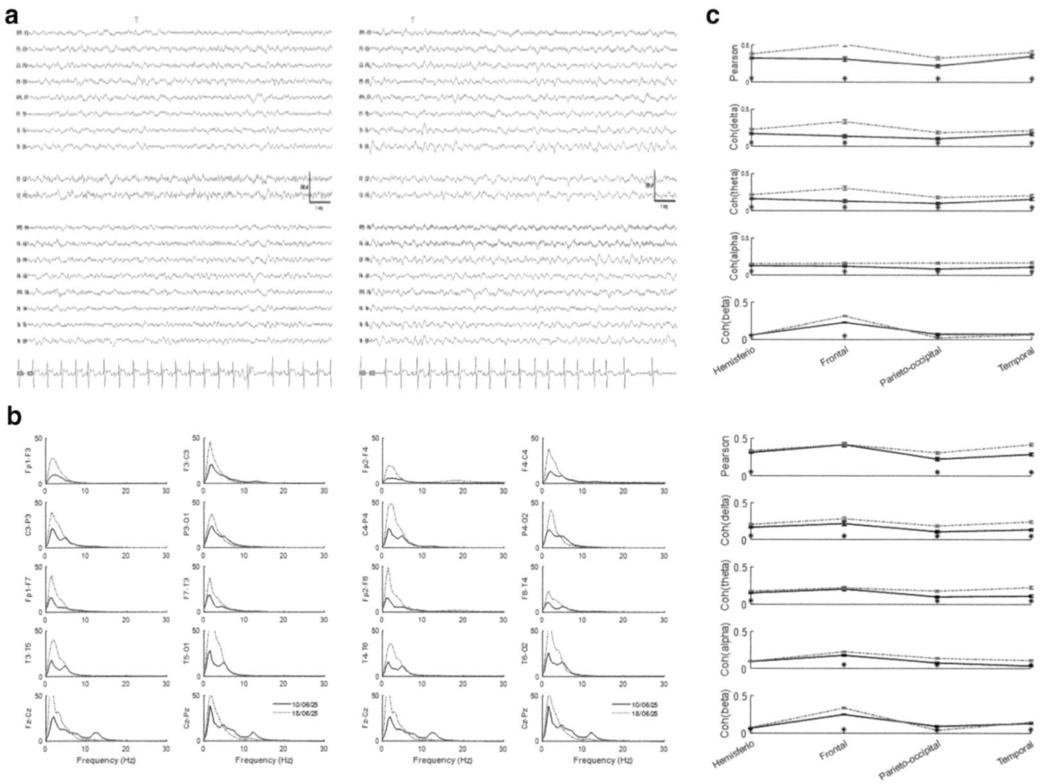

Fig. 9.6 Evolution of the encephalopathy under study and its correlation with the alert level. (**a**) Initial raw recording (left) and 5 days after recording (right). (**b**) Average spectra per channel for the initial recording (solid line) and 5 days after recording (dashed line). (**c**) Synchronization for correlation and coherence by hemisphere and lobe for the left (top) and right (bottom) hemispheres. Asterisks indicate $p < 0.05$. Initial recording = solid line; recording on day 5 = dashed line

gradient or dominant posterior rhythm, with the generalized presence of delta and theta rhythms. After 5 days, the patient showed a decrease in the level of alertness, so a new EEG recording was repeated (Fig. 9.6a, right). This record presents visual characteristics very similar to the previous one, which does not correspond very well with the decrease in the alert level. However, comparing the analyses of both recordings reveals very different characteristics. The averaged spectra by channel show a marked general increase in the delta band and a marked decrease in the faster rhythms (Fig. 9.6b). Furthermore, the new recording shows a marked increase in synchronization across the entire spectrum, especially in anterior regions (Fig. 9.6c). Taken together, these results demonstrate a marked worsening of the patient's preexisting encephalopathy and clearly explain the patient's altered level of alertness and tendency toward sleep, even though the raw record was similar.

To finish, we will discuss the specific properties of SARS-CoV2 encephalopathy, which were numerically different from conventional toxic-metabolic encephalopathy and severe hypoxic encephalopathy following cardiorespiratory arrest [95].

9.10.1 Encephalopathy in SARS-CoV2 Patients

Severe acute respiratory syndrome coronavirus disease (SARS-CoV2) causes acute highly lethal disease (COVID-19). The mechanisms of CNS infection and the pathophysiology by CoV2 are still debated and it has been proposed a direct invasion through blood-brain barrier (BBB), a neuronal pathway, hypoxia damage, immune-response mediated injury, or angiotensin-converter enzyme 2, among others. In fact, in long COVID syndrome has several neurological symptoms, including difficulty thinking or concentrating, headaches, sleep problems, depression, or anxiety [96] that show the persistence of CNS involvement.

We studied patients delivered from intensive care unit (ICU) after SARS-CoV2 infection (COVID) and we used two control groups from patients previously studied in our hospital i) patients with infectious toxic encephalopathy (ENC), taken as a kind of gold-standard encephalopathic pattern and ii) patients after cardio-respiratory arrest (CRA), as example of a severe hypoxic insult to the CNS. The main aim of our work was to describe the structure of EEG defined by the relative distribution of different bands (δ, θ, α, β) across the brain lobes, the global synchronization (measured as Pearson's correlation coefficient r), defined at lobes and hemispheres level and the complexity for power spectra, addressed by the SSE.

We showed that the structure of EEG was truly different for the three groups and for all the bands (except for ENC/COVID theta band, what was similar). Another interesting observation was that the value of χ^2 was always greater for the pair ENC/CRA, with the COVID value in the middle of both groups. Therefore, the structure of EEG for COVID patients is between the both extreme conditions. These results can be observed in the next table (Table 9.1).

From this table we can observe that EEG structure is truly different for the three groups and for all bands (except for ENC/COVID delta and theta bands, which are similar). Another interesting observation is that the value of χ^2 is always greater for ENC/CRA, with the COVID value in the middle of the two groups. Therefore, the structure of the EEG for COVID patients is between the two extreme conditions.

However, although overall the three groups were dissimilar, we assessed exactly what the differences were for every band. Then, we computed one-way ANOVA (ANOVA on ranks when normality failed) for every lobe and band. These results are shown in Fig. 9.7.

Significant difference below $p < 0.05$ by ANOVA is indicated by asterisks. The *post hoc* results indicating pairs of groups are presented in a color code. The pattern of ENC and CRA is clearly different for all bands, except for the parieto-occipital theta band. However, the COVID group is not completely different from ENC and CRA, although it is clearly observed that the distribution is between the two extreme groups. Nonetheless, the behavior of temporal lobes clearly differs for ENC and COVID groups for δ, α, and β bands.

The comparison between pairs of groups (COVID/ENC, COVID/CRA and ENC/CRA), showed that all the three groups were different as regards complexity and synchronization. Interestingly, we plotted the values of synchronization and SSE, comparing the three groups for every lobe (and hemisphere, in the case of r) using ANOVA/ANOVA on ranks (Fig. 9.8).

On the contrary to EEG bands, SSE was higher for COVID group and lower for

Table 9.1 Comparison by paired groups of patients at EEG bands using χ^2

Paired groups	δ	p	θ	p	α	p	β	p
ENC/COVID	14.0	0.05	1.4	n.s	21.8	0.001	17.7	0.01
ENC/CRA	156.2	0.001	20.9	0.001	54.2	0.001	73.2	0.001
COVID/CRA	68.9	0.001	15.3	0.01	17.5	0.01	34.1	0.001

Statistically significance for 5 degrees of freedom is indicated by p in column placed at the right of every band
n.s not-significant

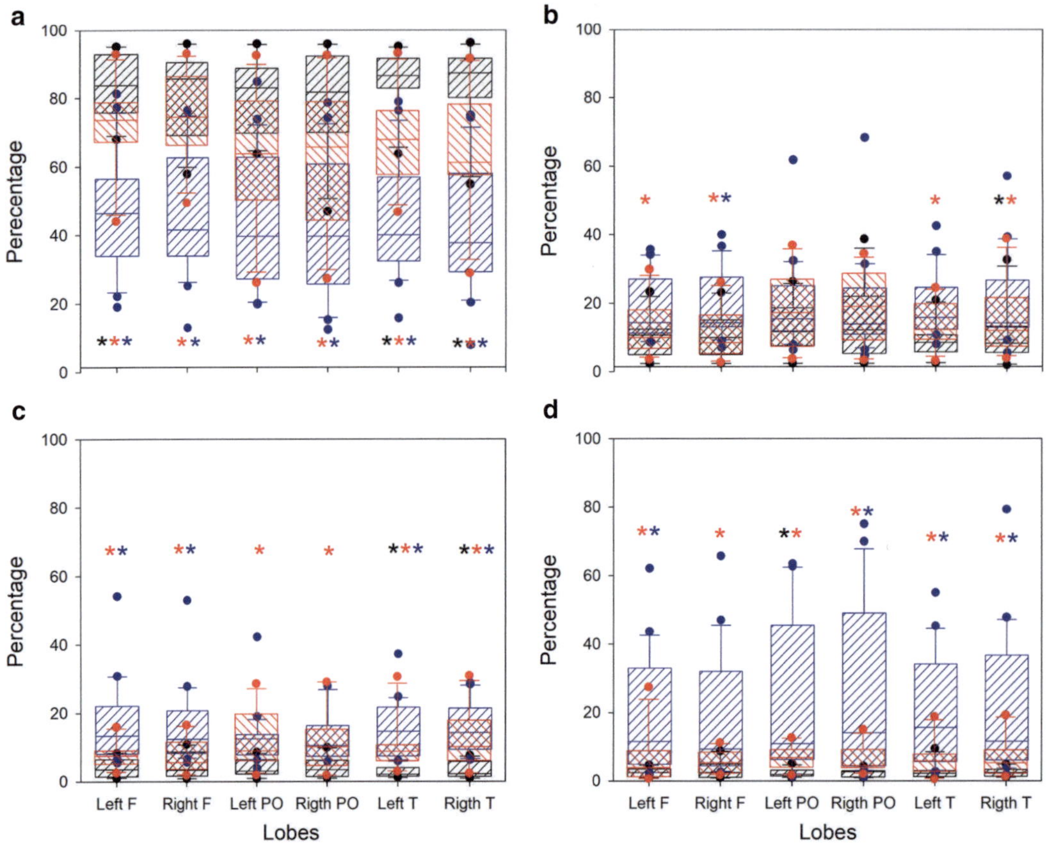

Fig. 9.7 Box plots showing comparison of EEG structure for different bands: (**a**) delta, (**b**) theta, (**c**) alpha, and (**d**) beta. Striped black box: ENC; striped red box: COVID; striped blue box: CRA; black asterisk: difference between ENC and COVID; red asterisk: difference between ENC and CRA; blue asterisk: difference between COVID and CRA. (Pastor et al. [95])

ENC. From Fig. 9.8a, we can observe that SSE was different for all the groups at all the lobes.

In this paper, we showed for first time that severely affected COVID patients developed an encephalopathy with specific EEG features with spectral and connectivity alterations, meanwhile raw traces appear near to physiology in some patients. The fundamental fact in this type of patient, therefore, was that the diagnosis of a specific type of encephalopathy is numerical, so that without this method, it was not easy (and even not possible at all) to identify that the brain involvement suffered by the patient had its origin in COVID infection.

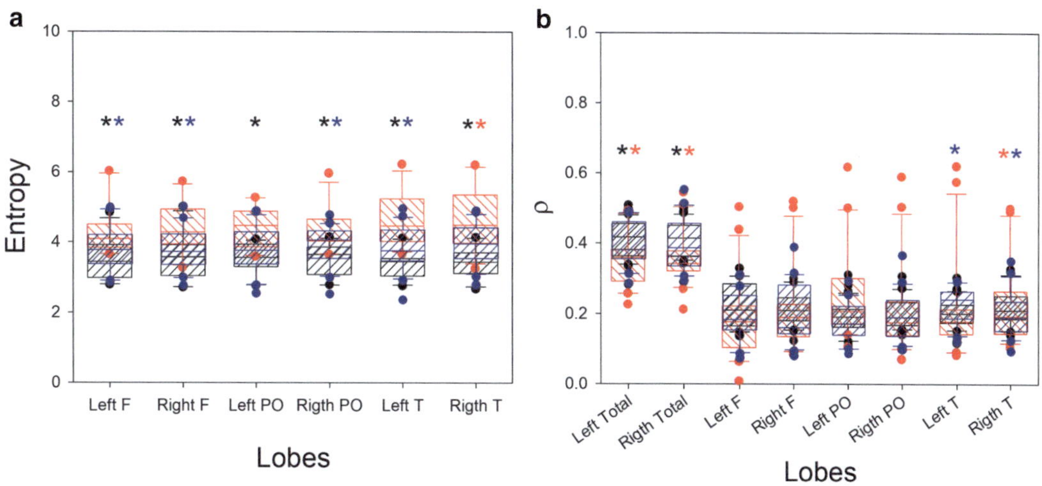

Fig. 9.8 Box plot showing comparison of the structure of entropy and synchronization for different groups. (**a**) SSE; (**b**) *r*. Stripped black box = ECN, stripped red box = COVID and stripped blue box = CRA. Black asterisk = difference between ENC and COVID; Red asterisk = difference between ENC and CRA and blue asterisk = difference between COVID and CRA [95]

9.11 Concluding Remarks

In general, encephalopathies present nonspecific characteristics. However, in some cases, the numerical changes are specific, as we described in COVID-19 encephalopathy.

Usually, the clinical course is typically slow, so visual EEG changes are imperceptible. However, qEEG can detect subtle changes, thus establishing a prognosis before clinical modifications are observed.

References

1. Kaplan PW, Rossetti AO. EEG patterns and imaging correlations in encephalopathy: encephalopathy part II. J. Clin. Neurophysiol. 2011;28:233–51. https://doi.org/10.1097/WNP.0b013e31821c33a0.
2. Erkkinen MG, Berkowitz AL. A clinical approach to diagnosing encephalopathy. Am J Med. 2019;132(10):1142–7.
3. Mizuguchi M, Yamanouchi H, Ichiyama T, Shiomi M. Acute encephalopathy associated with influenza and other viral infections. Acta Neurol Scand Suppl. 2007;186:45–56.
4. Kim Y, Kim JW. Toxic encephalopathy. Saf Health Work. 2012;4:243–56. https://doi.org/10.5491/SHAW.2012.3.4.243.
5. Young GB. Encephalopathy of infection and systemic inflammation. J. Clin. Neurophysiol. 2013;30:454–61. https://doi.org/10.1097/WNP.0b013e3182a73d83.
6. Tauber SC, Eiffert H, Bruck W, Nau R. Septic encephalopathy and septic encephalitis. Expert Rev Anti Infect Ther. 2017;15:121–32. https://doi.org/10.1080/14787210.2017.1265448.
7. Sutter R, Kaplan PW. Uncovering clinical and radiological associations of triphasic waves in acute encephalopathy: a case-control study. Eur J Neurol. 2014;21(4):660–6. https://doi.org/10.1111/ene.12372. Epub 2014 Feb 10. PMID: 24506269.
8. Kim KT, Roh YN, Cho NH, Jeon JC. Clinical correlates of frontal intermittent rhythmic delta activity without structural brain lesion. Clin EEG Neurosci. 2021;52(1):69–73. https://doi.org/10.1177/1550059420922741. Epub 2020 May 15. PMID: 32412802.
9. Thomas C, Hestermann U, Walther S, Pfueller U, Hack M, Oster P, Mundt C, Weisbrod M. Prolonged activation EEG differentiates dementia with and without delirium in frail elderly patients. J Neurol Neurosurg Psychiatry. 2008;79(2):119–25. https://doi.org/10.1136/jnnp.2006.111732. Epub 2007 May 22. PMID: 17519320.
10. Sutter R, Pang T, Kaplan PW. EEG in metabolic disorders, intoxications and epileptic encephalopathies. In: Shomer DL, Lopes da Silva F, editors. Niedermeyer's electroencephalography: basic principles, clinical applications and related fields. 7th ed. Oxford University Press; 2018. ISBN 9780190228484.
11. Marchetti P, D'Avanzo C, Orsato R, Montagnese S, Schiff S, Kaplan PW, Piccione F, Merkel C, Gatta A, Sparacino G, Toffolo GM, Amodio

P. Electroencephalography in patients with cirrhosis. Gastroenterology. 2011;141(5):1680-9.e1-2. https://doi.org/10.1053/j.gastro.2011.06.085. Epub 2011 Jul 18. PMID: 21763244.
12. Amodio P, Gatta A. Neurophysiological investigation of hepatic encephalopathy. Metab Brain Dis. 2005;20(4):369–79. https://doi.org/10.1007/s11011-005-7921-5. PMID: 16382347.
13. Silverman D. Some observations on the EEG in hepatic coma. Electroencephalogr Clin Neurophysiol. 1962;14(1):53–9. https://doi.org/10.1016/0013-4694(62)90006-8.
14. Kaplan PW. Nonconvulsive status epilepticus in the emergency room. Epilepsia. 1996;37(7):643–50. https://doi.org/10.1111/j.1528-1157.1996.tb00628.x. PMID: 8681896.
15. Hirsch LJ, Fong MWK, Leitinger M, LaRoche SM, Beniczky S, Abend NS, Lee JW, Wusthoff CJ, Hahn CD, Westover MB, Gerard EE, Herman ST, Haider HA, Osman G, Rodriguez-Ruiz A, Maciel CB, Gilmore EJ, Fernandez A, Rosenthal ES, Claassen J, Husain AM, Yoo JY, So EL, Kaplan PW, Nuwer MR, van Putten M, Sutter R, Drislane FW, Trinka E, Gaspard N. American Clinical Neurophysiology Society's Standardized Critical Care EEG Terminology: 2021 version. J Clin Neurophysiol. 2021;38(1):1–29. https://doi.org/10.1097/WNP.0000000000000806. PMID: 33475321; PMCID: PMC8135051.
16. Amodio P, Pellegrini A, Ubiali E, Mathy I, Piccolo FD, Orsato R, Gatta A, Guerit JM. The EEG assessment of low-grade hepatic encephalopathy: comparison of an artificial neural network-expert system (ANNES) based evaluation with visual EEG readings and EEG spectral analysis. Clin Neurophysiol. 2006;117(10):2243–51. https://doi.org/10.1016/j.clinph.2006.06.714. Epub 2006 Aug 22. PMID: 16931145.
17. Kuppachi S, Lwin L, Yoo J, Suneja M. Bilateral basal ganglia lesions in end-stage kidney disease presenting as acute chorea. Clin Kidney J. 2013;6(4):450–1. https://doi.org/10.1093/ckj/sft073. PMID: 27293580; PMCID: PMC4898345.
18. Hughes JR. Correlations between EEG and chemical changes in uremia. Electroencephalogr Clin Neurophysiol. 1980;48(5):583–94. https://doi.org/10.1016/0013-4694(80)90293-X.
19. Chuang CL, Chen KP, Kwan SY, Yang WC. Creutzfeldt-Jakob-like EEG in a patient with end-stage renal failure. Nephrol Dial Transplant. 2004;19(1):252–4. https://doi.org/10.1093/ndt/gfg476. PMID: 14671069.
20. Lacerda G, Krummel T, Sabourdy C, Ryvlin P, Hirsch E. Optimizing therapy of seizures in patients with renal or hepatic dysfunction. Neurology. 2006;67(12 Suppl 4):S28–33. https://doi.org/10.1212/wnl.67.12_suppl_4.s28. PMID: 17190918.
21. Markand ON. Electroencephalography in diffuse encephalopathies. J Clin Neurophysiol. 1984;1(4):357–407. PMID: 6242404.
22. Pastor J. Fundamentos biofísicos de la actividad neuronal [Biophysical foundations of the neuronal activity]. Rev Neurol. 2000;30(8):741–55. Spanish. PMID: 10893739.
23. van der Helm-van Mil AH, van Vugt JP, Lammers GJ, Harinck HI. Hypernatremia from a hunger strike as a cause of osmotic myelinolysis. Neurology. 2005;64(3):574–5. https://doi.org/10.1212/01.WNL.0000150730.80325.11. PMID: 15699405.
24. Zwang HJ, Cohn D. Electroencephalographic changes in acute water intoxication. Clin Electroencephalogr. 1981;12(1):35–40. https://doi.org/10.1177/155005948101200107. PMID: 7226563.
25. Azuma H, Akechi T, Furukawa TA. Absence status associated with focal activity and polydipsia-induced hyponatremia. Neuropsychiatr Dis Treat. 2008;4(2):495–8. https://doi.org/10.2147/ndt.s2554. PMID: 18728738; PMCID: PMC2518380.
26. Allen EM, Singer FR, Melamed D. Electroencephalographic abnormalities in hypercalcemia. Neurology. 1970;20(1):15–22. https://doi.org/10.1212/wnl.20.1.15. PMID: 5460769.
27. Gupta M, Lehl SS, Sachdev A. Idiopathic hypoparathyroidism: an unusual cause of acute encephalopathy. BMJ Case Rep. 2011;2011:bcr1220103593. https://doi.org/10.1136/bcr.12.2010.3593. PMID: 22699480; PMCID: PMC3070391.
28. Shao Y, Cai WJ, Wang XL, Chen K, Du J, Zang L, Ba JM, Yan WH, Pei Y, Guo QH, Gu WJ, Dou JT, Lyu ZH, Mu YM. Clinical data analysis of 22 cases with hypoparathyroidism misdiagnosed as epilepsy. Neuro Endocrinol Lett. 2022;43(2):113–8. PMID: 35933617.
29. Kretsch MJ, Sauberlich HE, Newbrun E. Electroencephalographic changes and periodontal status during short-term vitamin B-6 depletion of young, nonpregnant women. Am J Clin Nutr. 1991;53(5):1266–74. https://doi.org/10.1093/ajcn/53.5.1266. PMID: 2021133.
30. Sechi G, Serra A. Wernicke's encephalopathy: new clinical settings and recent advances in diagnosis and management. Lancet Neurol. 2007;6(5):442–55. https://doi.org/10.1016/S1474-4422(07)70104-7. PMID: 17434099.
31. de la Monte SM, Kril JJ. Human alcohol-related neuropathology. Acta Neuropathol. 2014;127(1):71–90. https://doi.org/10.1007/s00401-013-1233-3. Epub 2013 Dec 27. PMID: 24370929; PMCID: PMC4532397.
32. Emre U, Erdal Y, Söylemez C, KaragöZ Y, Mahmutoğlu AS. Wernicke's encephalopathy: similar and different aspects of the three cases. Noro Psikiyatr Ars. 2019;57(2):165–8. https://doi.org/10.29399/npa.22854. PMID: 32550785; PMCID: PMC7285645.
33. Gibson D, Watters A, Bauschka M. Seizures in eating disorders. Int J Eat Disord. 2023;56(8):1650–60. https://doi.org/10.1002/eat.23969. Epub 2023 Apr 24. PMID: 37092766.

34. El Otmani H, Moutaouakil F, Fadel H, El Ouafi N, Abdoh Rafai M, El Moutaouakil B, Slassi I. Chorea-ballismus in acute non-ketotic hyperglycaemia. Funct Neurol. 2009;24(3):129–32. PMID: 20018139.
35. Odak M, Douedi S, Upadhyaya V, Fadhel M, Cosentino J. Focal Neurological Seizure due to Hyperglycemic Hyperosmolar Non-Ketotic Syndrome in Undiagnosed Diabetes Mellitus. Cureus. 2020;12(8):e9909. https://doi.org/10.7759/cureus.9909. PMID: 32968572; PMCID: PMC7505616.
36. Ozer F, Mutlu A, Ozkayran T. Reflex epilepsy and non-ketotic hyperglycemia. Epileptic Disord. 2003;5(3):165–8. PMID: 14684353.
37. Tribl G, Howorka K, Heger G, Anderer P, Thoma H, Zeitlhofer J. EEG topography during insulin-induced hypoglycemia in patients with insulin-dependent diabetes mellitus. Eur Neurol. 1996;36(5):303–9. https://doi.org/10.1159/000117277. PMID: 8864713.
38. Blaabjerg L, Juhl CB. Hypoglycemia-induced changes in the electroencephalogram: an overview. J Diabetes Sci Technol. 2016;10(6):1259–67. https://doi.org/10.1177/1932296816659744. PMID: 27464753; PMCID: PMC5094337.
39. Ruggeri RM, Giuffrida G, Campennì A. Autoimmune endocrine diseases. Minerva Endocrinol. 2018;43(3):305–22. https://doi.org/10.23736/S0391-1977.17.02757-2. Epub 2017 Oct 9. PMID: 28990742.
40. Raff H, Carroll T. Cushing's syndrome: from physiological principles to diagnosis and clinical care. J Physiol. 2015;593(3):493–506. https://doi.org/10.1113/jphysiol.2014.282871. Epub 2015 Jan 5. PMID: 25480800; PMCID: PMC4324701.
41. Reisch N, Peczkowska M, Januszewicz A, Neumann HP. Pheochromocytoma: presentation, diagnosis and treatment. J Hypertens. 2006;24(12):2331–9. https://doi.org/10.1097/01.hjh.0000251887.01885.54. PMID: 17082709.
42. Kollmannsberger A, Hochheuser W, Schwarz K, Scriba PC. Addison's encephalopathy. Electroencephalogr Clin Neurophysiol. 1969;26(4):448. s
43. Shipley JE, Schteingart DE, Tandon R, Pande AC, Grunhaus L, Haskett RF, Starkman MN. EEG sleep in Cushing's disease and Cushing's syndrome: comparison with patients with major depressive disorder. Biol Psychiatry. 1992;32(2):146–55. https://doi.org/10.1016/0006-3223(92)90017-t. PMID: 1330006.
44. Murray K, Amin U, Maciver S, Benbadis SR. EEG findings in posterior reversible encephalopathy syndrome. Clin EEG Neurosci. 2019;50(5):366–9. https://doi.org/10.1177/1550059419856968. Epub 2019 Jun 19.
45. Faigle R, Sutter R, Kaplan PW. Electroencephalography of encephalopathy in patients with endocrine and metabolic disorders. J Clin Neurophysiol. 2013;30(5):505–16. https://doi.org/10.1097/WNP.0b013e3182a73db9. PMID: 24084183; PMCID: PMC3826953.
46. Song TJ, Kim SJ, Kim GS, Choi YC, Kim WJ. The prevalence of thyrotoxicosis-related seizures. Thyroid. 2010;20(9):955–8. https://doi.org/10.1089/thy.2009.0276. PMID: 20718679.
47. River Y, Zelig O. Triphasic waves in myxedema coma. Clin Electroencephalogr. 1993;24(3):146–50. https://doi.org/10.1177/155005949302400313. PMID: 8403447.
48. Spatola L, Maringhini S, Canale C, Granata A, D'Amico M. Lithium poisoning and renal replacement therapy: pathophysiology and current clinical recommendations. Int Urol Nephrol. 2023;55(10):2501–5. https://doi.org/10.1007/s11255-023-03558-5. Epub 2023 Mar 20. PMID: 36940005.
49. Casanova B, de Entrambasaguas M, Perla C, Gómez-Siurana E, Benetó A, Burguera JA. Lithium-induced Creutzfeldt-Jakob syndrome. Clin Neuropharmacol. 1996;19(4):356–9. https://doi.org/10.1097/00002826-199619040-00009. PMID: 8828999.
50. Schulz C, Mavrogiorgou P, Schröter A, Hegerl U, Juckel G. Lithium-induced EEG changes in patients with affective disorders. Neuropsychobiology. 2000;42(Suppl 1):33–7. https://doi.org/10.1159/000054850. PMID: 11093069.
51. Varma S, Bishara D, Besag FM, Taylor D. Clozapine-related EEG changes and seizures: dose and plasma-level relationships. Ther Adv Psychopharmacol. 2011;1(2):47–66. https://doi.org/10.1177/2045125311405566. PMID: 23983927; PMCID: PMC3736902.
52. Joutsiniemi SL, Gross A, Appelberg B. Marked clozapine-induced slowing of EEG background over frontal, central, and parietal scalp areas in schizophrenic patients. J Clin Neurophysiol. 2001;18(1):9–13. https://doi.org/10.1097/00004691-200101000-00003. PMID: 11290933.
53. Alper K, Schwartz KA, Kolts RL, Khan A. Seizure incidence in psychopharmacological clinical trials: an analysis of Food and Drug Administration (FDA) summary basis of approval reports. Biol Psychiatry. 2007;62(4):345–54. https://doi.org/10.1016/j.biopsych.2006.09.023. Epub 2007 Jan 16. PMID: 17223086.
54. Yoshino A, Yoshimasu H. Nonconvulsive status epilepticus complicating neuroleptic malignant syndrome improved by intravenous diazepam. J Clin Psychopharmacol. 2000;20(3):389–90. https://doi.org/10.1097/00004714-200006000-00020. PMID: 10831032.
55. Maguire MJ, Marson AG, Nevitt SJ. Antidepressants for people with epilepsy and depression. Cochrane Database Syst Rev. 2021;4(4):CD010682. https://doi.org/10.1002/14651858.CD010682.pub3. PMID: 33860531; PMCID: PMC8094735.
56. Kugler J, Lorenzi E, Spatz R, Zimmermann H. Drug-induced paroxysmal EEG-activities. Pharmakopsychiatr Neuropsychopharmakol. 1979;12(2):165–72. https://doi.org/10.1055/s--0028-1094607. PMID: 461504.
57. Rumpl E, Hinterhuber H. Unusual 'spike-wave stupor' in a patient with manic-depressive psychosis treated with amitriptyline. J Neurol. 1981;226(2):131–5.

https://doi.org/10.1007/BF00313441. PMID: 6186783.
58. Taniguchi G, Miyajima M, Watanabe M, Murata Y, Sone D, Watanabe Y, Okazaki M, Kobayashi-Kimura M, Kato M, Onuma T. Nonconvulsive status epilepticus in the elderly associated with newer antidepressants used at therapeutic doses: a report of three cases. Epilepsy Behav Case Rep. 2014;3:8–11. https://doi.org/10.1016/j.ebcr.2014.10.003. PMID: 25737963; PMCID: PMC4338854.
59. Sand T, Bråthen G, Michler R, Brodtkorb E, Helde G, Bovim G. Clinical utility of EEG in alcohol-related seizures. Acta Neurol Scand. 2002;105(1):18–24. https://doi.org/10.1034/j.1600-0404.2002.00058.x. PMID: 11903104.
60. Leone M, Tonini C, Bogliun G, Monaco F, Mutani R, Bottacchi E, Gambaro P, Rocci E, Tassinari T, Cavestro C, Beghi E. Chronic alcohol use and first symptomatic epileptic seizures. J Neurol Neurosurg Psychiatry. 2002;73(5):495–9. https://doi.org/10.1136/jnnp.73.5.495. s
61. Fernández-Torre JL, Kaplan PW. Subacute encephalopathy with seizures in alcoholics syndrome: a subtype of nonconvulsive status epilepticus. Epilepsy Curr. 2019;19(2):77–82. https://doi.org/10.1177/1535759719835676. PMID: 30955427; PMCID: PMC6610417.
62. Lukas SE, Mendelson JH, Woods BT, Mello NK, Teoh SK. Topographic distribution of EEG alpha activity during ethanol-induced intoxication in women. J Stud Alcohol. 1989;50(2):176–85. https://doi.org/10.15288/jsa.1989.50.176. PMID: 2927132.
63. Van Stavern GP, Gorman M. Orbital infarction after cocaine use. Neurology. 2002;59(4):642–3. https://doi.org/10.1212/wnl.59.4.642. PMID: 12196670.
64. Chang TR, Kowalski RG, Caserta F, Carhuapoma JR, Tamargo RJ, Naval NS. Impact of acute cocaine use on aneurysmal subarachnoid hemorrhage. Stroke. 2013;44(7):1825–9. https://doi.org/10.1161/STROKEAHA.111.000749. Epub 2013 May 7. PMID: 23652270.
65. Herning RI, Jones RT, Hooker WD, Mendelson J, Blackwell L. Cocaine increases EEG beta: a replication and extension of Hans Berger's historic experiments. Electroencephalogr Clin Neurophysiol. 1985;60(6):470–7. https://doi.org/10.1016/0013-4694(85)91106-x. PMID: 2408845.
66. Majlesi N, Shih R, Fiesseler FW, Hung O, Debellonio R. Cocaine-associated seizures and incidence of status epilepticus. West. J Emerg Med. 2010;11(2):157–60. PMID: 20823966; PMCID: PMC2908651.
67. Vlcek P, Monkova I, Nerandzic Z, Lippert-Grüner M. Delayed encephalopathy after acute carbon monoxide poisoning: a case study. Brain Inj. 2024;38(5):331–6. https://doi.org/10.1080/02699052.2024.2311339. Epub 2024 Feb 3. PMID: 38308510.
68. Blume WT, Lüders HO, Mizrahi E, Tassinari C, van Emde BW, Engel J Jr. Glossary of descriptive terminology for ictal semiology: report of the ILAE task force on classification and terminology. Epilepsia. 2001;42(9):1212–8. https://doi.org/10.1046/j.1528--1157.2001.22001.x. PMID: 11580774.
69. Reynolds C, King MD, Gorman KM. The phenotypic spectrum of SCN2A-related epilepsy. Eur J Paediatr Neurol. 2020;24:117–22. https://doi.org/10.1016/j.ejpn.2019.12.016. Epub 2019 Dec 12. PMID: 31924505.
70. Raga S, Specchio N, Rheims S, Wilmshurst JM. Developmental and epileptic encephalopathies: recognition and approaches to care. Epileptic Disord. 2021;23(1):40–52. https://doi.org/10.1684/epd.2021.1244. PMID: 33632673.
71. Ahmad SF, Ahmad KA, Ng YT. Neonatal epileptic encephalopathies. Semin Pediatr Neurol. 2021;37:100880. https://doi.org/10.1016/j.spen.2021.100880. Epub 2021 Mar 4. PMID: 33892847.
72. Asadi-Pooya AA. Lennox-Gastaut syndrome: a comprehensive review. Neurol Sci. 2018;39(3):403–14. https://doi.org/10.1007/s10072-017-3188-y. Epub 2017 Nov 9. PMID: 29124439.
73. Nelson JA, Knupp KG. Lennox-gastaut syndrome: current treatments, novel therapeutics, and future directions. Neurotherapeutics. 2023;20(5):1255–62. https://doi.org/10.1007/s13311-023-01397-x. Epub 2023 Jun 23. PMID: 37353676; PMCID: PMC10480363.
74. Galovic M, Schmitz B, Tettenborg B. EEG in inflammatory disorders, cerebrovascular disease, trauma and migraine. In: Shomer DL, Lopes da Silva F, editors. Niedermeyer's electroencephalography: basic principles, clinical applications and related fields. 7th ed. Oxford University Press; 2018. ISBN 9780190228484.
75. Upton A, Gumpert J. Electroencephalography in the early diagnosis of herpes simplex encephalitis. Rev Electroencephalogr Neurophysiol Clin. 1971;1(1):81–3. https://doi.org/10.1016/s0370-4475(71)80016-3. PMID: 5173700.
76. Wieser HG, Schwarz U, Blättler T, Bernoulli C, Sitzler M, Stoeck K, Glatzel M. Serial EEG findings in sporadic and iatrogenic Creutzfeldt-Jakob disease. Clin Neurophysiol. 2004;115(11):2467–78. https://doi.org/10.1016/j.clinph.2004.05.032. PMID: 15465434.
77. Huang B, Shafiian N, Masi PJ, Gordon ML, Franceschi AM, Giliberto L. Creutzfeldt-Jakob disease presenting as psychiatric disorder: case presentation and systematic review. Front Neurol. 2024;15:1428021. https://doi.org/10.3389/fneur.2024.1428021. PMID: 39268068; PMCID: PMC11390469.
78. Keyrouz SG, Labib BT, Sethi R. MRI and EEG findings in Heidenhain variant of Creutzfeldt-Jakob disease. Neurology. 2006;67(2):333. https://doi.org/10.1212/01.wnl.0000208487.18608.41. PMID: 16864833.
79. Baiardi S, Capellari S, Ladogana A, Strumia S, Santangelo M, Pocchiari M, Parchi P. Revisiting the Heidenhain variant of Creutzfeldt-Jakob disease: evidence for prion type variability influencing clinical course and laboratory findings. J Alzheimers Dis.

2016;50(2):465–76. https://doi.org/10.3233/JAD-150668. s
80. Westmoreland BF, Sharbrough FW, Donat JR. Stimulus-induced EEG complexes and motor spasms in subacute sclerosing panencephalitis. Neurology. 1979;29(8):1154–7. https://doi.org/10.1212/wnl.29.8.1154. PMID: 572503.
81. Tinuper P, de Carolis P, Galeotti M, Baldrati A, Gritti FM, Sacquegna T. Electroencephalogram and HIV infection: a prospective study in 100 patients. Clin Electroencephalogr. 1990;21(3):145–50. https://doi.org/10.1177/155005949002100310. PMID: 2364556.
82. Michael BD, Solomon T. Seizures and encephalitis: clinical features, management, and potential pathophysiologic mechanisms. Epilepsia. 2012;53(Suppl 4):63–71. https://doi.org/10.1111/j.1528--1167.2012.03615.x. PMID: 22946723.
83. Kerling F, Blümcke I, Stefan H. Pitfalls in diagnosing limbic encephalitis a case report. Acta Neurol Scand. 2008;118(5):339–42. https://doi.org/10.1111/j.1600--0404.2008.01020.x. Epub 2008 Mar 27. PMID: 18384457.
84. Bartel P, Schutte CM, Becker P, van der Meyden C. Discrimination between viral and nonviral meningitis by visually analyzed and quantitative electroencephalography. Clin Electroencephalogr. 1999;30(2):35–8. https://doi.org/10.1177/155005949903000202. PMID: 10358780.
85. Konno S, Sugimoto H, Nemoto H, Kitazono H, Murata M, Toda T, Nakazora H, Nomoto N, Wakata N, Kurihara T, Fujioka T. Triphasic waves in a patient with tuberculous meningitis. J Neurol Sci. 2010;291(1–2):114–7. https://doi.org/10.1016/j.jns.2009.12.027. Epub 2010 Feb 8. PMID: 20116807.
86. Schmitt SE, Pargeon K, Frechette ES, Hirsch LJ, Dalmau J, Friedman D. Extreme delta brush: a unique EEG pattern in adults with anti-NMDA receptor encephalitis. Neurology. 2012;79(11):1094–100. https://doi.org/10.1212/WNL.0b013e3182698cd8. Epub 2012 Aug 29. PMID: 22933737; PMCID: PMC3525298.
87. Huang Q, Liao Y, Ma M, Wu Y. Delta brush variant: a novel ictal EEG pattern in anti-NMDAR encephalitis. Epilepsia Open. 2020;5(3):507–13. https://doi.org/10.1002/epi4.12423. PMID: 32913959; PMCID: PMC7469758.
88. Irani SR, Michell AW, Lang B, Pettingill P, Waters P, Johnson MR, Schott JM, Armstrong RJ, Zagami AS, Blcasel A, Somerville ER, Smith SM, Vincent A. Faciobrachial dystonic seizures precede Lgi1 antibody limbic encephalitis. Ann Neurol. 2011;69(5):892–900. https://doi.org/10.1002/ana.22307. Epub 2011 Mar 17. PMID: 21416487.
89. Pollak L, Klein C, Schiffer J, Flechter S, Rabey J. Electroencephalographic abnormalities in aseptic meningitis and noninfectious headache. A comparative study. Headache. 2001;41(1):79–83. https://doi.org/10.1046/j.1526-4610.2001.111006079.x. PMID: 11168607.
90. Waterloo K, Omdal R, Jacobsen EA, Kløw NE, Husby G, Torbergsen T, Mellgren SI. Cerebral computed tomography and electroencephalography compared with neuropsychological findings in systemic lupus erythematosus. J Neurol. 1999;246(8):706–11. https://doi.org/10.1007/s004150050436. PMID: 10460449.
91. Bertsias GK, Ioannidis JP, Aringer M, Bollen E, Bombardieri S, Bruce IN, Cervera R, Dalakas M, Doria A, Hanly JG, Huizinga TW, Isenberg D, Kallenberg C, Piette JC, Schneider M, Scolding N, Smolen J, Stara A, Tassiulas I, Tektonidou M, Tincani A, van Buchem MA, van Vollenhoven R, Ward M, Gordon C, Boumpas DT. EULAR recommendations for the management of systemic lupus erythematosus with neuropsychiatric manifestations: report of a task force of the EULAR standing committee for clinical affairs. Ann Rheum Dis. 2010;69(12):2074–82. https://doi.org/10.1136/ard.2010.130476. Epub 2010 Aug 19. PMID: 20724309.
92. Lampropoulos CE, Koutroumanidis M, Reynolds PP, Manidakis I, Hughes GR, D'Cruz DP. Electroencephalography in the assessment of neuropsychiatric manifestations in antiphospholipid syndrome and systemic lupus erythematosus. Arthritis Rheum. 2005;52(3):841–6. https://doi.org/10.1002/art.20957. PMID: 15751063.
93. Aykutlu E, Baykan B, Serdaroglu P, Gökyigit A, Akman-Demir G. Epileptic seizures in Behçet disease. Epilepsia. 2002;43(8):832–5. https://doi.org/10.1046/j.1528-1157.2002.01102.x. PMID: 12181001.
94. Muthaffar OY, Alyazidi AS. Epilepsia partialis continua: A review. Neurosciences (Riyadh). 2024;29(2):71–6. https://doi.org/10.17712/nsj.2024.2.20230074. s
95. Pastor J, Vega-Zelaya L, Martin AE. Specific EEG encephalopathic pattern in SARS-CoV-2 patients. J. Clin. Med. 2020;91:545. https://doi.org/10.3390/jcm9051545.
96. Koc HC, Xiao J, Liu W, Li Y, Chen G. Long COVID and its Management. Int J Biol Sci. 2022;18(12):4768–80. https://doi.org/10.7150/ijbs.75056. PMID: 35874958; PMCID: PMC9305273.

10 Dementias and Psychiatric Pathologies

10.1 Introduction

Conventional EEG recordings in psychiatric disorders generally show nonspecific or apparently normal results, so for some time it was thought that qEEG could be a good diagnostic method for psychiatric disorders. However, methodological inconsistencies and overinterpretation of the results have led to this hope not being borne out by the results. However, in recent years, interest in the use of qEEG as a research method in psychiatric pathology has been revived [1], although it has hardly any clinical use.

One of the major clinical problems in patients with psychiatric symptoms is the possible existence of an organic pathology that justifies them [1]. Routine EEG studies in patients with psychiatric disorders usually show benign variants or diffuse slowing [2]. Therefore, an EEG study is often requested to check for the possibility of an organic (neurological) pathology that could explain these symptoms, such as epilepsy, focal disorders, or moderate encephalopathies. When no abnormalities are found in the EEG, the symptoms are classified as psychiatric.

10.2 EEG in Dementias

There is no scientific evidence that rsEEG can be useful in the diagnosis of patients with dementia [3], so its use in clinical practice is not recommended.

When studying patients with dementia, it is important to keep in mind that these patients are generally elderly, so it is crucial to consider the pattern of physiological aging in order to distinguish pathological states [4–6]. It is important to remember the following facts:

- The posterior dominant rhythm may be between 7 and 8 Hz, especially in patients over 80 years of age.
- Reduction in total alpha rhythm time, which is also less reactive to eye opening and closing.
- Appearance of subharmonic variants of the theta and delta frequencies.

Synchronization studies of rsEEG studies have described a decrease in the presence of alpha rhythm [7–9] and connectivity in the posterior region compared to young adults [10]. Changes in other bands have also been described, but inconclusively. A reduction in left frontal and central hemispheric connectivity has also been described [11].

Despite the inconclusive results, numerous findings have been found in qEEGs of patients with different types of dementia. The main findings are briefly summarized below (see [6] for a detailed description).

1. Alzheimer's disease (AD). Generally it's observed a generalized decrease in alpha and beta power density, primarily in posterior regions, along with a generalized increase in

J. Pastor, *Quantitative EEG Practical Guide*, https://doi.org/10.1007/978-3-032-05457-9_10

delta and theta power. Furthermore, a reduction in the complexity of brain dynamics and a decrease in alpha band connectivity have been observed, with a reduction in the flow of information between frontal and parietal lobes [12–16].
2. Vascular dementia (VD). Elevated delta/theta power values and low alpha/beta power values, along with a generalized increase in theta power [17–20]. Complexity values are higher than in AD [21], although coherence in central regions is decreased [22].
3. Frontotemporal dementia (FTD). A decrease in alpha power density has been described in the frontal and temporal poles, with preservation of delta, alpha, and beta power, compared to AD [23, 24] and relative preservation of connectivity [25].
4. Dementia with Lewy bodies (DLB). Increased power density in the theta and delta bands has been observed, especially compared to AD, as well as increased variability of delta power [26–28], which also shows increased coherence.
5. Parkinson's disease dementia (PDD). Compared with AD, it has been described an increased delta and theta power density, along with increased central delta and posterior theta power [29, 30], has been described, which is accompanied by higher alpha and beta coherence values than in AD.

10.3 qEEG in Psychiatric Pathologies

As we observed in the case of dementia, there is no indication for rsEEG in psychiatric disorders. It is generally performed to rule out the presence of organic pathology (e.g., epilepsy or encephalopathy). However, numerous authors have shown a larger proportion of abnormal EEGs in individuals with psychopathology. In summary, they have pointed out the following data:

- A significant percentage of nonepileptic adults have qEEG epileptiform discharges (EDs) [31] that are attributable to original brain anomalies, medications, and psychiatric disorders themselves.
- In up to 68% of qEEGs in psychiatric patients display evidence of pathophysiology as spectral composition [32].
- In up to 30% of panic attack patients it has been demonstrated the presence of qEEG abnormalities, and the incidence of atypical qEEG findings in mood disorders reaches up to 40% [33].
- A significant amount of youngsters (up to 69%) with behavior disorders, with a predominance of aggressiveness have EEG deviations from normality [34].

In view of these results, the EEG has been used for the classification of various psychiatric pathologies. For example, EEG utility in discrimination/differentiation between depression and dementia (sensitivity 72–93% and specificity 75–88%) [32], between affective disorders and schizophrenia [35], between schizophrenia and unipolar and bipolar depression, and between panic disorder and depression [36] have been demonstrated. However, the results have not been conclusive.

However, the main use given to qEEG in practice is the difference between states of organic cause and those that can present with psychiatric symptoms [37]. It is not, therefore, a positive diagnosis of psychiatric pathology, but the exclusion of an organic pathology. Among these situations, we can find diagnosis of nonepileptic psychogenic seizures (NEPS [38]), subclinical ictal patterns, encephalopathic patterns in dementias or the degree of brain maturation [39, 40].

Nevertheless, new approach in the field of psychiatric pathologies has been addressed recently. In fact, there are promising data in the pathophysiology and/or diagnosis of schizophrenia, e.g., by the anomalies of theta/gamma coupling [41], theta power [42], or functional connectivity [43, 44]. It has also been postulated that affective disorder [45], depression, autistic spectrum disorder, or attention deficit hyperactivity disorder can be addressed by specific changes of qEEG [46–48].

However, most of these achievements of qEEG occur in the scientific field and, subsequently, do not have practical and daily continuity in clinical use. Therefore, the main use of EEG continues to be to evaluate the presence of organic pathology. However, although we are not yet in a position to clinically use the diagnostic tools listed for different pathologies above, we are in a position to go a step further in the positive diagnosis of these pathologies and not only make a diagnosis by exclusion.

Therefore, we are going to show below some examples of patients with cognitive psychiatric symptoms who presented alterations in the qEEG and that can serve as a positive diagnosis in the daily practice of dementia or psychiatric pathology.

10.3.1 Two Practical Examples

We will show an example of the information provided by qEEG in a patient with insidious language impairment. This is a 64-year-old right-handed male who has reported hypofluent speech, with blockages, circumlocutions, and semantic paraphasias, for the past 2 years. He does not present anomia. At the time of the recording, he was not receiving treatment except for his high blood pressure. The rsEEG recording (Fig. 10.1a) shows a pattern of physiological characteristics, with an anteroposterior gradient and the presence of a posterior dominant rhythm to 9 Hz, symmetrical (13.9/16.4 μV) and reactive. The average spectra (Fig. 10.1b) exhibit symmetrical and physiological characteristics. The average power of the bands in the different lobes (Fig. 10.1c) shows an increase in left frontal delta power, with minimal decreases in left hemispheric alpha and beta power. However, the anomalies in the synchronization analysis were very striking. Fig. 10.1d shows the connectogram for Pearson's correlation. The important interhemispheric asymmetry is evident, with loss of synchronization links in the left frontoparietal region. Figure 10.1e shows the dynamics of the different coherence bands. In addition to the very important asynchrony of all bands in the left frontal and temporal lobes, another anomaly is frequently observed in patients with cognitive and/or psychiatric alterations. In both frontal regions, but especially in the left, a significant instability is observed in the dynamics of Pearson's correlation and the delta and theta bands (Fig. 10.1e, left column). These synchronization alterations, which are directly related to brain connectivity, probably serve as a justification for cognitive alterations in a patient without morphological alterations and without electrogenesis abnormalities, estimated from the spectral power of the different EEG bands.

The MRI study demonstrated the existence of cortico-subcortical atrophy (grade 2 on the Pasquier Global Atrophy scale), with atrophy of the mesial regions of the temporal lobes, without signs of microhemorrhage. Genetic evaluation showed the existence of a mutation in the C9ORF72 gene.

Finally, an example of objective alteration in the qEEG analysis in a patient with psychiatric symptoms is shown. A 50-year-old right-handed female patient was referred to a v-EEG study for 2–3 min episodes of behavioral arrest that sometimes associated upper limbs movements, inability to speak, and cephalic and eye movements without loss of consciousness. These episodes take place from the age of 35. She also reported episodes of falling to the ground and movements of the 4 limbs without cephalic version that may be associated with urinary incontinence. On treatment with clonazepam (0.5 mg/8 h) and oxcarbazepine (300 mg/12 h)..

During the study, a paroxysmal episode similar to those commonly reported was recorded. It occurred during wakefulness and lasted approximately 4 min. Clinical manifestations consisted of a sudden flexion of the head, with closed eyes, and subsequent lateral head movements (denial). There was no evidence of impaired consciousness or associated subjective sensations. The patient was apparently unable to speak and made laryngeal movements, although she obeyed commands. The episode ended with a deep inspiration. Bioelectrical manifestations only showed movement, muscle, and EOG artifacts. There was no evidence of an ictal or post-ictal pattern.

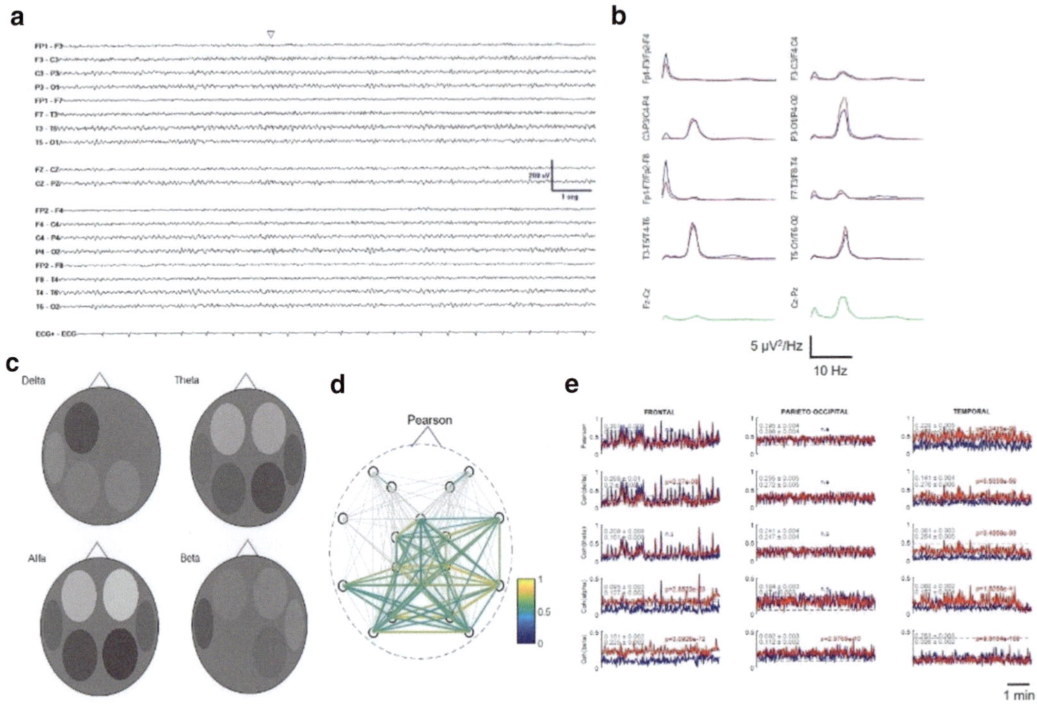

Fig. 10.1 Numerical alterations in a patient with primary aphasia. (**a**) EEG recording of physiological characteristics. (**b**) Average spectra per channel. Red = right hemisphere; blue = left hemisphere; green = mid-line. (**c**) Maps of average spectral power by lobes. (**d**) Channel connectogram for Pearson correlation. (**e**) Band dynamics for lobe synchronization measurements

The rsEEG shows a pattern of low amplitude, with neither clear antero-posterior gradient nor posterior dominant rhythm and without epileptiform waveforms (Fig. 10.2a). The spectral pattern is symmetrical and with excess of delta activity and slowing of faster frequencies (Fig. 10.2b), compatible with mild encephalopathy. Therefore, no alterations were observed in the voltage functions of the differential channels. However, the connectogram (Fig. 10.2c) was abnormal, with a more homogeneous pattern than normal for the patient's age. However, both the hemispheric synchronization dynamics (Fig. 10.2d) and the lobe dynamics (Fig. 10.2e) show a significant decrease in synchronization across the entire spectrum, especially in the posterior half. In the frontal lobes, very striking, high-amplitude oscillations in synchronization are observed. This alteration in basal synchronization is shown in detail in the analysis of frontoparietal synchronization in both hemispheres. Figure 10.2f shows, in the top row, the dynamics of the Pearson correlation for the F3-C3/P3-O1 and F4-C4/P4-O2 channels, and in the bottom row, the cross-correlation analysis. In both hemispheres, the Pearson correlation is zero. Cross-correlation analysis shows a pattern very different from the frontoparietal synchronization typically observed, even in epileptic patients.

These examples are significant in proving the contribution of qEEG in clinical practice. In these cases, the visual EEG recording was completely normal and without associated pathological manifestations, such as epileptiform waveforms or any other paroxysmal activity. There are still many doubts about the diagnostic interpretation of alterations in the recording, such as synchronization or interlobar asymmetries. However, the point to highlight is that a recording that is physiological on the surface, after a simple numerical analysis, shows alterations that can explain the patient's symptoms. In other words, what could be considered a false negative diagnosis (i.e., a physiological recording in a patient with symptoms), is actually a

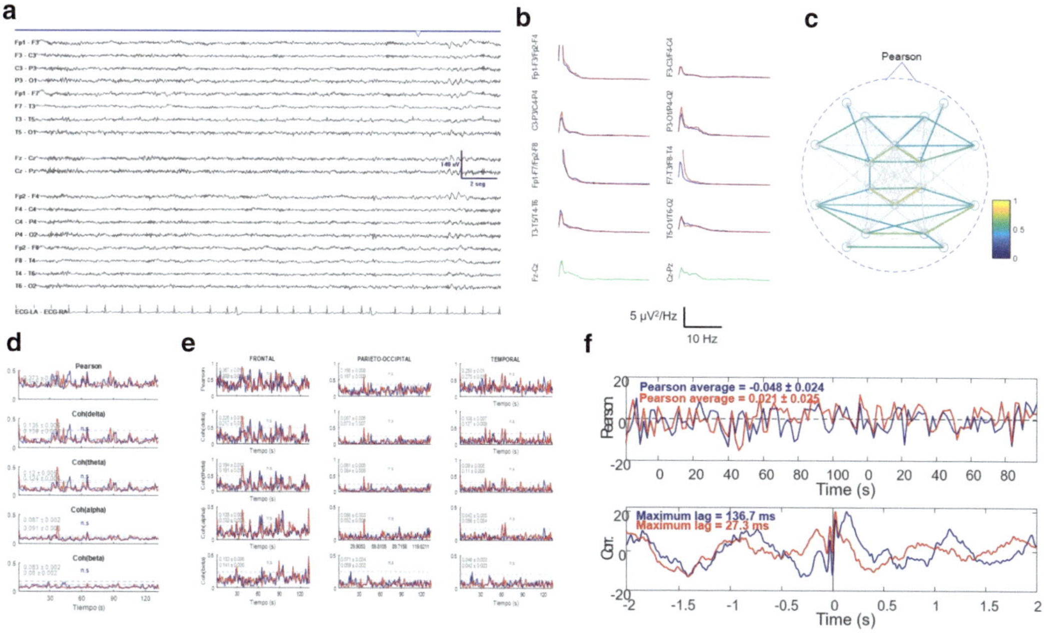

Fig. 10.2 Analysis rsEEG of a patient with NEPS. See text for details. (**a**) Raw recording of physiological characteristics de visu. (**b**) Average power spectra per channel. (**c**) Connectogram. (**d**) Average dynamics by hemisphere (**e**) and by lobes of the Pearson correlation and banded coherence measures. The dashed horizontal lines show the confidence interval for age. (**f**) Analysis of frontoparietal synchronization. The upper row shows the dynamics of the Pearson correlation, and the lower row shows the cross-correlation. The mean values ± SEM are shown. Red = right hemisphere; blue = left hemisphere

true positive, since it shows the existence of an organic alteration in numerical analysis.

We believe it is very important to remember that the presumed diagnosis in the second case, as in most similar situations, was epilepsy, so the doctor who requested the registration actually expected to find the presence of epileptiform activity or seizures.

These facts help to reconsider qEEG from a radically new perspective [49], thus expanding the field of application of electroencephalographic diagnosis.

10.4 Concluding Remarks

Conventional rsEEG is a highly useful tool for studying numerous brain pathologies, such as epilepsy, but not exclusively. However, there are many questions about its usefulness in diagnosing psychiatric disorders or differentiating between specific types of dementia. Nonetheless, objective measurement of parameters such as scalp spectral composition or synchronization offers potential improvements in these fields.

In fact, numerous changes in these magnitudes of spectral composition and synchronization have been demonstrated in these types of pathologies, which encourages further investigation into the numerical characterization of psychiatric pathologies and dementias.

Furthermore, the demonstration of numerically measurable organic alterations seems to us of extraordinary clinical value since, firstly, it identifies an anomalous brain structure associated with the clinical symptoms and, secondly, it can be of great value for the evolutionary study of the patient.

References

1. Kanner AM, Faan FF, Bermeo-Ovalle A. EEG in psychiatric disorders: does it have a role in their evaluation? In: Shomer DL, Lopes da Silva F, editors. Niedermeyer's electroencephalography: basic principles, clinical applications and related fields. 7th ed. Oxford University Press; 2018. ISBN 9780190228484.
2. Inui K, Motomura E, Okushima R, Kaige H, Inoue K, Nomura J. Electroencephalographic findings in patients with DSM-IV mood disorder, schizophrenia, and other psychotic disorders. Biol Psychiatry. 1998;43(1):69–75. https://doi.org/10.1016/s0006-3223(97)00224-2. PMID: 9442346.
3. Jelic V, Kowalski J. Evidence-based evaluation of diagnostic accuracy of resting EEG in dementia and mild cognitive impairment. Clin EEG Neurosci. 2009;40(2):129–42. https://doi.org/10.1177/155005940904000211. PMID: 19534305.
4. Klass DW, Brenner RP. Electroencephalography of the elderly. J Clin Neurophysiol. 1995;12(2):116–31. https://doi.org/10.1097/00004691-199503000-00002. PMID: 7797626.
5. Babiloni C, Carducci F, Lizio R, Vecchio F, Baglieri A, Bernardini S, Cavedo E, Bozzao A, Buttinelli C, Esposito F, Giubilei F, Guizzaro A, Marino S, Montella P, Quattrocchi CC, Redolfi A, Soricelli A, Tedeschi G, Ferri R, Rossi-Fedele G, Ursini F, Scrascia F, Vernieri F, Pedersen TJ, Hardemark HG, Rossini PM, Frisoni GB. Resting state cortical electroencephalographic rhythms are related to gray matter volume in subjects with mild cognitive impairment and Alzheimer's disease. Hum Brain Mapp. 2013;34(6):1427–46. https://doi.org/10.1002/hbm.22005. Epub 2012 Feb 14. PMID: 22331654; PMCID: PMC6869852.
6. Babiloni C, Del Percio C, Buján A. EEG in dementing disorders. In: Shomer DL, Lopes da Silva F, editors. Niedermeyer's electroencephalography: basic principles, clinical applications and related fields. 7th ed. Oxford University Press; 2018. ISBN 9780190228484.
7. Gaál ZA, Boha R, Stam CJ, Molnár M. Age-dependent features of EEG-reactivity–spectral, complexity, and network characteristics. Neurosci Lett. 2010;479(1):79–84. https://doi.org/10.1016/j.neulet.2010.05.037.
8. Klimesch W. EEG alpha and theta oscillations reflect cognitive and memory performance: a review and analysis. Brain Res Brain Res Rev. 1999;29(2–3):169–95. https://doi.org/10.1016/s0165-0173(98)00056-3. PMID: 10209231.
9. Müller V, Lindenberger U. Lifespan differences in nonlinear dynamics during rest and auditory oddball performance. Dev Sci. 2012;15(4):540–56. https://doi.org/10.1111/j.1467-7687.2012.01153.x. PMID: 22709403.
10. Vysata O, Kukal J, Prochazka A, Pazdera L, Simko J, Valis M. Age-related changes in EEG coherence. Neurol Neurochir Pol. 2014;48(1):35–8. https://doi.org/10.1016/j.pjnns.2013.09.001. Epub 2014 Jan 23. PMID: 24636768.
11. Zhu C, Guo X, Jin Z, Sun J, Qiu Y, Zhu Y, Tong S. Influences of brain development and ageing on cortical interactive networks. Clin Neurophysiol. 2011;122(2):278–83. https://doi.org/10.1016/j.clinph.2010.06.016. Epub 2010 Jul 15. PMID: 20637691.
12. Sorbi S, Hort J, Erkinjuntti T, Fladby T, Gainotti G, Gurvit H, Nacmias B, Pasquier F, Popescu BO, Rektorova I, Religa D, Rusina R, Rossor M, Schmidt R, Stefanova E, Warren JD, Scheltens P, EFNS Scientist Panel on Dementia and Cognitive Neurology. EFNS-ENS guidelines on the diagnosis and management of disorders associated with dementia. Eur J Neurol. 2012;19(9):1159–79. https://doi.org/10.1111/j.1468-1331.2012.03784.x. PMID: 22891773.
13. Dauwels J, Vialatte F, Cichocki A. Diagnosis of Alzheimer's disease from EEG signals: where are we standing? Curr Alzheimer Res. 2010;7(6):487–505. https://doi.org/10.2174/156720510792231720. PMID: 20455865.
14. Dauwels J, Vialatte F, Musha T, Cichocki A. A comparative study of synchrony measures for the early diagnosis of Alzheimer's disease based on EEG. NeuroImage. 2010;49(1):668–93. https://doi.org/10.1016/j.neuroimage.2009.06.056. Epub 2009 Jun 30. PMID: 19573607.
15. Babiloni C, Triggiani AI, Lizio R, Cordone S, Tattoli G, Bevilacqua V, Soricelli A, Ferri R, Nobili F, Gesualdo L, Millán-Calenti JC, Buján A, Tortelli R, Cardinali V, Barulli MR, Giannini A, Spagnolo P, Armenise S, Buenza G, Scianatico G, Logroscino G, Frisoni GB, Del Percio C. Classification of single Normal and Alzheimer's disease individuals from cortical sources of resting state EEG rhythms. Front Neurosci. 2016;10:47. https://doi.org/10.3389/fnins.2016.00047. PMID: 26941594; PMCID: PMC4763025.
16. Vecchio F, Miraglia F, Marra C, Quaranta D, Vita MG, Bramanti P, Rossini PM. Human brain networks in cognitive decline: a graph theoretical analysis of cortical connectivity from EEG data. J Alzheimer's Dis. 2014;41(1):113–27. https://doi.org/10.3233/JAD-132087. PMID: 24577480.
17. Sloan EP, Fenton GW. EEG power spectra and cognitive change in geriatric psychiatry: a longitudinal study. Electroencephalogr Clin Neurophysiol. 1993;86(6):361–7. https://doi.org/10.1016/0013-4694(93)90131-e. PMID: 7686470.
18. Bennett D. Public health importance of vascular dementia and Alzheimer's disease with cerebrovascular disease. Int J Clin Pract Suppl. 2001;120:41–8.
19. Gorelick PB, Scuteri A, Black SE, Decarli C, Greenberg SM, Iadecola C, Launer LJ, Laurent S, Lopez OL, Nyenhuis D, Petersen RC, Schneider JA, Tzourio C, Arnett DK, Bennett DA, Chui HC,

Higashida RT, Lindquist R, Nilsson PM, Roman GC, Sellke FW, Seshadri S, American Heart Association Stroke Council, Council on Epidemiology and Prevention, Council on Cardiovascular Nursing, Council on Cardiovascular Radiology and Intervention, and Council on Cardiovascular Surgery and Anesthesia. Vascular contributions to cognitive impairment and dementia: a statement for healthcare professionals from the american heart association/american stroke association. Stroke. 2011;42(9):2672–713. https://doi.org/10.1161/STR.0b013e3182299496. Epub 2011 Jul 21. PMID: 21778438; PMCID: PMC3778669.

20. van Straaten EC, de Haan W, de Waal H, Scheltens P, van der Flier WM, Barkhof F, Koene T, Stam CJ. Disturbed oscillatory brain dynamics in subcortical ischemic vascular dementia. BMC Neurosci. 2012;13:85. https://doi.org/10.1186/1471-2202-13-85. PMID: 22827860; PMCID: PMC3425137.
21. Jeong J, Chae JH, Kim SY, Han SH. Nonlinear dynamic analysis of the EEG in patients with Alzheimer's disease and vascular dementia. J Clin Neurophysiol. 2001;18(1):58–67. https://doi.org/10.1097/00004691-200101000-00010.
22. Leuchter AF, Dunkin JJ, Lufkin RB, Anzai Y, Cook IA, Newton TF. Effect of white matter disease on functional connections in the aging brain. J Neurol Neurosurg Psychiatry. 1994;57(11):1347–54. https://doi.org/10.1136/jnnp.57.11.1347. PMID: 7964810; PMCID: PMC1073185.
23. Caso F, Cursi M, Magnani G, Fanelli G, Falautano M, Comi G, Leocani L, Minicucci F. Quantitative EEG and LORETA: valuable tools in discerning FTD from AD? Neurobiol Aging. 2012;33(10):2343–56. https://doi.org/10.1016/j.neurobiolaging.2011.12.011. Epub 2012 Jan 13. PMID: 22244088.
24. Nishida K, Yoshimura M, Isotani T, Yoshida T, Kitaura Y, Saito A, Mii H, Kato M, Takekita Y, Suwa A, Morita S, Kinoshita T. Differences in quantitative EEG between frontotemporal dementia and Alzheimer's disease as revealed by LORETA. Clin Neurophysiol. 2011;122(9):1718–25. https://doi.org/10.1016/j.clinph.2011.02.011. Epub 2011 Mar 10. PMID: 21396882.
25. de Haan W, Pijnenburg YA, Strijers RL, van der Made Y, van der Flier WM, Scheltens P, Stam CJ. Functional neural network analysis in frontotemporal dementia and Alzheimer's disease using EEG and graph theory. BMC Neurosci. 2009;10:101. https://doi.org/10.1186/1471-2202-10-101. PMID: 19698093; PMCID: PMC2736175.
26. Andersson M, Hansson O, Minthon L, Rosén I, Londos E. Electroencephalogram variability in dementia with lewy bodies, Alzheimer's disease and controls. Dement Geriatr Cogn Disord. 2008;26(3):284–90. https://doi.org/10.1159/000160962. Epub 2008 Oct 8. PMID: 18841014.
27. Kai T, Asai Y, Sakuma K, Koeda T, Nakashima K. Quantitative electroencephalogram analysis in dementia with Lewy bodies and Alzheimer's disease. J Neurol Sci. 2005;237(1–2):89–95. https://doi.org/10.1016/j.jns.2005.05.017. PMID: 16019033.
28. Bonanni L, Thomas A, Tiraboschi P, Perfetti B, Varanese S, Onofrj M. EEG comparisons in early Alzheimer's disease, dementia with Lewy bodies and Parkinson's disease with dementia patients with a 2-year follow-up. Brain. 2008;131(Pt 3):690–705. https://doi.org/10.1093/brain/awm322. Epub 2008 Jan 17. PMID: 18202105.
29. Babiloni C, Del Percio C, Lizio R, Marzano N, Infarinato F, Soricelli A, Salvatore E, Ferri R, Bonforte C, Tedeschi G, Montella P, Baglieri A, Rodriguez G, Famà F, Nobili F, Vernieri F, Ursini F, Mundi C, Frisoni GB, Rossini PM. Cortical sources of resting state electroencephalographic alpha rhythms deteriorate across time in subjects with amnesic mild cognitive impairment. Neurobiol Aging. 2014;35(1):130–42. https://doi.org/10.1016/j.neurobiolaging.2013.06.019. Epub 2013 Jul 30. PMID: 23906617.
30. Fonseca LC, Tedrus GM, Carvas PN, Machado EC. Comparison of quantitative EEG between patients with Alzheimer's disease and those with Parkinson's disease dementia. Clin Neurophysiol. 2013;124(10):1970–4. https://doi.org/10.1016/j.clinph.2013.05.001. Epub 2013 Jun 5. PMID: 23746496.
31. Sam MC, So EL. Significance of epileptiform discharges in patients without epilepsy in the community. Epilepsia. 2001;42(10):1273–8. https://doi.org/10.1046/j.1528-1157.2001.17101.x. PMID: 11737162.
32. Hughes JR. The EEG in psychiatry: an outline with summarized points and references. Clin Electroencephalogr. 1995;26(2):92–101. https://doi.org/10.1177/155005949502600206. PMID: 7781196.
33. Shelley BP, Trimble MR, Boutros NN. Electroencephalographic cerebral dysrhythmic abnormalities in the trinity of nonepileptic general population, neuropsychiatric, and neurobehavioral disorders. J Neuropsychiatry Clin Neurosci. 2008;20(1):7–22. https://doi.org/10.1176/jnp.2008.20.1.7. PMID: 18305281.
34. Fingelkurts AA, Fingelkurts AA. Quantitative electroencephalogram (qEEG) as a natural and non-invasive window into living brain and mind in the functional continuum of healthy and pathological conditions. Appl Sci. 2022;12:9560. https://doi.org/10.3390/app12199560.
35. Abrams R, Taylor MA. Differential EEG patterns in affective disorder and schizophrenia. Arch Gen Psychiatry. 1979;36(12):1355–8. https://doi.org/10.1001/archpsyc.1979.01780120085010. PMID: 496554.
36. Hughes JR, John ER. Conventional and quantitative electroencephalography in psychiatry. J Neuropsychiatry Clin Neurosci. 1999;11(2):190–208. https://doi.org/10.1176/jnp.11.2.190. PMID: 10333991.
37. Kanner AM. Management of psychiatric and neurological comorbidities in epilepsy. Nat Rev Neurol.

2016;12(2):106–16. https://doi.org/10.1038/nrneurol.2015.243. Epub 2016 Jan 18. PMID: 26782334.
38. Vega-Zelaya L, Alvarez M, Ezquiaga E, Nogeiras J, Toledo M, Sola RG, Pastor J. Chapter 10: Psychogenic non-epileptic seizures in a surgical epilepsy unit: experience and a comprehensive review. In: Holmes MD, editor. Epilepsy topics. InTech; 2014. p. 193–224. https://doi.org/10.5772/57439. ISBN 978-953-51-1630-1.
39. Drislane FW. The clinical use of ambulatory EEG. In: Chang BS, Schachter SC, Schomer DL, editors. Atlas of ambulatory EEG. Amsterdam: Elsevier; 2005. p. 17–25.
40. Hegerl U, Hensch T. The vigilance regulation model of affective disorders and ADHD. Neurosci Biobehav Rev. 2014;44:45–57. https://doi.org/10.1016/j.neubiorev.2012.10.008. Epub 2012 Oct 22. PMID: 23092655.
41. Lisman J, Buzsáki G. A neural coding scheme formed by the combined function of gamma and theta oscillations. Schizophr Bull. 2008;34(5):974–80. https://doi.org/10.1093/schbul/sbn060. Epub 2008 Jun 16. PMID: 18559405; PMCID: PMC2518638
42. Cao Y, Han C, Peng X, Su Z, Liu G, Xie Y, Zhang Y, Liu J, Zhang P, Dong W, Gao M, Sha S, Zhao X. Correlation between resting theta power and cognitive performance in patients with schizophrenia. Front Hum Neurosci. 2022;16:853994. https://doi.org/10.3389/fnhum.2022.853994. PMID: 35529780; PMCID: PMC9074816.
43. Krukow P, Jonak K, Karakuła-Juchnowicz H, Podkowiński A, Jonak K, Borys M, Harciarek M. Disturbed functional connectivity within the left prefrontal cortex and sensorimotor areas predicts impaired cognitive speed in patients with first-episode schizophrenia. Psychiatry Res Neuroimaging. 2018;275:28–35. https://doi.org/10.1016/j.pscychresns.2018.03.001. Epub 2018 Mar 4. PMID: 29526598.
44. Perrottelli A, Giordano GM, Brando F, Giuliani L, Mucci A. EEG-based measures in at-risk mental state and early stages of schizophrenia: a systematic review. Front Psych. 2021;12:653642. https://doi.org/10.3389/fpsyt.2021.653642. PMID: 34017273; PMCID: PMC8129021.
45. Kesebir S, Yosmaoğlu A. QEEG in affective disorder: about to be a biomarker, endophenotype and predictor of treatment response. Heliyon. 2018;4(8):e00741. https://doi.org/10.1016/j.heliyon.2018.e00741. PMID: 30148219; PMCID: PMC6106696.
46. Ippolito G, Bertaccini R, Tarasi L, Di Gregorio F, Trajkovic J, Battaglia S, Romei V. The role of alpha oscillations among the Main neuropsychiatric disorders in the adult and developing human brain: evidence from the last 10 years of research. Biomedicine. 2022;10(12):3189. https://doi.org/10.3390/biomedicines10123189. PMID: 36551945; PMCID: PMC9775381.
47. Livint Popa L, Dragos H, Pantelemon C, Verisezan Rosu O, Strilciuc S. The role of quantitative EEG in the diagnosis of neuropsychiatric disorders. J Med Life. 2020;13(1):8–15. https://doi.org/10.25122/jml-2019-0085. PMID: 32341694; PMCID: PMC7175442
48. Kopańska M, Ochojska D, Dejnowicz-Velitchkov A, Banaś-Ząbczyk A. Quantitative electroencephalography (QEEG) as an innovative diagnostic tool in mental disorders. Int J Environ Res Public Health. 2022;19(4):2465. https://doi.org/10.3390/ijerph19042465. PMID: 35206651; PMCID: PMC8879113.
49. Pastor J, Vega-Zelaya M-AE. Necessity of quantitative EEG in daily clinical practice. In: Nakano H, editor. Electroencephalography. InTech; 2021; ISBN 978-1-83968-289-6.

11 Sleep

11.1 Introduction

Sleep in human beings is a behavioral state that alternates with waking, relative to which it is characterized by attenuation of motor output, an increased threshold to sensory input, typical changes in central and peripheral physiology, stereotypic postures, relatively easy reversibility (distinguishing it from coma), and diminished conscious awareness. Although some kind of consciousness can be obtained during the rapid eye movement state (REM), in which dreaming takes place [1, 2].

However, sleep is not a uniform state. On the contrary, it is composed of different well-coordinated stages and regulated by different brain structures placed at brainstem, diencephalon, thalamus, and cerebral cortex. In fact, this is an active state promoted by complex neural networks finely tuned. The states forming the sleep are organized in cycles of awake-sleep and non-REM-REM stages.

In the next, we shall briefly summarize the main brain cycles involved.

11.2 Circadian Awake-Sleep Cycle

Human sleep occurs with circadian periodicity (from the Latin circa = about; dia = day). Hence, humans (and many other animals) have an internal "clock" that operates even in the absence of external information about the time of day. The main hypothalamic nucleus regulating the circadian rhythm is the suprachiasmatic nucleus (SCN). Many of the genes and proteins responsible for circadian rhythms have now been discovered in mammals [3, 4]. At the start of the day, the transcription of specific genes into proteins (CLOCK and BMAL1) increases and, when the concentration of both proteins is enough, they act as transcriptional enhancers of other kinds of genes. As a result, the proteins PER1, PER2, CRY, or VP are synthesized. These proteins are modified in the cytoplasm. When the cytoplasmic concentrations of PER2 and CRY increase, both associate as CRY–PER2, diffuse back into the nucleus, increasing the synthesis of CLOCK and BMAL1, which binds to CRY, inhibiting their ability to stimulate the synthesis of the other genes [2, 5]. The total time course of these cycles takes about 24 h.

To synchronize physiological processes with the day-night cycle (a process named photo entrainment), the biological clock must detect changes in light levels as night approaches [6]. The receptors that sense these light changes are in the ganglionar layer of the outer retina, projecting through the retinohypothalamic tract to the SCN. The SCN also governs other functions that are synchronized with the sleep–wake cycle, including body temperature, hormone secretion (e.g., cortisol), blood pressure, and urine production, which in adults is reduced at night because of the circadian regulation of antidiuretic hormone (ADH or vasopressin) production [4, 7–9].

J. Pastor, *Quantitative EEG Practical Guide*, https://doi.org/10.1007/978-3-032-05457-9_11

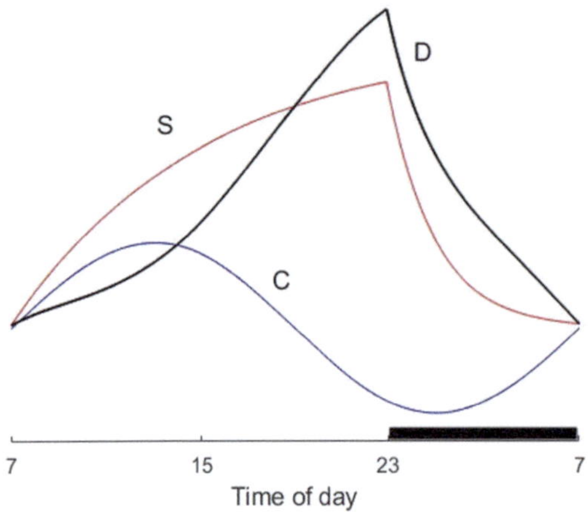

Fig. 11.1 Hypothetical model of sleep and the drive to sleep along a 24 h. The two mechanisms, circadian (C, blue) and S (red) are depicted. The difference between both is the drive of sleep (D). Horizontal bar represents the time in bed

The hypothetical model of the drive to sleep depends on two interacting mechanisms [10] one dependent on circadian rhythm (C). During a normal sleep-wake cycle, the drive to sleep (probably adenosine) accumulates during the daytime, declines during sleep, and is obviously associated with the sleep deficit [11]. This second process is termed S. Therefore, the highest drive to sleep is exactly before going to the bed. The difference between both processes reflects the sleep drive (Fig. 11.1).

11.3 The Physiological Structure of Sleep

Although we can use the same distribution of EEG bands described above for the wakefulness, the neural regulation of sleep and, therefore the features of EEG recording, is specific and very different from awake [12, 13]. Besides, there are specific waveforms that only appear during the sleep.

Sleep is not a uniform state. In contrast, it is composed of different well-coordinated stages and is regulated by different brain structures located at the brainstem, diencephalon, thalamus, and cerebral cortex. In fact, this is an active state promoted by complex neural networks that are finely tuned. Normal human sleep comprises two distinct states: non-rapid eye movement (NREM) and REM sleep.

The detailed neural substrate of the sleep is out of the scope of this book, and we refer to excellent works dedicated to that topic [2, 14–18].

The non-rapid eye movement (NREM) sleep is regulated by brainstem-hypothalamic inputs via the reciprocal antagonistic arousal (ARAS) and the sleep-promoting system (ventrolateral preoptic nucleus—VLPO—and melanin-concentrating hormone neurons). The hypothalamic structure inhibits (by GABA and galaninergic fibers) the brainstem ARAS, which reciprocally inhibits VLPO in a kind of "flip-flop" regulating circuit [19, 20], creating a bistable feedback loop with two alternate and stable firing patterns, excluding intermediate states.

Slow oscillation is a prevailing pattern during part of the NREM sleep, mainly during the stages known as N2 and N3 which differs by the amount of slow wave present at the background [13, 21]. Delta rhythm of the sleep is very well known. The depolarization of reticularis nucleus (a nutshell-like nucleus partially covering the thalamus) by acetylcholine (Ach) hyperpolarizes ThC cells in interstitial thalamus. The change in resting membrane potential, from a tonic to a phasic state, allows the interplay of ionic currents inhibited at more depolarized potentials (high voltage calcium current) that brings the neurons to a completely different state, eliciting slow calcium-dependent depolarization at 1–3 Hz [2, 22–24].

The regulation of the NREM-REM cycle depends heavily on the activity of different brainstem nuclei. Very briefly, noradrenergic, serotonergic, and histaminergic from ARAS are maximally active during waking, progressively decrease with deepening of NREM sleep, and almost completely stop during REM sleep. In contrast, monoaminergic neurons in the pedunculopontine and laterodorsal tegmental nuclei increase during REM sleep [2, 15, 25–27].

11.4 Sleep Stages

NREM sleep is composed of three different stages: stage 1 NREM (N1), stage 2 NREM (N2), and stage 3 NREM (N3). Overall, NREM sleep accounts for 75–80% of sleep time. Briefly, the main features of the different stages are [17, 18]:

N1 sleep comprises 3–8% of sleep time. This phase occurs most frequently in the transition from wakefulness to sleep stages or following arousals. In stage N1 sleep, alpha activity diminishes, and a low-voltage, mixed-frequency pattern appears. Vertex sharp waves (VSW, 50–200 ms) are prominent toward the end of this phase [28].

The N2 stage comprises 45–55% of the total sleep time. The typical findings include sleep spindles (SSs) and K-complexes (KCs). An SS is a 12–14 Hz waveform lasting at least 0.5 s, having a wax and waning appearance with a main location generalized with a maximum at frontocentral regions [29, 30], and it is generated within thalamic reticular neurons, eliciting inhibitory sequences on ThC cells [31, 32]. There are two types of spindles [33]: (i) Fast frequency spindles, with a maximum peak frequency at 14 Hz and placed near Cz. These kinds of spindles are the first to appear during the sleep and (ii) slow spindles, with a maximum frequency of about 12 Hz and maximum near Fz, this activity appears when the NREM sleep deepens. At the transition between N2 and N3 phases even slower spindles, around 10 Hz can be observed. Generation of spindles needs the interplay between thalamic reticular neuron and thalamus-cortex-thalamus network. The burst firing of afferents from the brain stem to reticular thalamic cells induces strong inhibitory postsynaptic potentials in the ThC neurons. This strong hyperpolarization removes the inactivation of the low-threshold Ca^{2+} current. This current gives rise to a high amplitude and long-duration calcium spike, eliciting a burst of spikes at the highest peak. These spikes travel to the cortex and, meanwhile, by collaterals, excite the reticular nucleus again [33, 34]. A KC is a waveform with two components (a negative wave followed by a positive one), both of which last more than 0.5 s and are generated intracortically [15]. KCs are generalized with a maximum at the vertex [35]. Delta waves (1–4 Hz) in the EEG may first appear at N2 sleep but in small amounts.

Stage N3 accounts for 15–20% of the total sleep time and constitutes slow-wave sleep, characterized by moderate to large amounts of high-amplitude, slow-wave activity.

REM sleep accounts for 20–25% of sleep time. EEG tracings are characterized by low-voltage, mixed-frequency activity with slow alpha (defined as 1–2 Hz slower than wake alpha) and theta waves. REM sleep can be divided into tonic (desynchronized EEG, atonia, and suppression of reflexes) and phasic sleep, characterized by rapid eye movements in all directions [15, 17]. Moreover, transient changes in heart rate, tongue movements, myoclonic twitching of the chin and limb muscles, and penis and clitoris erection are observed. Sawtooth waves, with a frequency in the theta range, involve the appearance of teeth on the cutting edge of a saw blade and often occur in conjunction with rapid eye movements. Although brain-stem nuclei are involved in eyeball movements and muscle tone, it is important to note that the latter can be considered epiphenomena and do not regulate the state of sleep.

There are a lot of numerical methods for sleep staging with different degrees of complexity. The use of CRANE simplifies tremendously the identification of the sleep stage, even in patients without additional electrodes, such as EOG or EMG (chin).

11.4.1 Specificity of Sleep Stages

The definition of the different phases of sleep is based on the spectral composition of scalp recordings, although in clinical practice the use of additional channels (EOG, EMG) is used in polysomnography (PSG). Usually, the number of channels used in PSG is reduced, between 4–6 channels and the montage can be monopolar or

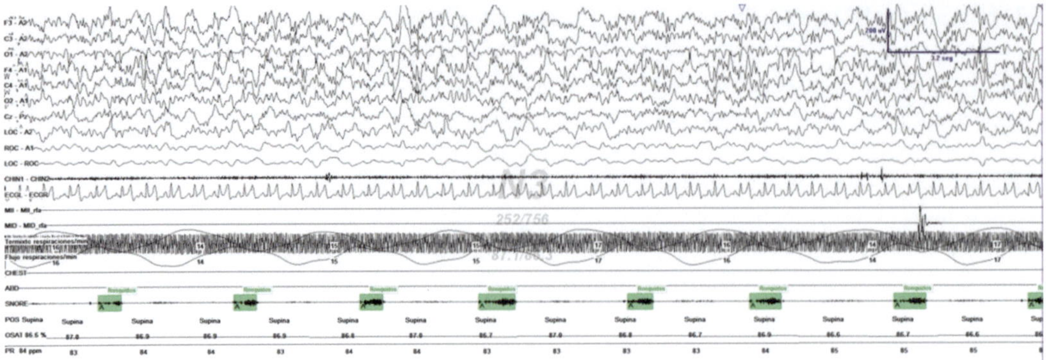

Fig. 11.2 PSG recording. Three electrodes are placed at each hemisphere and 2 at midline. LOC and ROC trace the lateral movements of EOG. EMG pickups activity from chin and muscle tibialis anterior. Other channels are also visible but not used to stage the sleep

differential. The use of EOG (electrodes placed 1 cm near the eye and referenced to mastoid electrodes), EMG (electrodes placed at the chin in a differential montage), and ECG is mandatory [21] (Fig. 11.2).

The spectral morphology, power bands composition, and synchronization properties are different for the wake and all the different sleep stages (see [13] for more details). The point to be noted is that basic numerical analysis on a conventional EEG recording may be sufficient for the identification of sleep phases. Figure 11.3 shows the different phases of sleep, including rsEEG (wakeful) and the power spectra obtained from those records from a person without organic pathology.

The example in Fig. 11.3 is taken from an individual person. We can see that the spectra are indeed apparently different, but one example is not enough. Therefore, we also analyzed the average of the spectra of all the people analyzed. We found that the spectra are clearly different for each sleep phase, so they can be used for sleep staging.

The average values of the power spectra for all scalp channels are symmetrical, with no differences observed between the right and left sides. Therefore, we will analyze the average spectra of a single hemisphere, to describe in detail the characteristics of each of the phases of sleep and wakefulness. Since the absolute values of the power spectra vary from person to person, we have obtained the average values from spectra normalized to the Cz-Pz channel during wakefulness. In this way, the differences between phases are preserved and the variance in the group is minimized.

In Fig. 11.4a, the progressive increase in delta power for the NREM sleep phases is evident, being maximum for N3. Also notable in this figure is the presence of an alpha component at 9.5 Hz in the posterior scalp region during wakefulness, which corresponds to the dominant posterior rhythm.

In Fig. 11.4b, which corresponds to the average normalized spectra of the right hemisphere, at higher gain, very interesting details are shown. During wakefulness, we can observe the presence of a harmonic of the dominant posterior rhythm around 19 Hz that is more striking in those channels where the amplitude of the alpha rhythm is greater at 9.5 Hz. During NREM sleep, especially in phases N2 and N3, the sigma component (which is the frequency band between 9

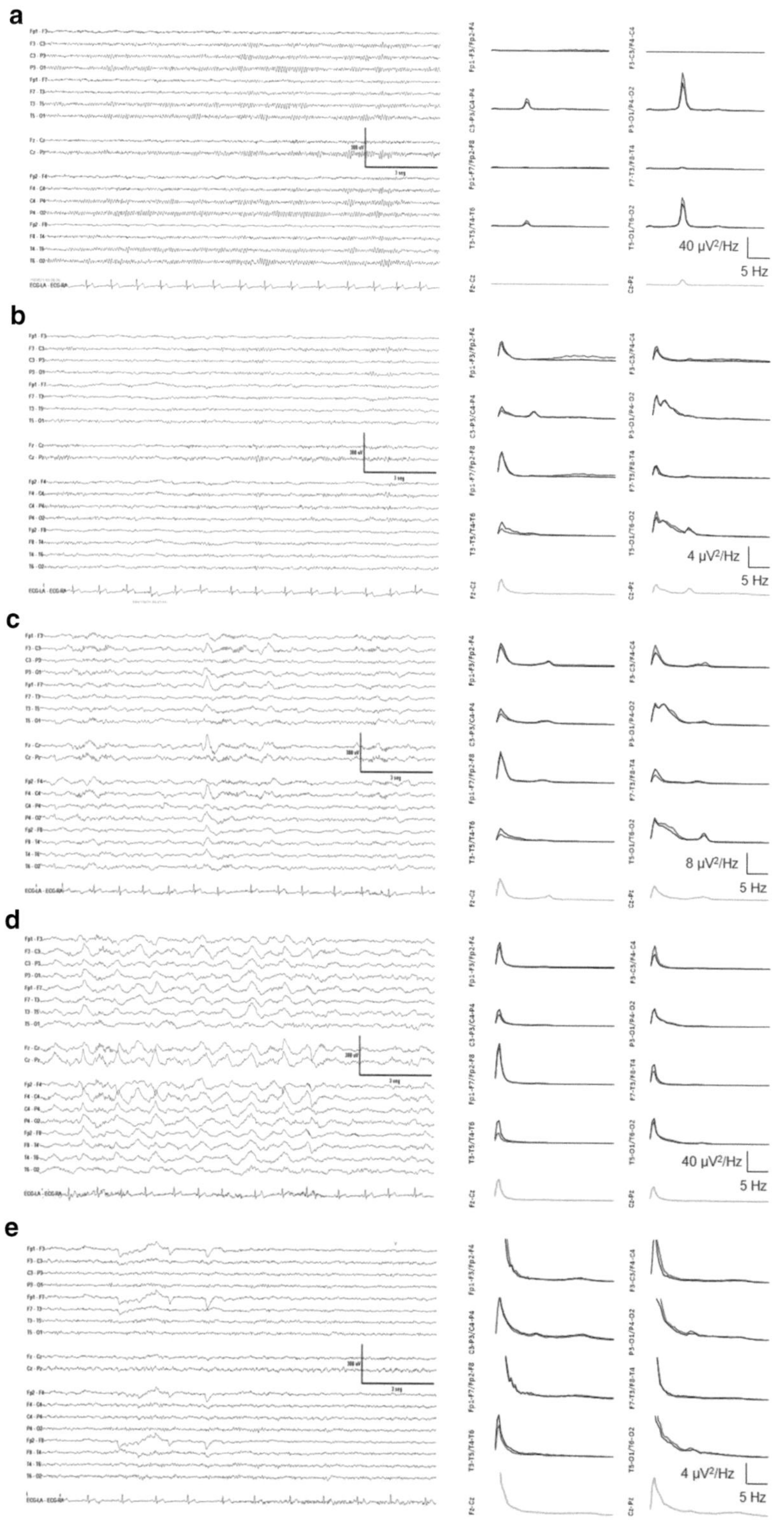

Fig. 11.3 Whole scalp recordings of the different phases of sleep from a person without organic pathology. (**a**) Wakeful, (**b**) N1, (**c**) N2, (**d**) N3, and (**e**) REM sleep stage. The left column shows the EEG recordings and the right column are the mean power spectra for all the channels. Red = right hemisphere; blue = left hemisphere; green = midline. Pay attention to the different ordinate scale

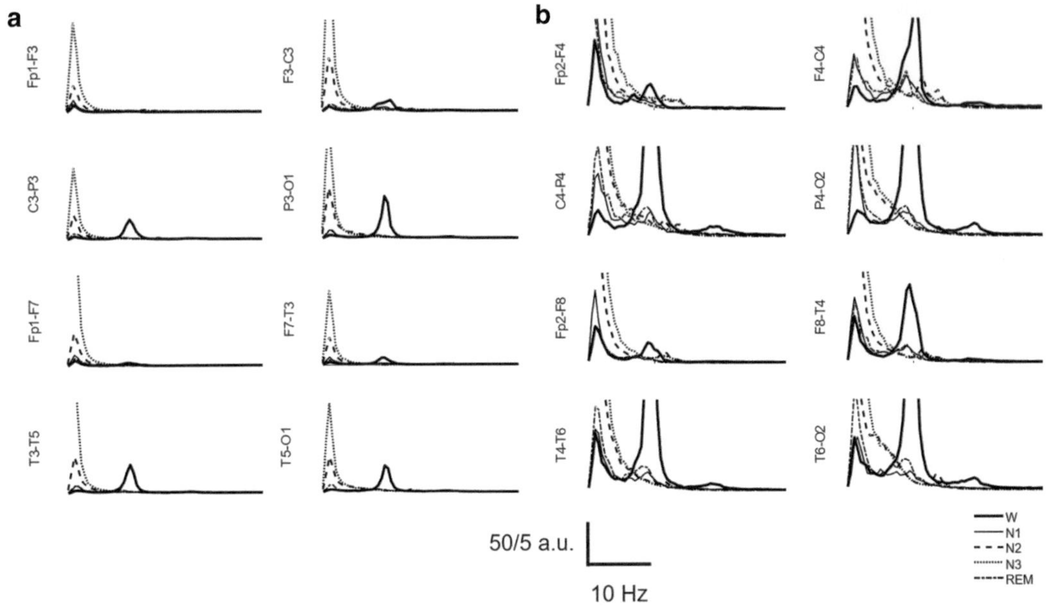

Fig. 11.4 Characteristics of the average spectra with two different gain values. (**a**) Left hemisphere, with a large gain. (**b**) Detail of the right hemisphere channels at higher gain. The ordinate has arbitrary units

and 16 Hz) is evident and corresponds to the sleep spindles. In addition to this, a theta component is observed, slower than the wake component (around 7 Hz) and which is maximum in parasagittal regions. Finally, during REM sleep there is no evidence of the sigma band, with a small delta component, except in frontopolar regions that show contamination by the rapid eye movements characteristic of this phase.

The record obtained from the PSG is called a hypnogram and has the characteristic of the discontinuous function showing sudden jumps between the different phases of sleep.

Except for the flip-plop mechanism that regulates the transition from NREM to REM sleep, the phase transitions are not abrupt at all, but have a continuous dynamic. Probably the technical impediments to a correct staging made it necessary to use the Rechstaffen and Kales or ASSA criteria with step-like transitions from stage to stage. But it is clear that the phase transitions, as we can see in Fig. 11.5, are not abrupt, but take a time of tens of minutes. Identifying N3 with a presence of delta waves >50% is only an arbitrary cut-off point, but probably does not represent a significant physiological value, i.e., different from what can be 40% or 60%.

This smooth and gradual transition takes place in NREM sleep, since it is the same neuronal pro-

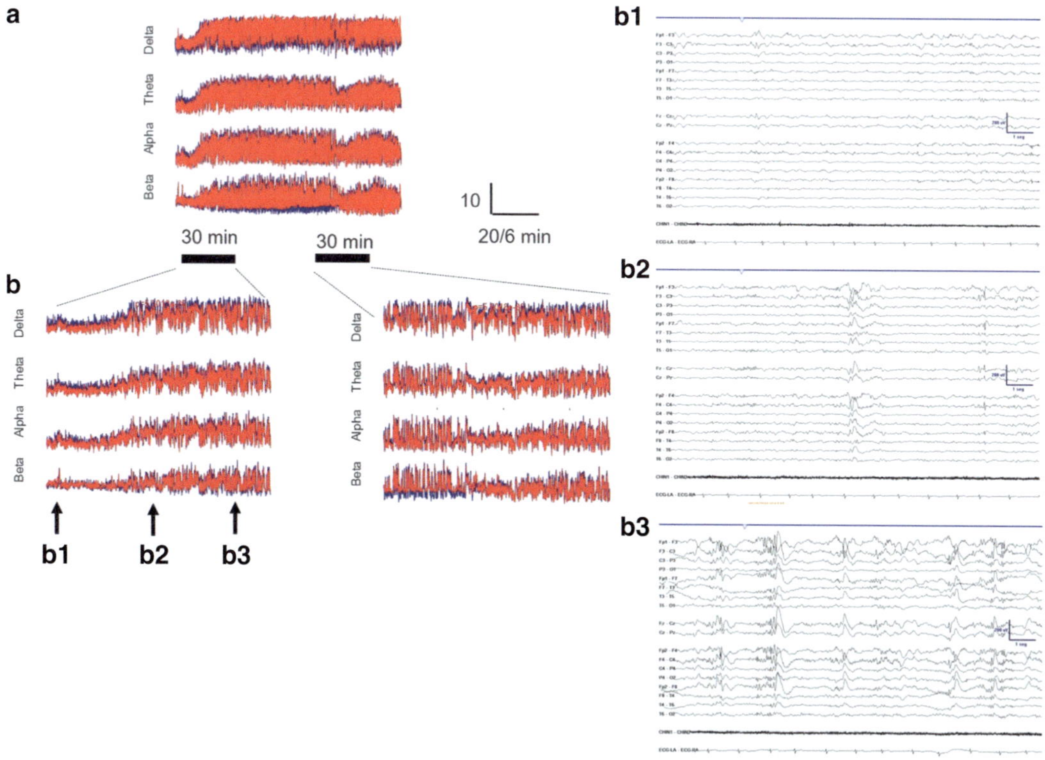

Fig. 11.5 Smooth transition during NREM sleep. (**a**) 110-min recording showing logPS dynamics of the frontal bands. Increased power is evident in all bands, but especially in the delta band. (**b**) Detail of 30-min duration during deepening of sleep from N2 to N3. (**b**) Transient superficialization during N3 sleep. Right column: recordings from the times indicated by the arrows. Red = right hemisphere; blue = left hemisphere

cess, whose substrate is thalamocortical hyperpolarization. However, the transition from NREM to REM sleep involves the activation of a different neural network, whose dynamics are described as a flip-flop transition system. Figure 11.6 shows the sudden transition from NREM to REM sleep in an 18-year-old patient on antiepileptic drugs. As can be seen, the power of the bands changes abruptly (Fig. 11.6b), with the disappearance of sleep spindles (Fig. 11.6c). The figure does not show any modification in the muscle tone of the chin because atony had already occurred 1 min before.

It is clear that the hypnogram, which is a discrete function, is an approximation of the real dynamics of sleep. It would therefore be necessary to debate whether it is necessary to modify this approximation or, on the contrary, since the technical tools allow it, to replace it with the real determination of the dynamics of sleep. The long-duration record shown in Fig. 11.7 exemplifies, even more clearly, how the N2 and N3 phases of sleep are artificial constructions, since, in reality, it is a continuous process. On the other hand, the transition from NREM to REM is a very rigidly defined and abrupt process.

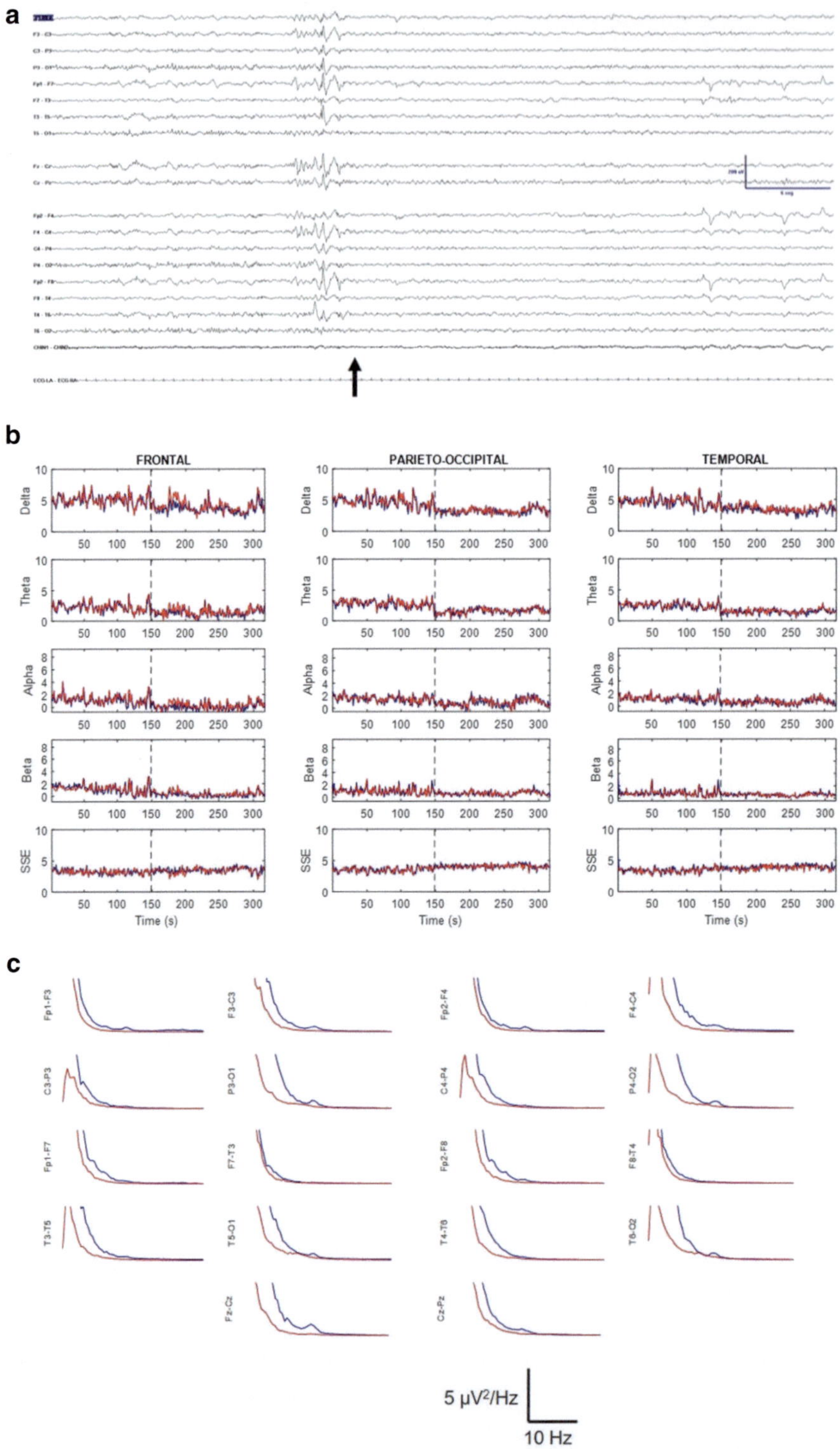

Fig. 11.6 Abrupt transition from NREM to REM sleep. (**a**) EEG recording during the transition. The arrow indicates the moment when this state change takes place. (**b**) LogPS dynamics. The vertical dashed line corresponds to the arrow in the upper recording. Blue = left hemisphere; red = right hemisphere. (**c**) Average spectra of both states. The presence of the sigma band during NREM sleep is observed. Blue = NREM; red = REM

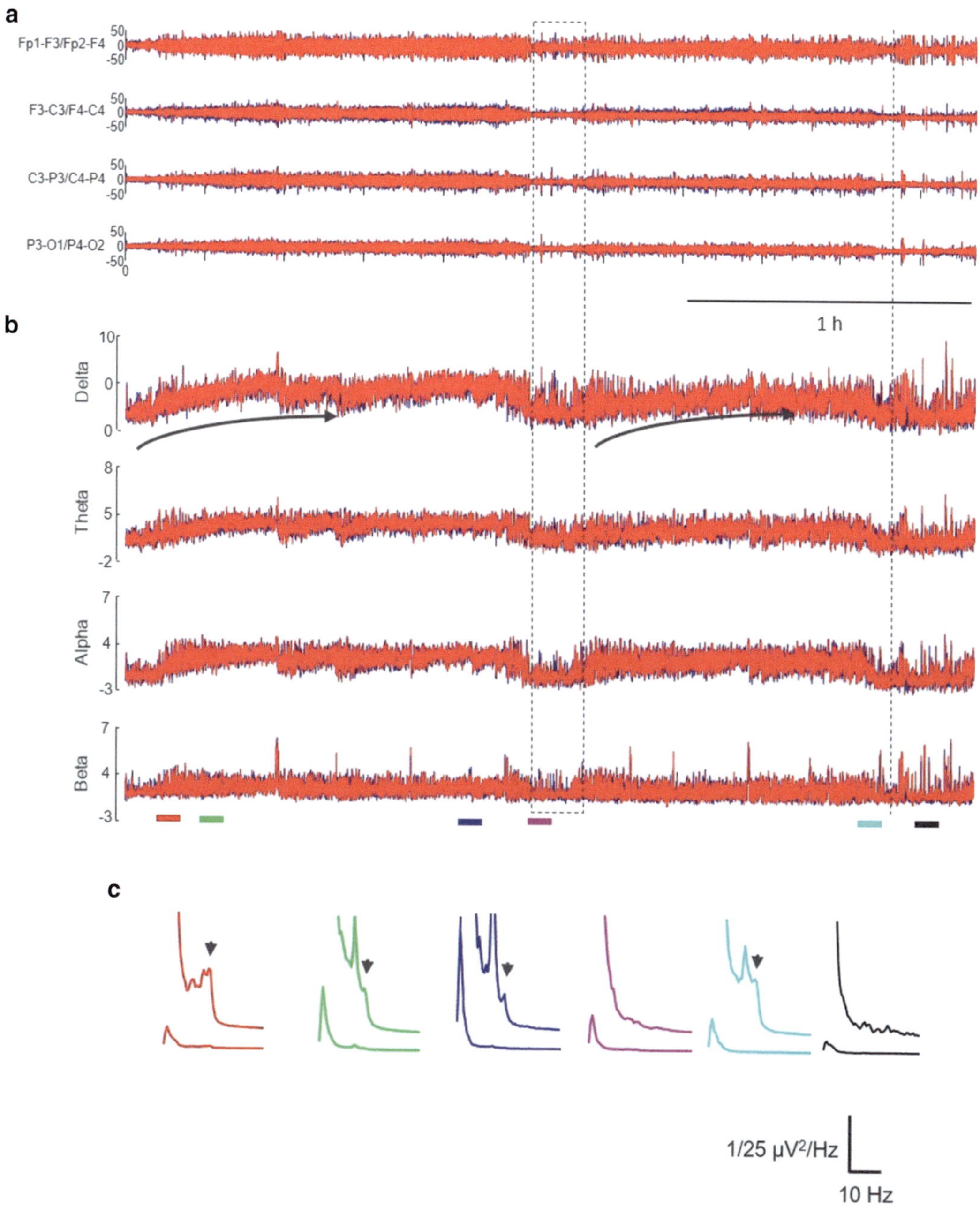

Fig. 11.7 Long-duration recording over two sleep cycles. (**a**) Raw recordings of the frontosagittal canals. (**b**) Dynamics of the logPS for all bands of the frontal channels. Dashed lines indicate REM sleep periods. (**c**) Average spectra of the periods indicated by the colored horizontal bars. They are shown with two different gains for frontopolar channels, to observe the gradual increase in delta power, which is the parameter defining the percentage of slow waves in the record. Arrowhead indicates the presence of a sigma component at 14 Hz, characteristic of NREM sleep

11.5 KC and SS Distributions Across the Scalp

We have analyzed the properties of the KCs and SSs across the scalp. Figure 11.8a shows an average KC (aKC) obtained from 32 individual waveforms. The amplitude is higher in frontal regions and lower in the occipital lobe. The amplitude from the aKC for both phases is shown in Fig. 11.8b. The error bars are the confidence intervals; therefore, taking into account that the distribution fit well to a Gaussian distribution, the amplitude was significant when the confidence intervals did not cross the 0 level. Interestingly (Fig. 11.8b), the aKC did not reach statistical significance at frontopolar regions (Fp1-F3/Fp2-F4), with the maximum amplitude at frontal lobes (F3-C3/F4-C4, 58.0 ± 3.8 and 55.3 ± 3.8 μV, respectively, peak-to-peak) and the lower amplitude at parieto-occipital channels (P3-O1/P4-O2, 26.2 ± 2.7 and 22.4 ± 2.1 μV, respectively, statistically significant). However, the KCs not only showed an antero-posterior gradient in the parasagittal region but also on the lateral side. In this case, the highest value was observed in the frontopolar lateral region (Fp1-F7/Fp2-F8, 65.0 ± 4.6 and 62.5 ± 36. μV, respectively), with the lowest value in the temporo-occipital region (T5-O1/T6-O2, 7.4 ± 1.4 and 11.7 ± 1.5 μV, respectively, both statistically significant). At the midline, the gradient was not as evident, with practically the same value in the frontal as in the parietal regions (61.6 ± 5.0 for Fz-Cz and 56.0 ± 3.8 for Cz-Pz). The total duration of aKC (Fig. 11.8c) was similar to the amplitude, with a maximum duration in the lateral frontopolar regions (Fp1-F7/Fp2-F8, 0.92 ± 0.07 and 0.95 ± 0.04 s, respectively), with a minimum duration in the temporo-occipital channels (T5-O1/T6-O2, 0.30 ± 0.04 and 0.48 ± 0.05 s, respectively). The correlations between both phases were highly significant (Student's t test), as can be observed for V1/V2 (Fig. 11.9d; $r = 0.965$, $p < 10^{-4}$), d1/d2 (Fig. 11.7e; $r = 0.898$, $p < 10^{-4}$), V1/d1 (Fig. 11.8f; $r = 0.675$, $p = 0.002$) and V2/d2 (Fig. 11.8g; $r = 0.753$, $p = 0.0003$).

From Fig. 11.8d, e, we can obtain the constitutive equations describing the canonical structure of the KCs. These properties can be useful for identifying whether a waveform is a true KC.

Sleep spindles were distributed across the entire scalp (Fig. 11.9a), but in some channels, they were difficult to distinguish from the background. Nevertheless, the presence of SSs (i.e., the σ band) can be easily identified in the PS by a well-defined σ component at approximately 13 Hz (Fig. 11.9b). These are the reasons why we choose to analyze their distribution and properties instead of measuring durations and amplitudes. We did not observe any difference in the symmetric channels between the right and left hemispheres for maximum peak frequency (pF, Fig. 11.9c, upper row) or maximum power (mP, Fig. 11.9c, lower row). The σ component was always present in the frontopolar region (Fp1-F3/Fp2-F4, 19/19 patients), followed by the occipital region (P3-O1/P4-O2 and T5-O1/T6-O2, 18/19), while the midtemporal areas presented a lower σ band (F7-T3/F8-T4 13/19 and T3-T5/T4-T6 12/19). We addressed the average value at the N2 sleep stage for pF and mP at the F, PO, and T lobes (considering both hemispheres). We obtained a lower value (median and 25–75 interpercentile range) for the F lobe (13.0, [12.5, 14.0] Hz and 1.05, [0.77, 1.62] μV2/Hz, respectively) and a higher value for the PO lobe (13.5, [13.0, 14.0] Hz and 1.07, [0.63, 3.47] μV2/Hz, respectively). The difference between the two regions was statistically significant for frequency ($p = 0.020$, for the Mann–Whitney Rank sum test; for pF Fig. 11.9d, upper row). No differences were observed regarding either the T lobe or the average mP (1.30, [0.83, 2.47] for F and 1.07 [0.63, 3.47] for PO, $p = 0.566$ for the Mann–Whitney Rank sum test; see Fig. 11.8d, lower row). Finally, we addressed the average pF and mP for the N2 and N3 sleep

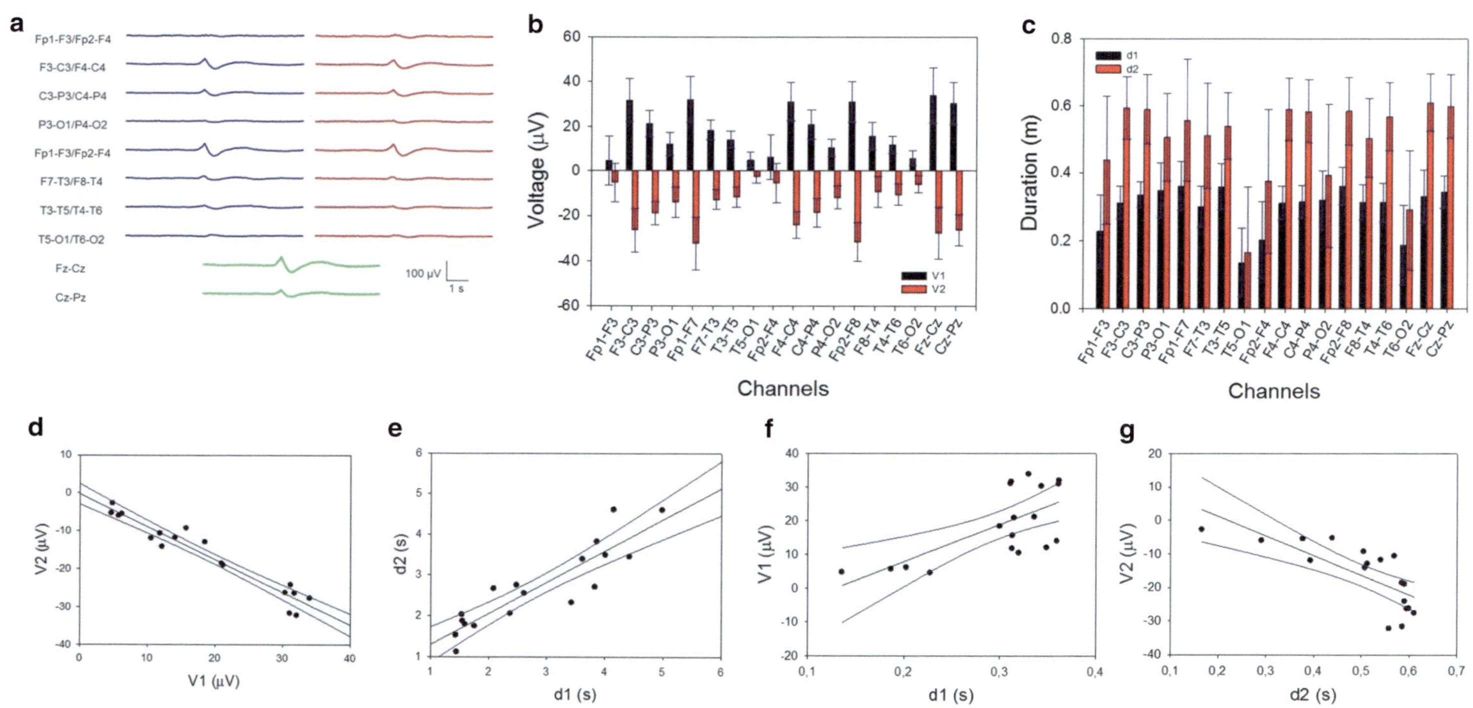

Fig. 11.8 Numerical properties of KCs. (**a**) Average KCs from a patient; blue lines = left hemisphere, red lines = right hemisphere, and green lines = midline. (**b**) Bar graph of the amplitudes of the first (V1, black) and second phases (V2, red). Error bars correspond to the coefficient interval, not the SEM. (**c**) Bar graph for the duration of both phases: d1 (black) and d2 (red). Errors are the CI. (**d**) Linear regression ±95% confidence band for the amplitude of the first and second components. (**e**) Linear regression ±95% confidence intervals for the durations of the first and second components. (**f**) Linear regression ±95% confidence band for the amplitude (V1) and duration (d1) of the first component. (**g**) Linear regression ±95% confidence band for the amplitude (V2) and duration (d2) of the second component. (Taken from [13])

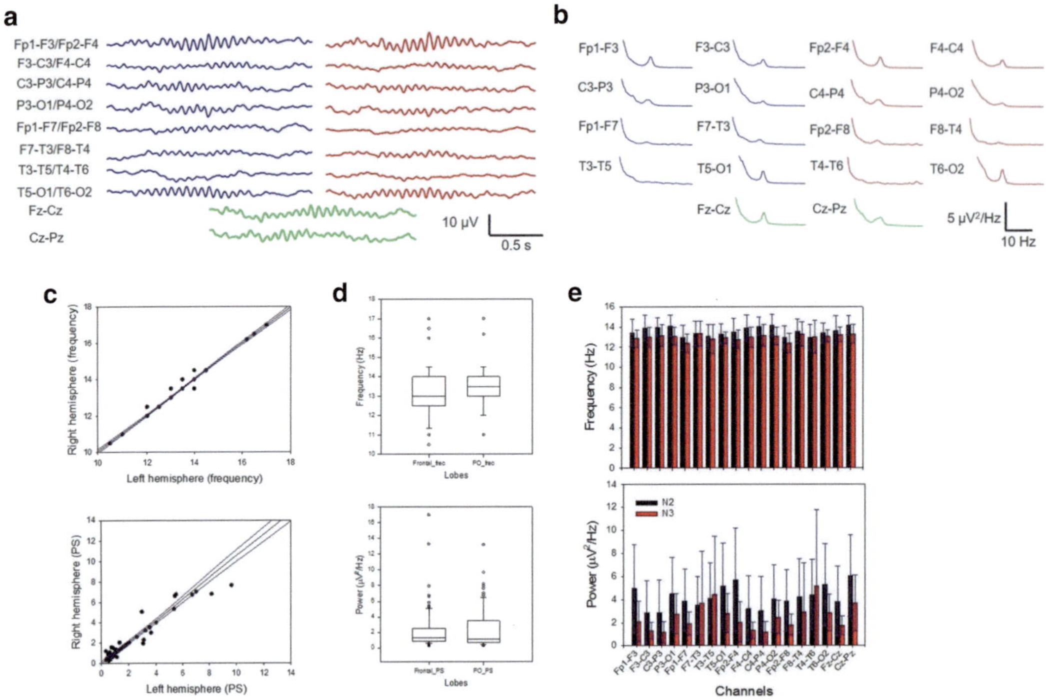

Fig. 11.9 Numerical properties of SSs. (**a**) Example of SSs from a patient. (**b**) aPS for all the channels in the same patient. A prominent peak 13 Hz (σ band) is clearly observed in 14 graphs; blue lines = left hemisphere, red lines = right hemisphere, and green lines = midline. (**c**) Upper row. Linear regression ±95% confidence band for the average frequency of symmetric channels in both hemispheres (r = 0.9961; $p < 10^{-4}$, function $y = -0.03 + 1.06x$). Lower row. Linear regression ±95% confidence band for the average power of symmetric channels of both hemispheres (r = 0.9924; $p < 10^{-4}$, function $y = 0.15 + 0.99x$) (**d**) Bar graph for the frequency (upper row) and the power (lower row) for all the channels at the N2 and N3 sleep stages. (**e**) Bar graph showing the frequency (upper row) and power (lower row) for all the channels. Black = N2 stage, red = N3 stage. The error bars are the CIs. (Taken from [13])

stages (Fig. 11.9e). The frequency was lower during the N3 stage than during the N2 stage (12.9 [12.7, 13.2] and 13.5 [13.3, 14.0] Hz, respectively, $p < 0.001$ for the Mann–Whitney rank sum test), and a similar result was obtained for the average mP, which was lower for N3 than for N2 (2.5 ± 0.3 and 4.2 ± 0.2 µV²/Hz, respectively, $p < 0.001$ for the Student-t test).

In a subset of 32% patients, the presence of the σ band of well-defined bimodal components during the N2 sleep stage was observed, as shown in Fig. 11.10, where bimodal distributions at 12.5 and 14.5 Hz can be observed at several channels, e.g., Fp1-F3/Fp2-F4, F7-T3/F8-T4, or Fz-Cz. All of these bimodal σ components were obtained during the N2 sleep stage and therefore cannot be confused with the slow SS.

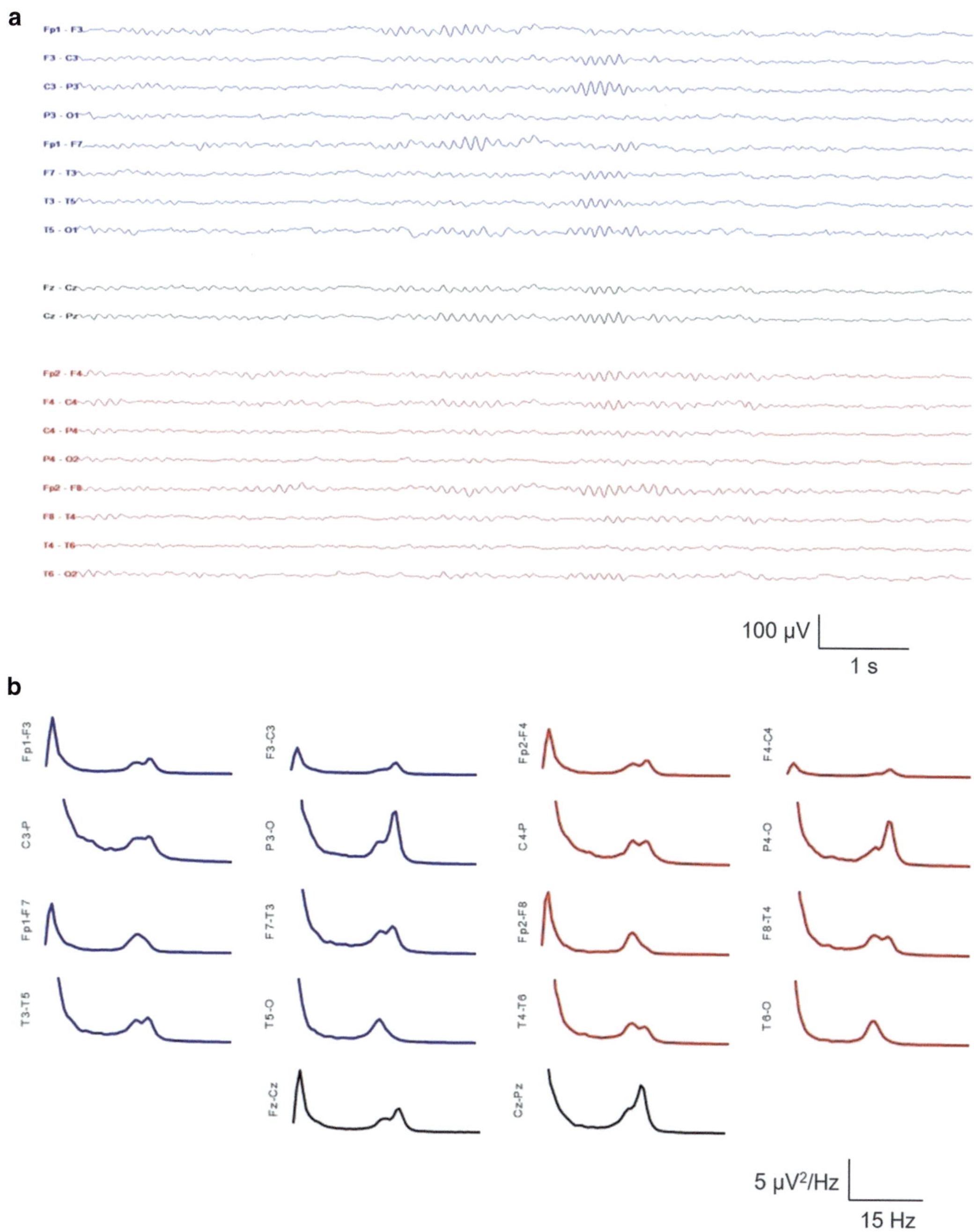

Fig. 11.10 Example of a bimodal σ band in the same patient during the N2 sleep stage. (**a**) Raw trace showing the coexistence of SSs at two different frequencies (12.5 and 14.5 Hz). (**b**) Average spectra for all the channels. (Taken from [13])

11.6 Some Practical Examples

Although it can be used to characterize various pathologies, probably the most direct use of qEEG in sleep studies is the identification of the different phases without the need for additional channels or even when these channels do not provide adequate information.

An example of this situation is shown next. It is 19-year-old, right-handed patient, referred from school to evaluation for a language disorder without significant intellectual delay, and difficulty with reading, writing, and calculation since childhood. Currently without treatment. The characteristics of the sleep are physiological (Fig. 11.11), except for a significant alteration during REM sleep. During this phase of sleep, the patient was able to maintain an anti-gravity posture with the right lower limb without it being supported anywhere, which indicates the presence of muscle tone.

In patients with impairment of brain physiology, in many cases it is not easy to identify the phases of sleep. For example, when there are no eye movements during possible REM sleep. In these cases, spectral analysis is essential for correct staging. In this example (Fig. 11.12), we show a right-handed male 22-year-old patient with a history of LGS and extensive bilateral polymicrogyria. He reports almost daily episodes of disconnection with cephalic version to left, and sometimes secondary generalization. On other occasions, he presents seizures with tonic posture and subsequent fall. Vagus nerve stimulator carrier since 2022. Anti-epileptic treatment with cannabidiol (2 ml/12 h), levetiracetam (500 mg/8 h), carbamazepine (400 mg/8 h), topiramate (200 mg/8 h), phenytoin (150 mg/8 h), and clobazam (5 mg/8 h). The alteration of brain physiology was so marked that it was very difficult to identify the different phases of sleep. Furthermore, although EOG artifacts were observed in wakefulness, during sleep it was not possible to identify these waveforms. However, spectral analysis demonstrated the existence of a large amplitude delta component during NREM sleep, which was associated with the presence of sigma band. However, these features disappeared during REM sleep, allowing adequate sleep staging (Fig. 11.12).

Finally, let's look at one last example of how numerical analysis allows us to identify states that are not accessible in visual studies. In this case, the patient is a 23-year-old, right-handed male with a history of meningoencephalitis at age 8. The patient presented severe cognitive impairment. He was referred for diagnosis due to the presence of paroxysmal events compatible with epilepsy. At the time of the study, the patient was being treated with levetiracetam (1000 mg/12 h). The severe impairment of background activity made it impossible to determine the existence of sleep phases, especially because characteristic elements such as vertex waves, K complexes, or sleep spindles were absent. Although vertical eye movements (EOG) were observed during wakefulness, no vertical or horizontal movements were observed during sleep. Therefore, visual analysis of the recording suggested that there were no distinct phases during behavioral sleep. This fact suggests an even more intense involvement of the CNS, also affecting subcortical structures such as the thalamus and brainstem. This was the conclusion obtained in previous v-EEG studies in which numerical analysis was not used. However, using average spectra across different sleep periods (Fig. 11.13a–c) allows us to demonstrate the existence of different sleep phases. Figure 11.13d shows the channel-averaged spectra for the right hemisphere (data from a single hemisphere are shown for simplicity) for the different phases shown in the top row. It is evident that the power of the delta band is greater in the green spectrum (corresponding to Fig. 11.13b) than in the blue (Fig. 11.13a) and red (Fig. 11.13c) spectra. But it's not just the structure of the slower bands (delta and theta) that is different for the different channels in the three states. There are also significant differences for the faster bands (alpha and beta), even though the power of these bands in this patient is abnormally low. In Fig. 11.13e, we can observe the spectra in greater detail. In each of these channels, the blue and green spectra are very similar, while the red spectra have a distinctly different shape. In the first two, no sigma

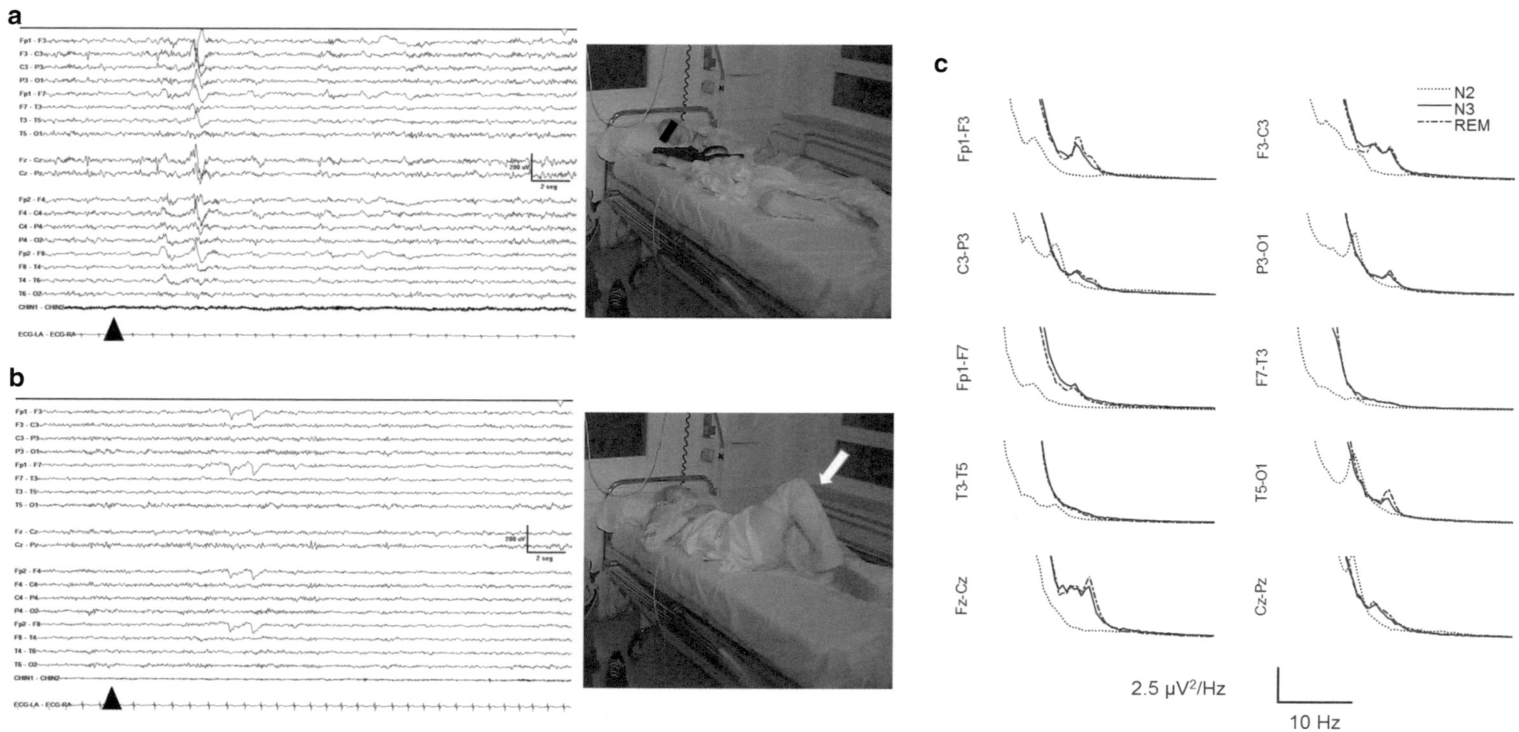

Fig. 11.11 Staging of the sleep. (**a**) NREM N2 stage with presence of K complex, sleep spindles, and muscle tone in chin (arrowhead). Patient remains supine. (**b**) Absence of NREM waveforms, presence of vertical eye movements, and absence of chin muscle tone. The patient keeps the right lower limb in an anti-gravitation posture (white arrow) that can only be maintained while preserving muscle tone. (**c**) Average power spectra showing the presence of sigma band in N2 (dotted) and N3 (solid) stages, but absent during REM sleep (dashed)

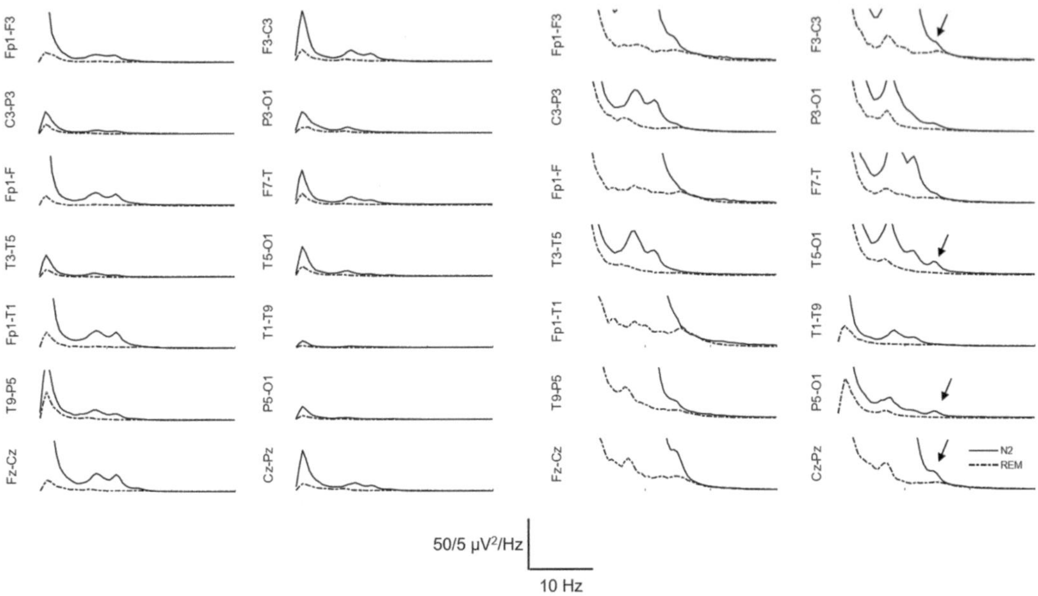

Fig. 11.12 Average spectra for all scalp channels during NREM (solid line) and REM (dashed line) sleep. The spectra are shown with two different gains, in order to appreciate both the delta component and the presence of the sigma band. The arrows indicate the presence of the sigma band during NREM sleep

band is observed, while the red spectra show a component with a maximum at 15 Hz in the frontal, occipital, and midline regions. Synchronization analysis (Fig. 11.13f) by regions for the three phases shows that differences are observed between the blue and green spectra, with respect to the red, while no differences have been observed between the blue and green spectra in any region of scalp. It is clear that the three phases analyzed present a mixture of properties that do not fit the characteristics of sleep phases. In this case, we have seen that the spectra of **a** and **b** have a clear predominance of the delta and theta bands, with a clear increase in power in the case of the spectra (green) of recording **b**, compared to **a** (blue). Furthermore, there is no difference in synchronization between these states, so we could assume that both are states of NREM sleep. However, the presence of the sigma band (although with the absence of sleep spindles in the visual analysis) is observed in state **c** (red spectrum), which presents less power in the slow bands and a different synchronization than states A and B.

To help identify the REM sleep phase in a patient with severe impairment, such as this one, the heart rate variability characteristic of this phase can be used. Figure 11.14 shows the instantaneous heart rate graphs for the three sleep states indicated in the previous figure. As can be seen, the heart rate in cases a and b is highly stable, while in case c, the heart rate shows significant irregularity.

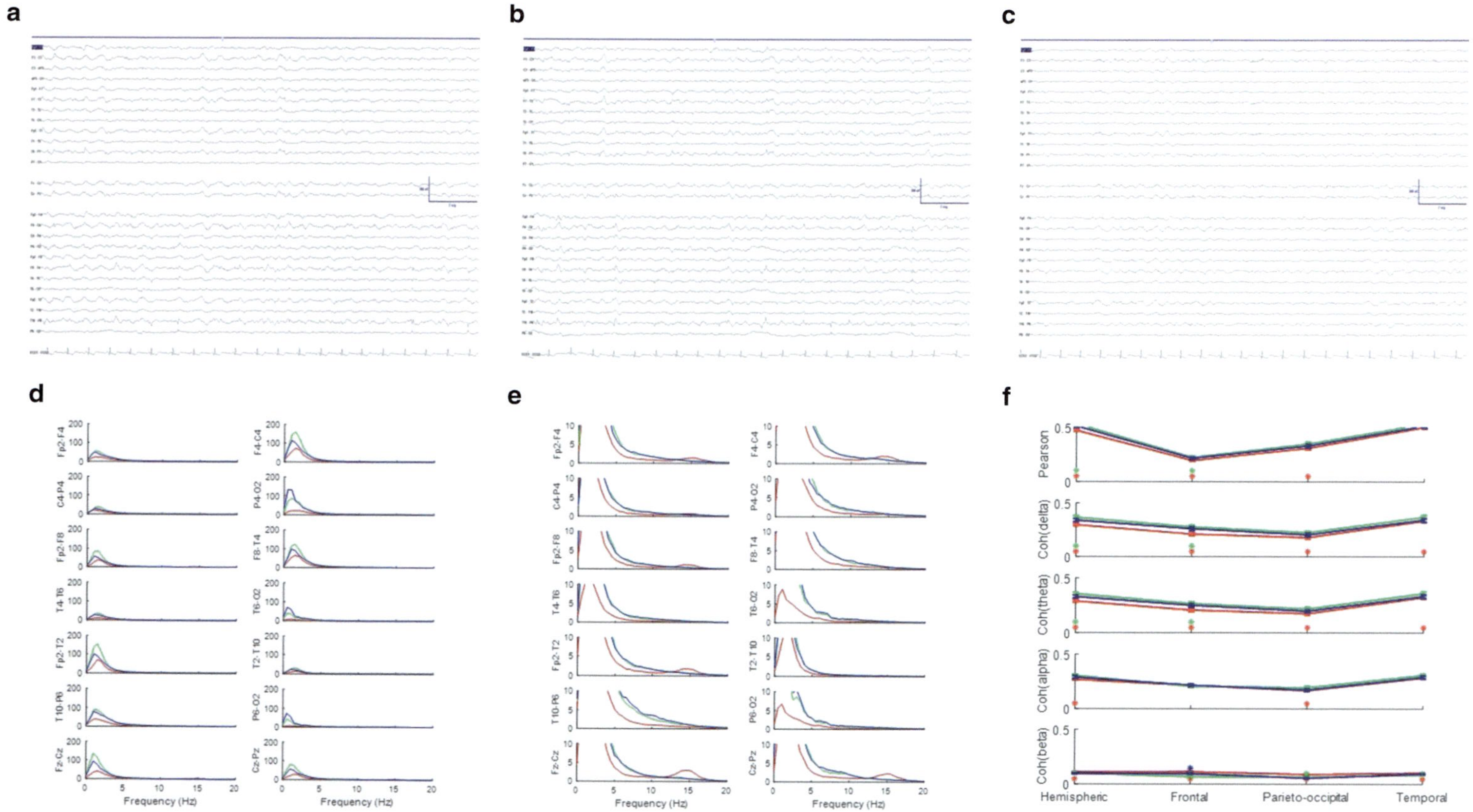

Fig. 11.13 Identification of sleep phases in a patient with severe encephalopathy. (**a–c**) sleep states analyzed. No sleep waveforms or eye movements that would allow identification are observed. (**d**) Average spectra for the right hemisphere for the three states. (**e**) The same spectra, shown with greater power detail. (**f**) Comparison of synchronization measures for different scalp regions of the right hemisphere. Record a = blue, record b = green, and record c = red. Red asterisk = significant difference between states a and c; green asterisk = difference between states b and c

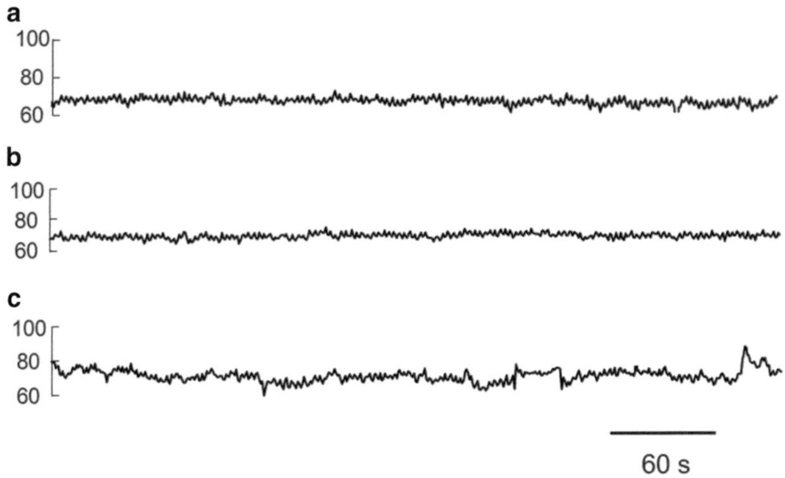

Fig. 11.14 Instantaneous heart rate variability identifies phases **a** and **b** as NREM sleep, while phase **c** corresponds to REM sleep

11.7 Concluding Remarks

Sleep staging through the analysis of average spectra is easy and reliable. The presence of a significant delta and sigma band allows for the identification of sleep stages, even in cases that are very difficult for conventional staging.

However, the information obtained through qEEG during sleep is not only clinically relevant; it also reveals a continuous process in the NREM phase or the presence of bimodal components of sleep spindles that are of scientific relevance.

References

1. Pace-Schott EF. Sleep architecture. In: Stickgold R, Walker MP, editors. The neuroscience of sleep. London: Academic; 2009. p. 11–7. ISBN 9780123750730.
2. McCormick DA, Westbrook GL. Sleep and dreaming. In: Kandel ER, Schwartz JH, Jessell TM, Siegelbaum SA, Hudspeth J, editors. Principles of neural science. 5th ed. New York: Ed. Elsevier; 2013. ISBN978-0-07-139011-8.
3. Patke A, Young MW, Axelrod S. Molecular mechanisms and physiological importance of circadian rhythms. Nat Rev Mol Cell Biol. 2020;21(2):67–84. https://doi.org/10.1038/s41580-019-0179-2. Epub 2019 Nov 25. PMID: 31768006.
4. Welsh DK, Takahashi JS, Kay SA. Suprachiasmatic nucleus: cell autonomy and network properties. Annu Rev Physiol. 2010;72:551–77. https://doi.org/10.1146/annurev-physiol-021909-135919. PMID: 20148688; PMCID: PMC3758475.
5. Namihira M, Honma S, Abe H, Tanahashi Y, Ikeda M, Honma K. Circadian rhythms and light responsiveness of mammalian clock gene, Clock and BMAL1, transcripts in the rat retina. Neurosci Lett. 1999;271(1):1–4. https://doi.org/10.1016/s0304-3940(99)00407-3. PMID: 10471199.
6. McCormick DA. Sleep and wakefulness. In: Purves D, Augustine GJ, Fitzpatrick D, Hall WC, AS LM, JO MN, Williams SM, editors. Neuroscience. 3rd ed. Sunderland: Sinauer Associates; 2004. ISBN 0-87893-725-0.
7. Moore RY, Silver R. Suprachiasmatic nucleus organization. Chronobiol Int. 1998;15(5):475–87. https://doi.org/10.3109/07420529808998703. PMID: 9787937.
8. Tsuno Y, Mieda M. Circadian rhythm mechanism in the suprachiasmatic nucleus and its relation to the olfactory system. Front Neural Circuits. 2024;18:1385908. https://doi.org/10.3389/fncir.2024.1385908. PMID: 38590628; PMCID: PMC11000122.
9. Rodríguez-Cortés B, Hurtado-Alvarado G, Martínez-Gómez R, León-Mercado LA, Prager-Khoutorsky M, Buijs RM. Suprachiasmatic nucleus-mediated glucose entry into the arcuate nucleus determines the daily rhythm in blood glycemia. Curr Biol. 2022;32(4):796–805.e4. https://doi.org/10.1016/j.cub.2021.12.039. Epub 2022 Jan 13. PMID: 35030330.
10. Daan S, Beersma DG, Borbély AA. Timing of human sleep: recovery process gated by a circadian pacemaker. Am J Phys. 1984;246(2 Pt 2):R161–83. https://doi.org/10.1152/ajpregu.1984.246.2.R161. PMID: 6696142.
11. Reichert CF, Deboer T, Landolt HP. Adenosine, caffeine, and sleep-wake regulation: state of the science and perspectives. J Sleep Res. 2022;31(4):e13597. https://doi.org/10.1111/jsr.13597. Epub 2022 May 16. PMID: 35575450; PMCID: PMC9541543.
12. Pastor J, Vega-Zelaya L. Normative structure of resting-state EEG in bipolar derivations for daily

clinical practice: a pilot study. Brain Sci. 2023;13:167. https://doi.org/10.3390/brainsci13020167.
13. Pastor J, Garrido Zabala P, Vega-Zelaya L. Structure of spectral composition and synchronization in human sleep on the whole scalp: a pilot study. Brain Sci. 2024;14:1007. https://doi.org/10.3390/brainsci14101007.
14. Jones BE. From waking to sleeping: neuronal and chemical substrates. Trends Pharmacol Sci. 2005;26(11):578–86. https://doi.org/10.1016/j.tips.2005.09.009. Epub 2005 Sep 23. Erratum in: Trends Pharmacol Sci. 2006 Apr;27(4):189. PMID: 16183137.
15. Rama AN, Cho SC, Kushida CA. Normal human sleep. In: Lee-Chiong T, editor. Sleep: a comprehensive handbook. Wiley; 2006. p. 3–10. ISBN-13 978-0-471-68371-1.
16. Zeitzer JM. Control of sleep and wakefulness in health and disease. Prog Mol Biol Transl Sci. 2013;119:137–54. https://doi.org/10.1016/B978-0-12-396971-2.00006-3. PMID: 23899597.
17. Carskado WC, Dement MA. Normal human sleep an overview. In: Kryger M, Roth T, Dement WC, editors. Principles and practice of sleep medicine. 6th ed. Amsterdam: Elsevier; 2017. ISBN 978-0-323-24288-2. https://doi.org/10.1016/C2012-0-03543-0.
18. Shin A, Park S, Shin W, Woo J, Jeong M, Kim J, Kim D. A brainstem-to-mediodorsal thalamic pathway mediates sound-induced arousal from slow-wave sleep. Curr Biol. 2023;33(5):875–885.e5. https://doi.org/10.1016/j.cub.2023.01.033. Epub 2023 Feb 7. PMID: 36754050.
19. Saper CB. The neurobiology of sleep. Continuum (Minneap Minn). 2013;19(1 Sleep Disorders):19–31. https://doi.org/10.1212/01.CON.0000427215.07715.73. PMID: 23385692.
20. Rahimi S, Joyce L, Fenzl T, Drexel M. Crosstalk between the subiculum and sleep-wake regulation: a review. J Sleep Res. 2024;33(5):e14134. https://doi.org/10.1111/jsr.14134. Epub 2024 Jan 9. PMID: 38196146.
21. American Academy of Sleep Medicine. AASM manual for the scoring of sleep and associated events: rules, terminology and technical specifications (Version 3). 2023. Available online: https://cir.nii.ac.jp/crid/1370004237604151044. Accessed on 18 July 2025.
22. Falup-Pecurariu C, Diaconu Ş, Ţînţ D, Falup-Pecurariu O. Neurobiology of sleep (Review). Exp Ther Med. 2021;21(3):272. https://doi.org/10.3892/etm.2021.9703. Epub 2021 Jan 25. PMID: 33603879; PMCID: PMC7851648.
23. Li Q, Song JL, Li SH, Westover MB, Zhang R. Effects of cholinergic neuromodulation on thalamocortical rhythms during NREM sleep: a model study. Front Comput Neurosci. 2020;13:100. https://doi.org/10.3389/fncom.2019.00100. PMID: 32038215; PMCID: PMC6990259.
24. Li Q, Song JL, Li SH, Westover MB, Zhang R. Effects of cholinergic neuromodulation on thalamocortical rhythms during NREM sleep: a model study. Front Comput Neurosci. 2020;12:100. https://doi.org/10.3389/fncom.2019.00100. PMID: 32038215; PMCID: PMC6990259.
25. Weber F. Modeling the mammalian sleep cycle. Curr Opin Neurobiol. 2017;46:68–75. https://doi.org/10.1016/j.conb.2017.07.009. Epub 2017 Aug 24. PMID: 28841438.
26. Le Bon O. Relationships between REM and NREM in the NREM-REM sleep cycle: a review on competing concepts. Sleep Med. 2020;70:6–16. https://doi.org/10.1016/j.sleep.2020.02.004. Epub 2020 Feb 15. PMID: 32179430.
27. Le Bon O. An asymmetrical hypothesis for the NREM-REM sleep alternation-what is the NREM-REM cycle? Front Neurosci. 2021;15:627193. https://doi.org/10.3389/fnins.2021.627193. PMID: 33897348; PMCID: PMC8060555.
28. Colrain IM, Webster KE, Hirst G, Campbell KB. The roles of vertex sharp waves and K-complexes in the generation of N300 in auditory and respiratory-related evoked potentials during early stage 2 NREM sleep. Sleep. 2000;23(1):97–106. PMID: 10678470.
29. Wendt SL, Welinder P, Sorensen HB, Peppard PE, Jennum P, Perona P, Mignot E, Warby SC. Inter-expert and intra-expert reliability in sleep spindle scoring. Clin Neurophysiol. 2015;126(8):1548–56. https://doi.org/10.1016/j.clinph.2014.10.158. Epub 2014 Nov 10. PMID: 25434753; PMCID: PMC4426257.
30. Warby SC, Wendt SL, Welinder P, Munk EG, Carrillo O, Sorensen HB, Jennum P, Peppard PE, Perona P, Mignot E. Sleep-spindle detection: crowdsourcing and evaluating performance of experts, non-experts and automated methods. Nat Methods. 2014;11:385–92.
31. Steriade M, Nuñez A, Amzica F. A novel slow (< 1 Hz) oscillation of neocortical neurons in vivo: depolarizing and hyperpolarizing components. J Neurosci. 1993;13(8):3252–65. https://doi.org/10.1523/JNEUROSCI.13-08-03252.1993. PMID: 8340806; PMCID: PMC6576541.
32. Steriade M, Contreras D, Curró Dossi R, Nuñez A. The slow (< 1 Hz) oscillation in reticular thalamic and thalamocortical neurons: scenario of sleep rhythm generation in interacting thalamic and neocortical networks. J Neurosci. 1993;13(8):3284–99. https://doi.org/10.1523/JNEUROSCI.13-08-03284.1993. PMID: 8340808; PMCID: PMC6576531.
33. Fernandez LMJ, Lüthi A. Sleep spindles: mechanisms and functions. Physiol Rev. 2020;100(2):805–68. https://doi.org/10.1152/physrev.00042.2018. Epub 2019 Dec 5. PMID: 31804897.
34. Mander BA, Winer JR, Walker MP. Sleep and human aging. Neuron. 2017;94(1):19–36. https://doi.org/10.1016/j.neuron.2017.02.004. PMID: 28384471; PMCID: PMC5810920.
35. Colrain IM. The K-complex: a 7-decade history. Sleep. 2005;28(2):255–73. https://doi.org/10.1093/sleep/28.2.255. PMID: 16171251.

12 Other Neurological Pathologies

12.1 Introduction

This chapter will review some cases not included in previous chapters, but which are of great interest and demonstrate the capability of qEEG to provide relevant clinical information that is not easy to observe visually.

We shall show examples of conditions such as migraine, normal pressure hydrocephalus, and medically unexplained physical symptoms. In all of these cases, qEEG provides highly relevant information.

12.2 Migraine

Migraine with aura comprises around one-third of migraine patients [1] and, usually, the aura precedes the headache attack. During the migraine aura, the patient experiences transient neurological deficits, such as visual (the most frequent), sensory, motor, speech, or brainstem symptoms. On the other hand, hemiplegic migraine is a severe subset of migraine including reversible hemiparesis in addition to other symptoms such as visual, sensory, or speech or vision alterations and headache [2].

During visual auras, EEG showed no consistent abnormalities related to aura, although transient focal slowing in occipital regions has been observed in quantitative studies. In contrast, in familial hemiplegic migraine, significant EEG abnormalities have been described consistently, including slowing over the affected hemisphere or bilaterally or suppression of EEG activity [3, 4].

This example shows the case of a 46-year-old right-handed woman with no significant medical history and no medication. Since the last 4 months, she had experienced episodes of visual disturbance in the left lower hemifield, associated with headache. An MRI scan revealed no abnormalities. A recording was performed during the occurrence of one of these visual disturbances and was maintained long enough for the symptoms to resolve. Figure 12.1a shows examples of the baseline recording (without symptoms) and the recording during the migraine event. In both cases, the recordings exhibit physiological characteristics. However, a comparison of the two appears to show a greater extent of the alpha component. Indeed, this change is observed in Fig. 12.1b1, which shows the average spectra by channel under baseline conditions (blue) and dur-

J. Pastor, *Quantitative EEG Practical Guide*, https://doi.org/10.1007/978-3-032-05457-9_12

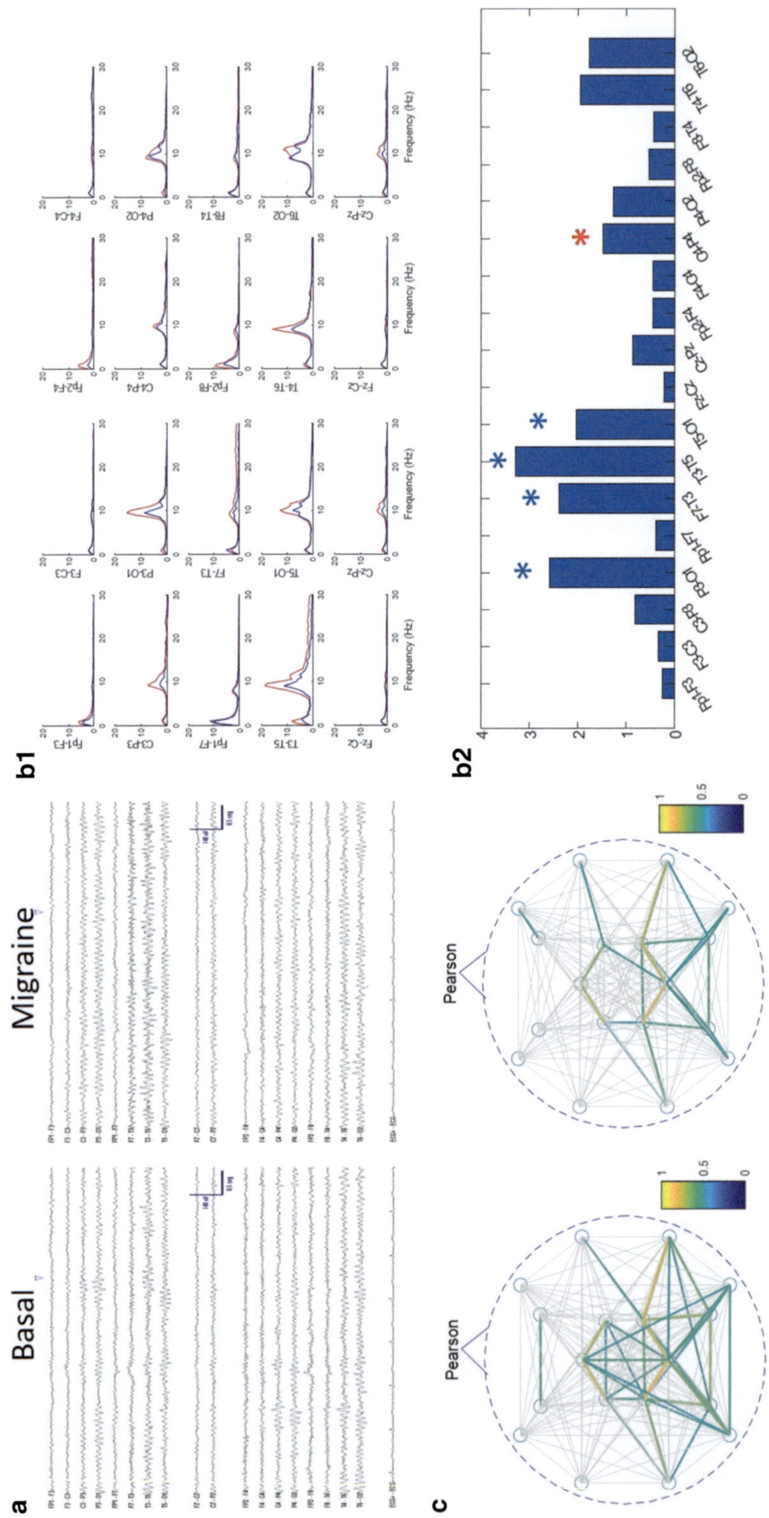

Fig. 12.1 rsEEG and recording during a migraine attack. (**a**) Raw recording during basal (left) and the migraine attack (right). (**b1**) Average spectra during basal (blue) and migraine attack (red) for all channels. (**b2**) Area difference between spectra per channel between both states. (**c**) Connectogram for the Pearson correlation coefficient during basal (left) and migraine (right)

ing the migraine episode (red), with higher components at parieto-temporo-occipital left regions. Figure 12.1b2 shows the difference in spectral area between the basal and migraine states by channel. Comparison of symmetrical channel pairs shows a marked increase during migraine in the channels marked with a blue asterisk (P3-O1, F7-T3, T3-T5, and T5-O1), while an increase was only observed in C4-P4 (red asterisk). During the migraine attack, the connectogram was also modified, which decreased significantly, as can be seen in Fig. 12.1c.

In some cases, we have also found changes in the rsEEG without the presence of symptoms. In this new example, a 39-year-old male with a history of hepatitis A and COVID-19 presented with episodes of unsteadiness, headache, and right visual field changes. These were not associated with nausea or vomiting. The patient improved with darkness and sleep. MRI and CT scans revealed no abnormalities. The recording shows a background of physiological characteristics and absence of epileptiform activity (Fig. 12.2a). However, the average spectra per channel (Fig. 12.2b) show the presence of anomalous theta components at 3.5 Hz in the right temporo-parietal region. In addition to this spectral alteration, a clear decrease in right parieto-temporal connectivity is observed. This alteration not only affects the Pearson correlation, but also affects the delta and theta coherences in the right parieto-occipital region (Fig. 12.2d).

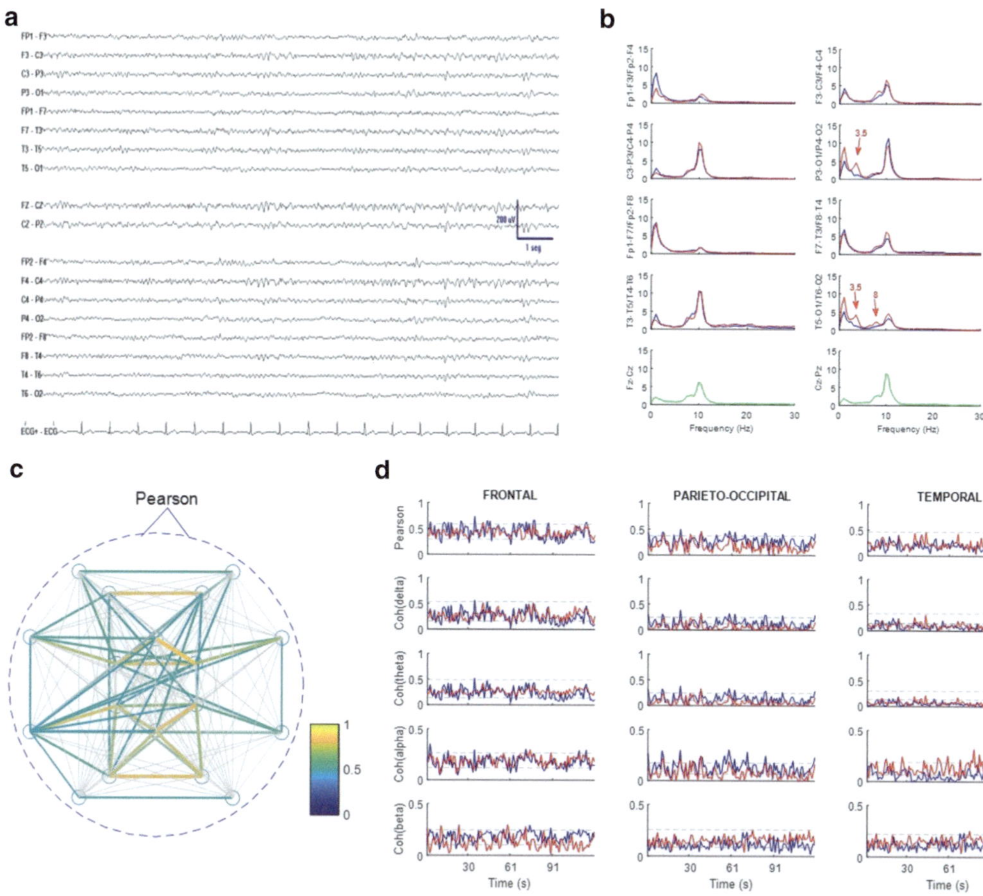

Fig. 12.2 Basal state of a migraine patient. (**a**) Raw recording. (**b**) Average spectra. Abnormal components are indicated. (**c**) Connectogram for Pearson correlation. (**d**) Dynamics of synchronization measurements by lobe. Red = right hemisphere; blue = left hemisphere

Although many questions remain to be resolved, it is interesting to note that not only during migraine attacks, but also under resting conditions, alterations in the qEEG can be observed in posterior regions that are not observed in visual analysis. It is very likely that these alterations are not constant, but rather are associated with the patient's pathophysiological state [4], but the truth is that, when they are found, they will help in a very robust way in the diagnosis.

12.3 Hydrocephalus

Idiopathic normal pressure hydrocephalus (INPH) is an adult-onset syndrome of undefined origin, with symptoms of cognitive deterioration, gait disturbance, and urinary incontinence, involving nonobstructive increase of the cerebral ventricles with normal cerebrospinal fluid pressure [5, 6]. However, INPH is largely underdiagnosed. Structural and functional changes of brain circuits in relation to symptoms and treatment response are expected to deliver major advances in this direction. Therefore, methods assessing connectivity such as, obviously, the EEG, can help to improve the diagnosis [7].

The underlying pathophysiological mechanisms involved in INPH are still debated, although subcortical white matter is the main locus of investigation [8–10]. Interestingly, subcortical dementias can exhibit relatively normal EEG patterns compared with cortical dementias [11, 12].

It has been described that among patients with INPH there are differences in the alpha band at right frontal and centrotemporal regions between patients responders and nonresponders to the cerebrospinal fluid trap test [13]. In this sense, it has been proposed that Neuronal Activity Topography analysis, which calculates normalized power variance of EEG bands, can detect cortical functional changes induced by CSF tapping in INPH [14].

Figure 12.3 shows two examples of patients diagnosed with INPH. In both cases, the cerebrospinal fluid trap test was positive. Both patients, male, aged 69 (left) and 74 (right), presented mild cognitive impairments, neither gait difficulty nor urinary incontinence. It is very interesting to observe the significant alteration of scalp connectivity (Fig. 12.3a), with an altered connectogram pattern, especially in the 74-year-old patient. The average spectra (Fig. 12.3b) show an almost physiological pattern, a good antero-posterior gradient, and the presence of a dominant posterior rhythm at 8.5 and 8 Hz, although an increase in delta power is observed.

It is clear, at least in these two examples, that the alteration in connectivity is much more intense and precede the alteration in the topographic distribution of EEG bands, which was almost normal. Therefore, anomalous connectivity patterns, with a relatively physiological structure of power bands should bring us to think in this pathology.

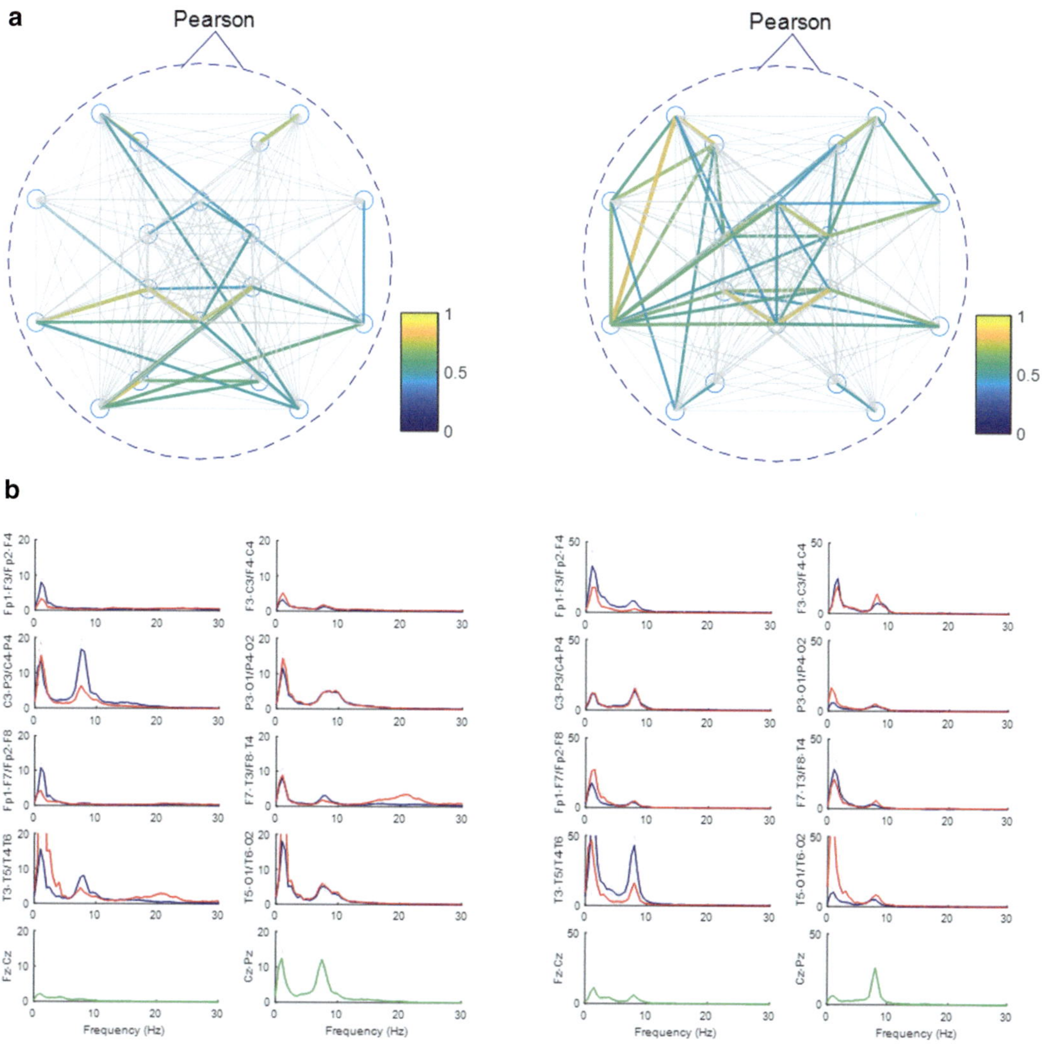

Fig. 12.3 qEEG in patients with INPH. (**a**) Connectogram and (**b**) Average power spectra during rsEEG. Red = right hemisphere; blue = left hemisphere

12.4 Highly Anomalous EEG. Medically Unexplained Physical Symptoms

To finish, it may be that both the symptomatology and the EEG pattern are related to the medically unexplained physical symptoms (MUPS), a kind of blackbox term to denote conditions defined by physical signs or subjective symptoms not better defined by another syndrome [15, 16]. MUPS are extremely common, accounting for 15–30% of primary care patients and around half of secondary care patients [17–19].

MUPS can be explained by a complex interaction of physiological and psychological factors [15] and they can cause disability persisting for at least 1 year in up to 30% of patients [17], affecting everyday functioning, interfering with work productivity, and resulting in use of healthcare resources than in other patient groups [20–22].

The most frequently reported symptoms by patients are pain, fatigue, neurological, gastrointestinal, dermatological, and rheumatological symptoms. Although in most cases these symptoms resolve spontaneously, a small but significant percentage may worsen over time [23].

Finally, we will show the example of a patient who fits well into the concept of MUP, since the EEG recording does not fit any previously described pattern and, although there is no unequivocal association between symptoms and EEG, the patient's subjective sensation appeared during the anomalous EEG pattern. This is a 41-year-old female, right-handed dominance with premorbid conditions of hypothyroidism, irritable bowel syndrome, chronic migraine, and supposed epilepsy. She was admitted to the v-EEG unit for presurgical evaluation. In a different center was diagnosed as having generalized epilepsy with fixation-off sensitivity [24]. Treatment performed with levetiracetam (1000 mg/12 h), oxcarbacepine (600 mg/12), and lamotrigine (150 mg/12 h). The symptoms consisted of paroxysmal episodes, lasting 3–10 min, in which the patient reported slowing of thought and speech.

The basal recording shows an apparently physiological pattern (Fig. 12.4a, top), with a well-defined antero-posterior gradient and a dominant posterior rhythm at 9 Hz. We called this state as desynchronized (EEGdesyn). This pattern alternated with a generalized high-voltage pattern, with no anteroposterior gradient, and an apparently generalized alpha rhythm (Fig. 12.4a, bottom), named as synchronized (EEGsyn). During the entire waking period recorded (>30 h), the EEGsyn state occurred 55% of the time. The patient reported a 95-second episode of slowed thinking, with no objective language disturbance. This episode occurred in EEGsyn, and no changes in this state were observed. The patient reported no subjective disturbances during the remainder of the period in EEGsyn.

The average spectral analysis (Fig. 12.4b) showed the presence of an anomalous pattern, with three generalized components at 5.5, 9, and 11 Hz, although in the fronto-polar region, the structure of the spectra was slightly different. Surprisingly, during the anomalous phase (Fig. 12.4a, below) no new spectral components appeared, but the power of the existing ones increases, especially the 9 Hz component.

Synchronization measures changed dramatically during the anomalous period (Fig. 12.4c), especially in frontal regions. The connectogram (Fig. 12.4d, EEGdesyn above and EEGsyn below) shows elevated, symmetrical, and widespread synchronization.

Anomalous patterns like this are rarely observed. Its description and study may be of great importance for a better understanding of brain dynamics in some poorly studied states.

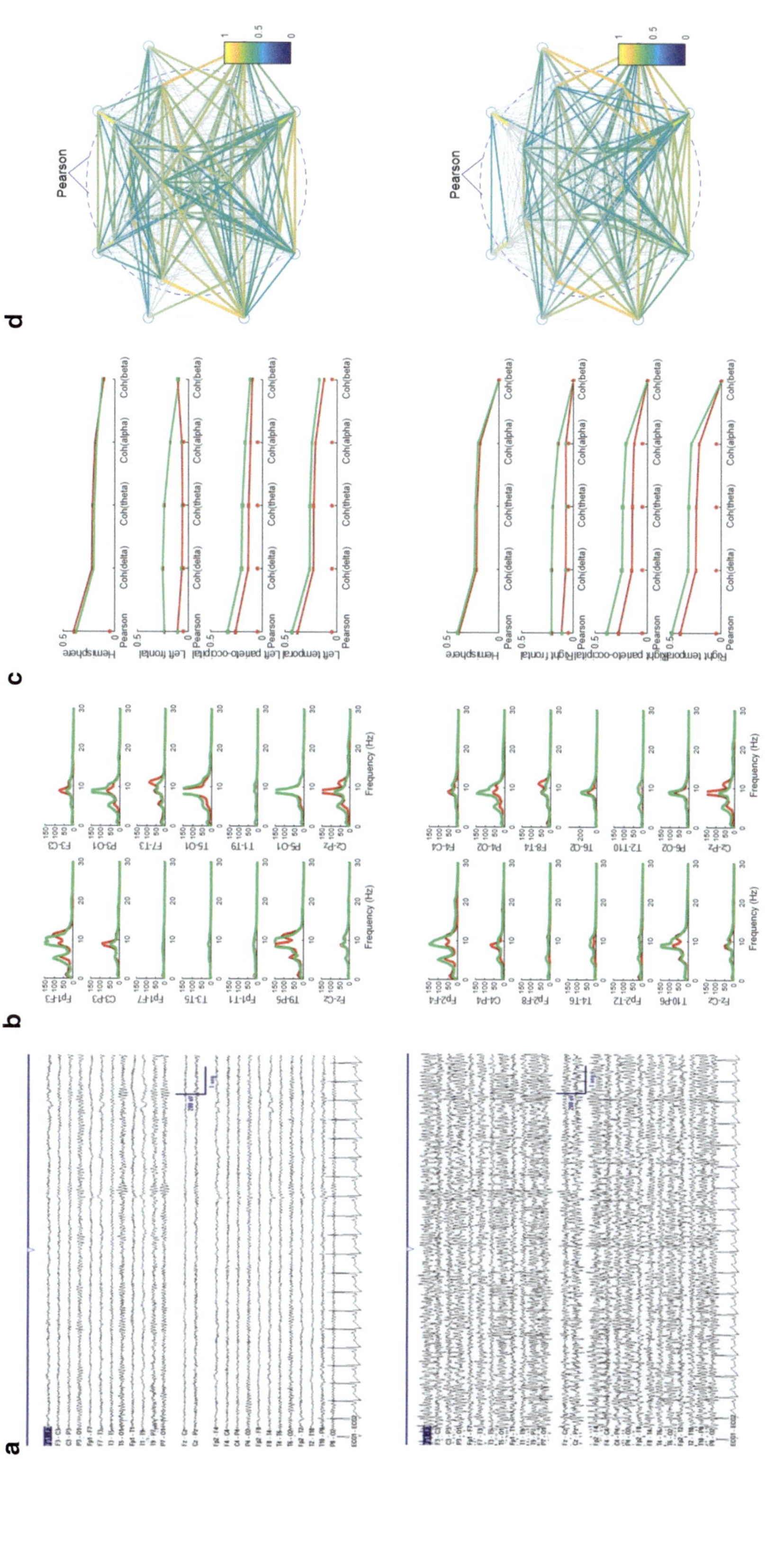

Fig. 12.4 EEG in a MUPS patient. (**a**) EEG recording during resting state (upper) and during EEGsyn (lower). (**b**) Mean average power spectra for the left (upper) and right hemispheres (lower). (**c**) Average synchronization measures for bands for the left (upper) and right hemisphere (lower). Red = EEGdesyn; Green = EEGsyn. (**d**) Connectogram during EEGdesyn (upper) and EEGsyn (lower)

12.5 Concluding Remarks

The contribution of qEEG in the first two pathological states was extremely valuable, as visual analysis proved nonspecific. However, in both cases, the alteration in numerical magnitudes was evident, which allowed for corroboration of the suspected diagnosis.

In the third case, the EEGsyn state is clearly abnormal de visu. Probably, it could be confused with a status epilepticus. Surprisingly, an anomalous pattern is already observed during the EEGdesyn state, with generalized components at 5.5 and 11 Hz. During the EEGsyn state, although the recording is highly anomalous, the spectral structure is very similar, with the same spectral components, except in the anterior region.

References

1. Akin M. Comparison of wavelet transform and FFT methods in the analysis of EEG signals. J Med Syst. 2002;26(3):241–7. https://doi.org/10.1023/a:1015075101937. PMID: 12018610.
2. Nandyala A, Shah T, Ailani J. Hemiplegic migraine. Curr Neurol Neurosci Rep. 2023;23(7):381–7. https://doi.org/10.1007/s11910-023-01277-z. Epub 2023 May 29. PMID: 37247170.
3. Riederer F, Beiersdorf J, Scutelnic A, Schankin CJ. Migraine Aura-catch me if you can with EEG and MRI-A narrative review. Diagnostics (Basel). 2023;13(17):2844. https://doi.org/10.3390/diagnostics13172844. PMID: 37685382; PMCID: PMC10486733.
4. Zhang N, Pan Y, Chen Q, Zhai Q, Liu N, Huang Y, Sun T, Lin Y, He L, Hou Y, Yu Q, Li H, Chen S. Application of EEG in migraine. Front Hum Neurosci. 2023;17:1082317. https://doi.org/10.3389/fnhum.2023.1082317. PMID: 36875229; PMCID: PMC9982126.
5. Gavrilov GV, Gaydar BV, Svistov DV, Korovin AE, Samarcev IN, Churilov LP, Tovpeko DV. Idiopathic normal pressure hydrocephalus (Hakim-Adams syndrome): clinical symptoms, diagnosis and treatment. Psychiatr Danub. 2019;31(Suppl 5):737–44. PMID: 32160166.
6. Shprecher D, Schwalb J, Kurlan R. Normal pressure hydrocephalus: diagnosis and treatment. Curr Neurol Neurosci Rep. 2008;8(5):371–6. https://doi.org/10.1007/s11910-008-0058-2. PMID: 18713572; PMCID: PMC2674287.
7. Griffa A, Van De Ville D, Herrmann FR, Allali G. Neural circuits of idiopathic Normal pressure hydrocephalus: a perspective review of brain connectivity and symptoms meta-analysis. Neurosci Biobehav Rev. 2020;112:452–71. https://doi.org/10.1016/j.neubiorev.2020.02.023. Epub 2020 Feb 20. PMID: 32088348.
8. Akai K, Uchigasaki S, Tanaka U, Komatsu A. Normal pressure hydrocephalus. Neuropathological study. Acta Pathol Jpn. 1987;37(1):97–110. PMID: 3577765.
9. Del Bigio MR. Neuropathological changes caused by hydrocephalus. Acta Neuropathol. 1993;85(6):573–85. https://doi.org/10.1007/BF00334666. PMID: 8337936.
10. Carswell C. Idiopathic normal pressure hydrocephalus: historical context and a contemporary guide. Pract Neurol. 2023;23(1):15–22. https://doi.org/10.1136/pn-2021-003291. Epub 2022 Sep 26. PMID: 36162853.
11. Jeong J. EEG dynamics in patients with Alzheimer's disease. Clin Neurophysiol. 2004;115(7):1490–505. https://doi.org/10.1016/j.clinph.2004.01.001. PMID: 15203050.
12. Verma NP, Greiffenstein MF, Verma N, King SD, Caldwell DL. Electrophysiologic validation of two categories of dementias--cortical and subcortical. Clin Electroencephalogr. 1987;18(1):26–33. PMID: 2951039.
13. Seo JG, Kang K, Jung JY, Park SP, Lee MG, Lee HW. Idiopathic normal pressure hydrocephalus, quantitative EEG findings, and the cerebrospinal fluid tap test: a pilot study. J Clin Neurophysiol. 2014;31(6):594–9. https://doi.org/10.1097/WNP.0000000000000105. PMID: 25462148.
14. Aoki Y, Kazui H, Tanaka T, Ishii R, Wada T, Ikeda S, Hata M, Canuet L, Musha T, Matsuzaki H, Imajo K, Yoshiyama K, Yoshida T, Shimizu Y, Nomura K, Iwase M, Takeda M. EEG and neuronal activity topography analysis can predict effectiveness of shunt operation in idiopathic normal pressure hydrocephalus patients. Neuroimage Clin. 2013;3:522–30. https://doi.org/10.1016/j.nicl.2013.10.009. PMID: 24273735; PMCID: PMC3830071.
15. McAndrew LM, Philips LA, Helmer DA, Maestro K, Engel CC, Greenberg LM, et al. High healthcare utilization near the onset of medically unexplained symptoms. J Psychosom Res. 2017;98:98–105. https://doi.org/10.1016/j.jpsychores.2017.05.001.
16. Jungwee P, Gilmour H. Medically unexplained physical symptoms (MUPS) among adults in Canada: comorbidity, health care use and employment. Health Matters. 2017;28:3–8.
17. Joyce E, Cowing J, Lazarus C, Smith C, Zenzuck V, Peters S. Training tomorrow's doctor to explain 'medically unexplained physical symptoms: an examination of UK medical educators' views of barriers and solutions. Patient Educ Couns. 2017;101:878–84. https://doi.org/10.1016/j.pec.2017.11.020.
18. Klaus K, Rief W, Brähler E, Martin A, Glaesmer H, Mewes R. The distinct between "medically unexplained" and "medically explained" in the context of somatoform disorders. Int J Behav

Med. 2013;20:161–71. https://doi.org/10.1007/s12529-012-9245-2.

19. Callegari C, Bortolaso P, Vender S. A single case report of recurrent surgery for chronic back pain and its implications concerning a diagnosis of Munchausen syndrome. Funct Neurol. 2006;6:103–8.
20. Ferrari S, Poloni N, Diefenbacher A, Barbosa A, Cosci F. From hysteria to somatic symptom disorders: searching for a common psychopathological ground. J Psychopathol. 2015;21:372–9.
21. Van Dessel NC, Van Der Wouden JC, Hoekstra T, Dekker J, Van Der Horst HE. The 2-year course of medically unexplained physical symptoms (MUPS) in terms of symptom severity and functional status: results of the PROSPECT cohort study. J Psychosom Res. 2018;104:76–87. https://doi.org/10.1016/j.jpsychores.2017.11.012.
22. Caselli I, Poloni N, Ielmini M, Diurni M, Callegari C. Epidemiology and evolution of the diagnostic classification of factitious disorder in DSM-5. Pyschol Res Behav Manag. 2017;10:387–94. https://doi.org/10.2147/PRBM.S153377.
23. Poloni N, Ielmini M, Caselli I, Ceccon F, Bianchi L, Isella C, Callegari C. Medically unexplained physical symptoms in hospitalized patients: a 9-year retrospective observational study. Front Psych. 2018;23(9):626. https://doi.org/10.3389/fpsyt.2018.00626. PMID: 30532714; PMCID: PMC6265408.
24. Pastor J, Vega-Zelaya L. Highly anomalous electroencephalographic synchronized pattern during wakefulness. Internal Med Res Open J. 2021;6(3):1–10. https://doi.org/10.31038/IMROJ.2021634.

13 The Future of qEEG in Clinics

13.1 Where Are We Now

The real interest in numerical techniques in the study of EEG in recent decades is beyond doubt. Without intending to carry out a rigorous and exhaustive exploration, but only as an estimate of the situation, we made a search in PubMed. Using the MesH *qEEG* and *quantified EEG*, we obtained the results shown in Fig. 13.1.

The data fit this exponential function paper(year) = $79.2 + 33.4e^{0.0786\text{year}}$ with extraordinary accuracy ($r^2 = 0.9801$). The relevant fact is that this is a growing function demonstrating the increasing interest in qEEG.

This crude approach does not differentiate between clinical and scientific applications, but certainly demonstrates that this is a rapidly expanding field. However, the title of this chapter refers to the application of qEEG to daily clinical practice. Of course, in our hospital, we can affirm that the clinical use of qEEG is daily and intensive. However, this observation cannot be generalized, at least to our country.

The general context of development of qEEG obviously depends on the use that is given to conventional EEG, and what is this use?

EEG recordings, in clinical practice, have been analyzed by electroencephalographers in a similar way since the beginning. Of course, there have been important milestones that have paved the way, such as the description of specific patterns in epilepsy [1–3] or sleep phases [4–6], or the introduction of synchronized video and EEG recording.

This form of analysis has influenced the development of the technique, making the EEG, for a large majority of physicians, a tool primarily oriented toward the study of epilepsy. Of course, it is easier to identify 100 µV sharp waves than to assess low-amplitude focal slowing in subcortical lesions or subtle changes in patients with dementia. These very small changes, many of them related to modifications in synchronization, cannot be assessed even by highly experienced neurophysiologists.

We have addressed superficially (e.g., in a non-rigorous manner) the burden of the different pathologies, to address the use of EEG in daily practice. To do that, we have searched in PubMed for the appearance of pathologies (see table 13.1) + EEG in the Title/Abstract (Table 13.1). Obviously, this is neither a rigorous nor exhaustive inquiry, but it's interesting.

Although this result can only be considered an approximation of the real situation, it indicates the excessive weight given to EEG studies of epilepsy. It is quite likely that a more serious study would have obtained similar results. But, obviously, the EEG is a technique for studying the brain including much more than epilepsy [7–10].

Therefore, we can see that we are at a time when most of the medical and scientific interest among EEG professionals is focused on epilepsy. This is likely reflected in the EEG studies per-

J. Pastor, *Quantitative EEG Practical Guide*, https://doi.org/10.1007/978-3-032-05457-9_13

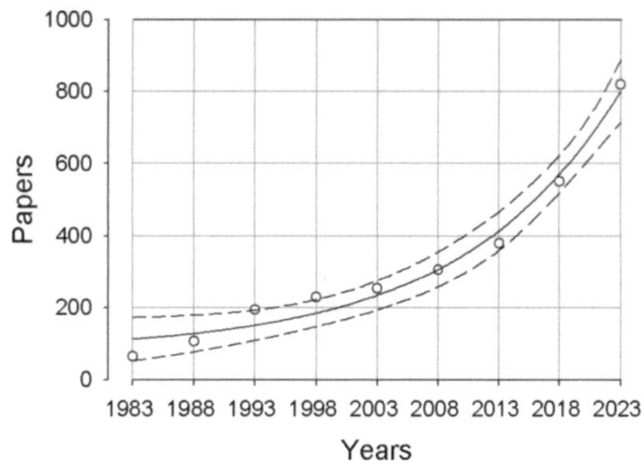

Fig. 13.1 Number of papers published (in 5-year periods) in the last 40 years. The solid line represents the fitted exponential function, and the dashed lines represent the 95% confidence interval

Table 13.1 Relevance of the study of different pathologies by EEG

Pathology+EEG	First year	Total papers	Percentage
Epilepsy	1946	20,051	50.8
Sleep	1949	13,385	33.9
Psychiatric	1951	2618	6.6
Dementia	1959	1965	5.0
Encephalopathies	1950	798	2.0
ICU	1977	662	1.7

Data obtained from PubMed

formed in most hospitals. On the other hand, we have seen a real explosion in the application of numerical tools to EEG analysis. Therefore, what can we expect from the interaction of these two conditions?

13.2 Looking to the Future

Quantified analysis allows the measurement of magnitudes that are impossible to assess with the naked eye, such as small changes in spectral composition, the frequency of a band, or synchronization. Moreover, it is precisely these types of magnitudes that are presumably implicated in psychiatric pathologies, dementias, or other neurological disorders whose study by EEG today results in a disappointing non-informative "physiological" result.

Quantitative EEG will likely allow not only the monitoring of ES/SE in the ICU, which is its main current use, but will also extend the real efficiency of EEG to many pathologies, improving not only its diagnostic capacity but also its specificity.

The development and application of EEG recordings is increasing daily. In addition to the methods reviewed in this book and many others like it, we also employ source determination techniques [11–14], machine learning, deep learning [15, 16], and even artificial intelligence.

I don't know if in the distant future medical diagnoses (and probably treatments) will rely on completely digital systems with no human involvement. Nevertheless, my prediction isn't that far off. I'm going to refer to the next 10 years. It's always uncertain to make a prediction about what will happen in the future. In general, the more significant the consequences of that prediction, the greater the risk. In this case, however, the consequences of an erroneous prediction won't be too severe (except for my reputation as a clairvoyant), so we can confidently take the risk.

This is the prediction: qEEG is the tool that will revolutionize clinical EEG practice.

This prediction presupposes several conditions. First, the manufacturers must develop packages for the analysis of conventional recordings, not just for cEEG in the ICU. Several manufacturers already have this option, such as Nihon Kohden® with a useful application, or Natus®, which have a simple analysis system, offering the possibility of incorporating a more complex system, such as Persyst®. Second, electroencephalographers responsible for these studies must familiarize themselves with the fundamentals of the methods used. A deep understanding of the system being studied and the tools being applied differentiates the technician from the specialist. And, finally, the numerical methods implemented in clinical practice should be pathology-directed, robust to artefact, and sufficiently easy to understand and manage by physicians.

To conclude, all of these methods must be normalized and validated to allow an interchangeable utility, also in clinical practice as with scientific purposes [17].

References

1. Goodwin JE. The significance of alpha variants in the EEG, and their relationship to an epileptiform syndrome. Am J Psychiatry. 1947;104(6):369–79. https://doi.org/10.1176/ajp.104.6.369. PMID: 18911621.
2. Jasper H, Flanigin H, Pertuisset B. Temporal lobe seizures: EEG and cortical electrographic studies. Trans Am Neurol Assoc. 1950;51:149–55. PMID: 14788097.
3. Lennox W. Ueber verschiedene, hauptsächlich statistische EEG-Untersuchungen bei Epileptikern [About various, mainly statistical EEG examinations in epileptics]. Praxis. 1949;38(36):789. German. PMID: 18140857.
4. Dement W, Kleitman N. Cyclic variations in EEG during sleep and their relation to eye movements, body motility, and dreaming. Electroencephalogr Clin Neurophysiol. 1957;9(4):673–90. https://doi.org/10.1016/0013-4694(57)90088-.
5. Jouvet M, Michel F, Courjon J. Sur un stade d'activité électrique cérébrale rapide au cours du sommeil physiologique. CR Soc Biol. 1959;153:1024–8.
6. Rechtschaffen A, Kales A. A manual of standardized terminology, tecniques and scoring system for sleep stage of human subjects. In: Brain information service/brain research institute. Los Angeles: University of Californian; 1968.
7. Pastor J, Vega-Zelaya L, Martin-Abad E. Necessity of quantitative EEG in daily clinical practice. In: Nakano H, editor. Electroencephalography. InTech; 2021. ISBN 978-1-83968-289-6.
8. Pastor J, Vega-Zelaya L. Normative structure of resting-state EEG in bipolar derivations for daily clinical practice: a pilot study. Brain Sci. 2023;13:167. https://doi.org/10.3390/brainsci13020167.
9. Pastor J, Garrido Zabala P, Vega-Zelaya L. Structure of spectral composition and synchronization in human sleep on the whole scalp: a pilot study. Brain Sci. 2024;14:1007. https://doi.org/10.3390/brainsci14101007.
10. Pastor J, Vega-Zelaya L. Titration of pharmacological responses in ICU patients by quantified EEG. Curr Neuropharmacol. 2022; https://doi.org/10.2174/1570159X20666220411083213.
11. Saletu B, Anderer P, Saletu-Zyhlarz GM. EEG topography and tomography (LORETA) in the classification and evaluation of the pharmacodynamics of psychotropic drugs. Clin EEG Neurosci. 2006;37(2):66–80. https://doi.org/10.1177/155005940603700205. PMID: 16733939.
12. Zeltser A, Ochneva A, Riabinina D, Zakurazhnaya V, Tsurina A, Golubeva E, Berdalin A, Andreyuk D, Leonteva E, Kostyuk G, Morozova A. EEG techniques with brain activity localization, specifically LORETA, and its applicability in monitoring schizophrenia. J Clin Med. 2024;13(17):5108. https://doi.org/10.3390/jcm13175108. PMID: 39274319; PMCID: PMC11395834.
13. Clemens B, Emri M, Fekete I, Fekete K. Epileptic diathesis: an EEG-LORETA study. Clin Neurophysiol. 2023 Jan;145:54–61. https://doi.org/10.1016/j.clinph.2022.11.004. Epub 2022 Nov 17. PMID: 36442376.
14. Piano C, Mazzucchi E, Bentivoglio AR, Losurdo A, Calandra Buonaura G, Imperatori C, Cortelli P, Della MG. Wake and sleep EEG in patients with Huntington disease: an eLORETA study and review of the literature. Clin EEG Neurosci. 2017 Jan;48(1):60–71. https://doi.org/10.1177/1550059416632413. Epub 2016 Apr 19. PMID: 27094758.
15. Sanz-García A, Pérez-Romero M, Pastor J, Sola RG, Vega-Zelaya L, Monasterio F, Torrecilla C, Vega G, Pulido P, Ortega G. Identifying causal relationships between EEG activity and intracranial pressure changes in neurocritical care patients. J Neural Eng. 15(6) https://doi.org/10.1088/1741-2552/aadeea.
16. Sanz-García A, Pérez-Romero M, Pastor J, Sola RG, Vega-Zelaya L, Monasterio F, Torrecilla C, Vega G, Pulido P, Ortega G. Potential EEG biomarkers of sedation doses in intensive care patients unveiled by using a machine learning approach. J Neural Eng. 2019;16:026031. https://doi.org/10.1088/1741-2552/ab039f.
17. Babiloni C, Barry RJ, Başar E, Blinowska KJ, Cichocki A, WHIM D, Klimesch W, Knight RT, Lopes da Silva F, Nunez P, Oostenveld R, Jeong J, Pascual-Marqui R, Valdes-Sosa P, Hallett M. International Federation of Clinical Neurophysiology (IFCN) – EEG research workgroup: recommendations on frequency and topographic analysis of resting state EEG rhythms. Part 1: applications in clinical research studies. Clin Neurophysiol. 2020;131(1):285–307. https://doi.org/10.1016/j.clinph.2019.06.234. Epub 2019 Sep 19. PMID: 31501011.

Part III

Mathematical Topics

14 Linear Algebra: Vector Analysis

14.1 Vectors and Cartesian Coordinates

Vector analysis is absolutely essential in electroencephalography. We only need to take into account that most of the electromagnetic magnitudes (e.g., the electric field or the current) are vectorial.

Each point in space (which we will call p) can be described in a unique way, by three values called coordinates (Fig. 14.1a), therefore we can write $p = (x_1, x_2, x_3)$. Additionally, we can describe a vector as a mathematical object defined as a directed line segment in $\mathbb{R}^3$, defined by its magnitude and direction. R is the set of real numbers. We can define a vector $\vec{v}$[1] between two points $a = (a_1, a_2, a_3)$ and $b = (b_1, b_2, b_3)$ as $\vec{v} = \overrightarrow{ab} = (b_1, b_2, b_3) - (a_1, a_2, a_3) = (b_1 - b_1, b_2 - a_2, b_3 - a_3)$ and means the position of b relative to a (Fig. 14.1b). The vector of any point a respect to the origin $O = (0, 0, 0)$ is called the position vector and allows to establish a mathematical relationship between coordinates and vectors, because if we have $a = (a_1, a_2, a_3)$, its position vector is $\vec{a} = (a_1, a_2, a_3) - (0,0,0) = (a_1 - 0, a_2 - 0, a_3 - 0) = (a_1, a_2, a_3)$.

Usually, the position vector is indicated by $\vec{r}$. Therefore, although coordinates and vectors are, strictly speaking, different mathematical objects, we can establish a direct relationship through the concept of vector position. Consequently, we can interchange both concepts.

Any vector can be described by the ordinal of coordinates, e.g., $\vec{v} = (v_1, v_2, v_3)$ where v_1 is the coordinate of axis x, v_2 the coordinate of axis y, and v_3 one of the axis z. Sometimes, axes x, y, and z are denoted as x_1, x_2, and x_3.

Besides direction, vectors are defined by their module or magnitude (v) and can be understood as the length of the arrow. It is defined as

$$v = \sqrt{v_1^2 + v_2^2 + v_1^2} = \sqrt{\sum_{i=1}^{3} v_i^2} \quad (14.1)$$

We can define a coordinate system which is a three-dimensional arrangement defined by three lines perpendicular to each other and crossed up at a point called the origin of coordinates. Every line represents a dimension and it's defined by a unitary vector, with origin at (0,0,0) and oriented exactly one unit in the positive direction. These unit vectors are usually designated as $\vec{i} = (1,0,0)$ for the x-axis, $\vec{j} = (0,1,0)$ for the y, and $\vec{k} = (0,0,1)$ for the z. Therefore, any point in space (a) will be described by the vectorial expression

$$a = (a_1, a_2, a_3) \equiv \vec{a} = a_1\vec{i} + a_2\vec{j} + a_3\vec{k} \quad (14.2)$$

The vectors $\vec{i}$, $\vec{j}$, and $\vec{k}$ form a canonical base or Cartesian coordinate system, meaning that all the vectors pertaining to the system can be expressed in terms of this base.

[1] Pay attention that vectors are denoted by an upper small arrow.

J. Pastor, *Quantitative EEG Practical Guide*, https://doi.org/10.1007/978-3-032-05457-9_14

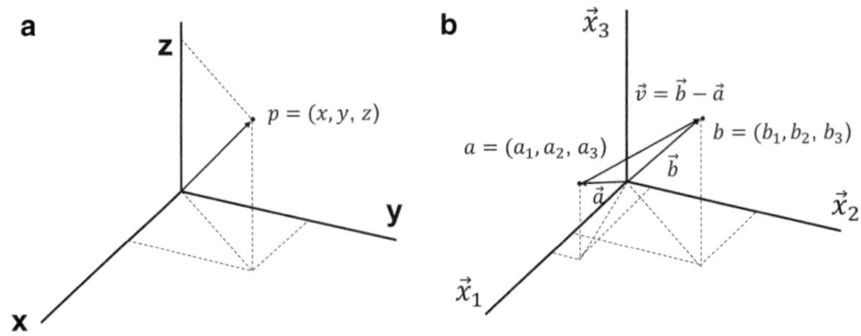

Fig. 14.1 Points at space and vectors. (**a**) Geometric representation of point p in Cartesian coordinates and its position vector. (**b**) Relative position of point $\vec{b}$ respect to point $\vec{a}$, defining the vector $\vec{v}$. Cartesian coordinates in **a** are expressed as $\vec{x}, \vec{y}, \vec{z}$, meanwhile in **b** the same axes are denoted as $\vec{x}_1, \vec{x}_2, \vec{x}_3$ to show that names are not relevant

Usually, we are interested only in the direction of the vector. In this case, we can compute the unitary vector (denoted as $\hat{r}$) according to

$$\hat{r} = \frac{\vec{r}}{\|\vec{r}\|} \tag{14.3}$$

This expression follows from the relationship $\vec{r} = \|\vec{r}\|\hat{r}$, where both concepts of module and direction are clearly separated.

Using r for the magnitude of vector $\vec{r}$, we can obtain the coordinates of point p by means of these expressions,

$$x = r\cos\alpha, \quad x = r\cos\beta, \quad x = r\cos\gamma, \tag{14.4}$$

where cosα, cosβ, and cosγ are called direction cosines, α being the angle between the given vector and the positive x-axis, β being the angle between the vector and the positive y-axis, and, finally, γ the angle between the vector and the positive z-axis.

A vector is a member of a three-dimensional vectorial space, denoted as $\mathbb{R}^3$ if the three elements are real numbers. The dimension of the system is the number of coordinates needed to define a point. In linear systems, a single coordinate is used. In a plane (e.g., in EEG voltage maps), two coordinates are used and when a field is calculated in electromagnetism (EM), it is done on three coordinates. There may be systems with more coordinates (in multivariate analysis of time series in EEG) and in these cases the abbreviated notation is much more useful than the vector notation with coordinates. We can change a row vector into a column vector by means of transposition, a concept developed in Chap. 15.

14.2 Operations with Vectors

There are several operations defined with vectors, like those defined with numbers, although not all the operations are allowed.

Suppose we have the vectors $\vec{a} = (a_1, a_2, a_3)$, $\vec{b} = (b_1, b_2, b_3)$, and $\vec{c} = (c_1, c_2, c_3)$.

A scalar is a magnitude that can be defined by a single real number. We can define the product of the scalar α by the vector $\vec{a}$ as $\alpha\vec{a} = (\alpha a_1, \alpha a_2, \alpha a_3)$.

14.2.1 Addition and Subtraction

We define (Fig 14.2a) $\vec{c} = \vec{a} + \vec{b}$

$$(a_1, a_2, a_3) + (b_1, b_2, b_3) = (a_1 + b_1, a_2 + b_2, a_3 + b_3) = (c_1, c_2, c_3)$$

So, we can use a more compact form as

$$c = \sum_{j=1}^{3} a_j + b_j ; j = 1,2,3$$

From now on, we will use this more compact notation.

Evidently, $\vec{c} = \vec{a} - \vec{b}$, Will be

$$c = \sum_{j=1}^{3} a_j - b_j ; j = 1,2,3$$

The vector addition and the product by a scalar (α, β) satisfy the following properties:
Associative: $(\alpha\beta)(a_1, a_2, a_3) = \alpha[\beta(a_1, a_2, a_3)]$

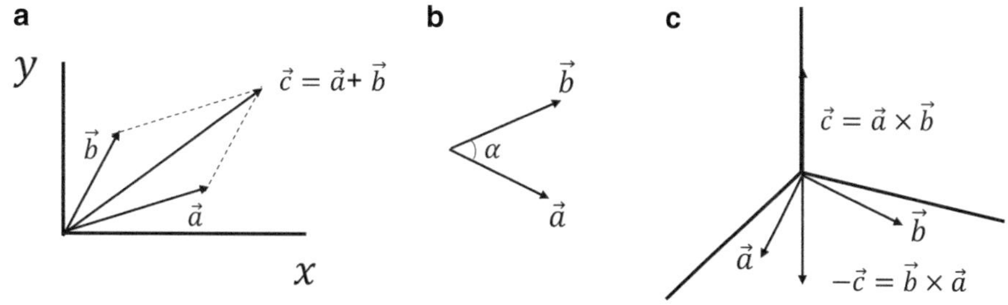

Fig. 14.2 Some operations with vectors. (**a**) Addition; (**b**) Dot product, and (**c**) Cross product

Distributive: $(\alpha + \beta)(a_1, a_2, a_3) = \alpha(a_1, a_2, a_3) + \beta(a_1,a_2, a_3)$ and
$\alpha[(a_1, a_2, a_3) + (b_1, b_2, b_3)] = \alpha(a_1, a_2, a_3) + \alpha(b_1, b_2, b_3)$

We can define two main ways to multiply vectors.

14.2.2 Dot or Scalar Product

Its result is not a vector but a scalar (Fig. 14.2b). Can be denoted by a dot (·), $\vec{a}\cdot\vec{b}$ or between brackets $\left(\vec{a},\vec{b}\right)$

$$\vec{a}\cdot\vec{b} = a_1b_1 + a_2b_2 + a_3b_3 = \sum_{i=1}^{3} a_ib_i$$

The dot product is very important for many functions. Among others, its main properties are:

- Commutative. $\vec{a}\cdot\vec{b} = \vec{b}\cdot\vec{a}$
- Let's find the norm of a vector. That is, $\|\vec{a}\| = \sqrt{(\vec{a},\vec{a})}$. So, if you want to normalize a vector, that is, find a vector with the same direction, but with a unit norm, you should do it like this $\hat{a} = \dfrac{\vec{a}}{\|\vec{a}\|}$
- Determination of the angle between two vectors. The angle formed between the vectors $\vec{a}$ and $\vec{b}$ is called α and can be calculated considering this definition of dot product

$$\vec{a}\cdot\vec{b} = ab\cos\alpha \rightarrow \cos\alpha = \frac{\vec{a}\cdot\vec{b}}{ab}$$

- Cauchy-Schwarz inequality. It is often very useful in problems of a geometric nature. For any pair of vectors, we have that $\left|\vec{a}\cdot\vec{b}\right|$ " $\vec{a}\vec{b}$. That is, the absolute value of the scalar product is less than or equal to the product of the modules of the vectors.
- Triangle inequality. For any two vectors it is verified that $\left\|\vec{a}+\vec{b}\right\| \le \|\vec{a}\| + \left\|\vec{b}\right\|$

The dot product defines the orientations of Cartesian coordinate systems because they must satisfy these relationships

$$\vec{i}\cdot\vec{i} = \vec{j}\cdot\vec{j} = \vec{k}\cdot\vec{k} = 1$$

$$\vec{i}\cdot\vec{j} = \vec{i}\cdot\vec{k} = \vec{j}\cdot\vec{k} = 0$$

The dot product has a straightforward statistical interpretation. Description of a variable can be achieved by its mean. In the case of a vector, we can consider the projection over a constant vector of unit norm. Let's see the constant n-vector with unitary norm,

$$\frac{1}{\sqrt{n}}\vec{1} = \frac{1}{\sqrt{n}}\begin{pmatrix}1\\ \vdots\\ 1\end{pmatrix}$$

The projection of vector $\vec{x}$ onto this vector is

$$\frac{1}{\sqrt{n}}\vec{1}'\vec{x} = \frac{1}{\sqrt{n}}\left(x_1,\ldots,x_n\right)\begin{pmatrix}1\\ \vdots\\ 1\end{pmatrix} = \sum_{i=1}^{n}\frac{x_i}{\sqrt{n}} = \frac{\overline{x}}{\sqrt{n}}$$

. The constant result from this projection is $\frac{1}{\sqrt{n}}\vec{1}\left(\overline{x}\sqrt{n}\right) = \overline{x}\vec{1}$

Therefore, the mean is the scalar obtained by projecting the data vector onto the constant direction. We can also define the variability of data by means of the standard deviation (SD), defined as

$$\mathrm{SD} = \frac{1}{\sqrt{n}}\left\|\vec{x} - \bar{x}\vec{1}\right\| = \sqrt{\frac{\sum_{i=1}^{n}\left(x_i - \bar{x}\right)^2}{n}}$$

The dot product can be useful in the assessment of linear dependency between variables $\vec{x}$ and $\vec{y}$. In fact, we can define covariance (cov) as the standardized dot product of both vectors measured as deviations from the mean

$$\mathrm{cov} = \frac{1}{n}\left(\vec{x} - \bar{x}\vec{1}\right)'\left(\vec{y} - \bar{y}\vec{1}\right) = \frac{\sum_{i=1}^{n}\left(x_i - \bar{x}\right)\left(y_i - \bar{y}\right)}{n} \tag{14.5}$$

In case of zero mean for both vectors, $\mathrm{cov} = \frac{1}{n}\vec{x}'\vec{y}$

It is interesting to observe that synchronization magnitudes can be stated in vector terms.

14.2.3 Vector or Cross Product

It's denoted as $\vec{a}\times\vec{b}$, hereby the name of cross product and the result is a vector orthogonal to the other two (Fig. 14.2c). We say that two vectors are orthogonal when they form a right angle and, therefore its dot product is 0.

$$\vec{a}\times\vec{b} = \begin{vmatrix} \vec{i} & \vec{j} & \vec{k} \\ a_1 & a_2 & a_3 \\ b_1 & b_2 & b_3 \end{vmatrix} = \vec{i}\begin{vmatrix} a_2 & a_3 \\ b_2 & b_3 \end{vmatrix} - \vec{j}\begin{vmatrix} a_1 & a_3 \\ b_1 & b_3 \end{vmatrix} + \vec{k}\begin{vmatrix} a_1 & a_2 \\ b_1 & b_2 \end{vmatrix}$$

In this definition, we have used the concept of determinant. Please, see the Chap. 15 Linear algebra: matrices, for more details about this concept.

Properties of the cross product

- It's not commutative $\vec{a}\times\vec{b} = -\left(\vec{b}\times\vec{a}\right)$
- $\vec{a}\times\left(\alpha\vec{b} + \beta\vec{c}\right) = \alpha\left(\vec{a}\times\vec{b}\right) + \beta\left(\vec{a}\times\vec{c}\right)$
- $\left\|\vec{a}\times\vec{b}\right\| = \left\|\vec{a}\right\|\left\|\vec{b}\right\|\sin\alpha \rightarrow \sin\alpha = \frac{\left\|\vec{a}\times\vec{b}\right\|}{\left\|\vec{a}\right\|\left\|\vec{b}\right\|}$

Triple product. It is the combination of dot and cross product and can be defined as $\vec{a}\cdot\left(\vec{b}\times\vec{c}\right)$. Considering that dot product is commutative, we have the identity $\vec{a}\cdot\left(\vec{b}\times\vec{c}\right) = \left(\vec{b}\times\vec{c}\right)\cdot\vec{a}$, but it's important to remember that the cross product is not commutative.

Another set of properties that define Cartesian coordinate systems relies in cross product and can be stated as

$$\vec{i}\times\vec{j} = \vec{k}$$

$$\vec{j}\times\vec{k} = \vec{i}$$

$$\vec{k}\times\vec{i} = \vec{j}$$

14.3 Other Coordinate Systems

In different fields of Physics and specially in EM, it is very useful to use other coordinate systems different from Cartesian. In fact, an adequate selection of the coordinate system can simplify the problem under consideration. We will briefly show two other systems of great relevance.

14.3.1 Circular Cylinder Coordinates

Every point can be defined in terms of coordinates ρ, φ, z. We define ρ as the perpendicular distance from the z-axis. The angle φ is formed between ρ and the positive x-axis (Fig 14.3a). The dominion for ρ, φ, and z are

$$0 \le \rho < \infty, \quad 0 \le \varphi \le 2\pi, \quad -\infty < z < \infty$$

The relationships with Cartesian coordinates are

$$\rho = \left(x^2 + y^2\right)^{1/2}, \quad \varphi = \tan^{-1}\left(\frac{y}{x}\right), \quad z = z \tag{14.6}$$

Inverting these equations for ρ and φ we obtain the transform relations,

$$x = \rho\cos\varphi, \quad y = \rho\sin\varphi, \quad z = z \tag{14.7}$$

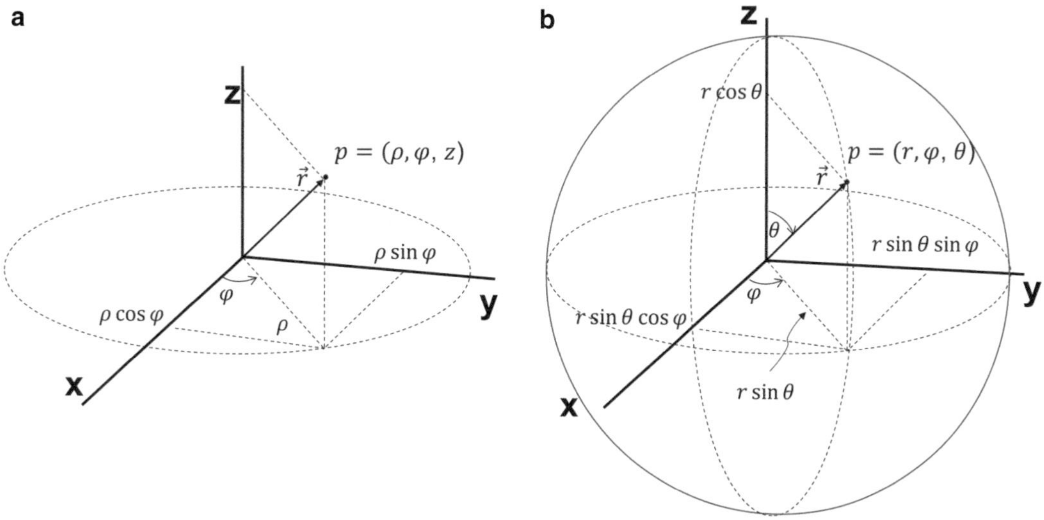

Fig. 14.3 Special coordinate systems. (**a**) Circular cylindrical coordinates system. (**b**) Spherical polar coordinates. The transformation for Coordinate systems is shown

14.3.2 Spherical Polar Coordinates

Every point in the space can be defined by means of these coordinates r, φ, θ. Being r the magnitude of $\vec{r}$. The projection of which in the xy-plane is $r \sin\theta$. The azimuth angle φ is defined between the $\vec{r}$ projection and the positive x-axis and θ is the polar angle between $\vec{r}$ and the positive z-axis (Fig 14.3b). The dominion of these coordinates are

$$0 < r < \infty, \quad 0 \le \varphi \le 2\pi, \quad 0 \le \theta \le \pi$$

The equations defining the coordinates in terms of Cartesian ones are

$$r = \left(x^2 + y^2 + z^2\right)^{1/2}, \quad \theta = \arccos\frac{z}{\left(x^2 + y^2 + z^2\right)^{1/2}}, \quad \varphi = \arctan\frac{y}{x} \tag{14.8}$$

The corresponding transformation equations are

$$x = r\sin\theta\cos\varphi, \quad y = r\sin\theta\sin\varphi, \quad z = r\cos\theta \tag{14.9}$$

14.4 Advanced Vector Analysis

This is a very relevant topic in EM and therefore we need a knowledge of some operators frequently used.

14.4.1 Derivative of a Vector by a Scalar

Let's suppose we have a continuous vectorial function that depends on a variable t, namely $\vec{v} = v_x(t)\vec{i} + v_y(t)\vec{j} + v_z(t)\vec{k}$. We can obtain its derivative according to this expression

$$\frac{d\vec{v}}{dt} = \frac{dv_x}{dt}\vec{i} + \frac{dv_y}{dt}\vec{j} + \frac{dv_z}{dt}\vec{k}$$

Then, we derive every function by t and construct a vector multiplying by the directional vectors and, obviously the result is a new vector.

14.4.2 Gradient of a Scalar

The concept of gradient is one of the most important in EM and related fields, for this reason it is extremely important to understand its meaning well. The gradient, denoted *grad* or ∇, indicates the direction in which the change of a scalar function is maximum.

The gradient is defined as the space partial derivative,

$$\nabla(x,y,z) = \frac{\partial}{\partial x}\vec{i} + \frac{\partial}{\partial y}\vec{j} + \frac{\partial}{\partial z}\vec{k} \quad (14.10)$$

We have explicitly stated that this operator is defined in Cartesian coordinates. Suppose a scalar (namely u) function, such that $u : \mathbb{R}^3 \to \mathbb{R}$. A typical example of a scalar field can be the temperature in a room, since for each point, given by (x, y, z), there will be a single temperature value. Suppose we know the value of the function at a point, p, which is $u(x, y, z)$. The value of the function will be slightly different for a small displacement $d\vec{s}$, that is, (dx, dy, dz). To find the value of the gradient, we will use the concept of exact differential, which is

$$du = \frac{\partial u}{\partial x}dx + \frac{\partial u}{\partial y}dy + \frac{\partial u}{\partial z}dz \quad (14.11)$$

But this expression can be written like this,

$$du = \nabla u \cdot d\vec{s} \quad (14.12)$$

Where, as we have indicated before $d\vec{s} = dx\vec{i} + dy\vec{j} + dz\vec{k}$

It is important to keep in mind that $du = u(x + dx, y + dy, z + dz) - u(x, y, z)$ which is a scalar magnitude, not a vectorial one. Therefore, for the above equality to hold, the following definition must be verified,

$$\nabla u = \frac{\partial u}{\partial x}\vec{i} + \frac{\partial u}{\partial y}\vec{j} + \frac{\partial u}{\partial z}\vec{k} \quad (14.13)$$

Now, this definition of gradient is valid only for Cartesian coordinates. For cylindrical or spherical coordinates, respectively, these expressions should have to be used,

$$\nabla(\rho,\varphi,z) = \frac{\partial}{\partial \rho}\hat{\rho} + \frac{1}{\rho}\frac{\partial}{\partial \varphi}\hat{\varphi} + \frac{\partial}{\partial z}\hat{z} \quad (14.14)$$

$$\nabla(r,\theta,\varphi) = \frac{\partial}{\partial r}\hat{r} + \frac{1}{r}\frac{\partial}{\partial \theta}\hat{\theta} + \frac{1}{r\sin\theta}\frac{\partial}{\partial \varphi}\hat{\varphi} \quad (14.15)$$

The gradient has the property of being orthogonal to the equipotential surfaces, in which its value does not change. This property is important, because it allows us to find the unit vector of that surface. To do this, keep in mind that the gradient is a vector, with all of the laws. Therefore, it can be written,

$$\nabla u = |\nabla u|\hat{n} \quad (14.16)$$

14.4.3 Divergence

Above, we have defined the ∇ as a vector operator. Then, we can apply the same operations defined for vectors. First, we can define the dot product with a vector $\vec{v}$

$$\nabla \cdot \vec{v} = \frac{\partial v_x}{\partial x} + \frac{\partial v_y}{\partial y} + \frac{\partial v_z}{\partial z} \quad (14.17)$$

This operation is known as divergence and the result is a scalar and states that the net flow out of a volume of a physical magnitude (e.g., current, air, probability, etc.) results in a decreased density inside the volume.

14.4.4 Curl

Another possible operation with the vector operator ∇ is the cross product. We obtain

$$\nabla \times \vec{v} = \begin{vmatrix} \vec{i} & \vec{j} & \vec{k} \\ \frac{\partial}{\partial x} & \frac{\partial}{\partial y} & \frac{\partial}{\partial z} \\ v_x & v_y & v_z \end{vmatrix} = \left(\frac{\partial v_z}{\partial y} - \frac{\partial v_y}{\partial z}\right)\vec{i} + \left(\frac{\partial v_x}{\partial z} - \frac{\partial v_z}{\partial x}\right)\vec{j} + \left(\frac{\partial v_y}{\partial x} - \frac{\partial v_x}{\partial y}\right)\vec{k} \quad (14.18)$$

Which is called the curl of $\vec{v}$ and the result is another vector orthogonal to the plane defined by ∇ and $\vec{v}$. And this vectorial nature is responsible that, in general, $\nabla \times \vec{v} \neq -\vec{v} \times \nabla$.

14.4.5 Laplacian

It is the other operator of great interest, defined as the dot product of a gradient by self, ∇ · ∇ = ∇^2:

$$\nabla^2 = \nabla\cdot\nabla = \frac{\partial^2}{\partial x^2} + \frac{\partial^2}{\partial y^2} + \frac{\partial^2}{\partial z^2} \quad (14.19)$$

We can apply the Laplacian onto a scalar function, namely $u(x,y,z)$, obtaining.

$$\nabla^2 u(\vec{r}) = \frac{\partial^2 u}{\partial x^2} + \frac{\partial^2 u}{\partial y^2} + \frac{\partial^2 u}{\partial z^2}$$

Nevertheless, the expression $\nabla^2\vec{v}$ represents the three equations in which Laplacian operates onto every component of $\vec{v}$, in this way,

$$\nabla^2 v_x = \frac{\partial^2 v_x}{\partial x^2} + \frac{\partial^2 v_x}{\partial y^2} + \frac{\partial^2 v_x}{\partial z^2}, \nabla^2 v_y$$
$$= \frac{\partial^2 v_y}{\partial x^2} + \frac{\partial^2 v_y}{\partial y^2} + \frac{\partial^2 v_y}{\partial z^2}, \nabla^2 v_z = \frac{\partial^2 v_z}{\partial x^2} + \frac{\partial^2 v_z}{\partial y^2} + \frac{\partial^2 v_z}{\partial z^2}$$

14.4.6 Some Vectorial Relations

We finally indicate some important vectorial results very frequently found in EM. Let's have u and v scalar functions, and $\vec{v}$ and $\vec{w}$ vectors.

Divergence of a curl is zero:

$$\nabla\cdot(\nabla\times\vec{v}) = 0 \quad (14.20)$$

Curl of a gradient is zero:

$$\nabla\times\nabla u = 0 \quad (14.21)$$

$$\nabla\cdot(\vec{v} + w) = \nabla\cdot\vec{v} + \nabla\cdot\vec{w} \quad (14.22)$$

$$\nabla\cdot(u + v) = \nabla\cdot u + \nabla\cdot v \quad (14.23)$$

$$\nabla\cdot(u\vec{v}) = \vec{v}(\nabla \mathrm{u}) + u(\nabla\cdot\vec{v}) \quad (14.24)$$

$$\nabla(\mathrm{R}^n) = -\nabla'(\mathrm{R}^n) = nR^{n-1}\hat{R} \quad (14.25)$$

and particularly important, when $n = -1$, we have

$$\nabla\left(\frac{1}{R}\right) = -\nabla'\left(\frac{1}{R}\right) = -\frac{\hat{R}}{R^2} = -\frac{\vec{R}}{R^3} \quad (14.26)$$

Suggested Reading

A rigorous and detailed description of these theories and concepts can be found in the next references.

Arfken GB, Weber HJ. Vector analysis. mathematical methods for physicists. 6th ed. Amsterdam: Elsevier; 2005. ISBN-13978-0-12-059876-2.

Fitzpatrick R. Vectors and vector fields. In: Maxwell's equations and the principles of electromagnetism. Hingham: Inifinty Science Press; 2008. ISBN: 978-1-934015-20-9.

Marsden JE, Tromba A. The geometry of Euclidean space. In: Vector calculus. 6th ed. New York: WH Freeman & Co Ltd.; 2011a. ISBN 978-1429215084.

Marsden JE, Tromba A. Vector valued functions. In: Vector calculus. 6th ed. New York: WH Freeman & Co Ltd.; 2011b. ISBN 978-1429215084.

Wangsness RK. Vectors. In: Electromagnetic fields. 2nd ed. New York: Wiley; 1986. ISBN 0-471-81186-6.

15 Linear Algebra: Matrices

15.1 Linear Dependency

We can say that a set of vectors $\vec{x}_1,\ldots,\vec{x}_n$ is linear dependent in case of exist scalars $c_1,\ldots,c_n$, not all of them null, such that

$$c_1\vec{x}_1 + \ldots + c_n\vec{x}_n = \vec{0} \tag{15.1}$$

Where

$$\vec{0} = \begin{pmatrix} 0 \\ \vdots \\ 0 \end{pmatrix} \tag{15.2}$$

If a set of vectors is not linear dependent, we can define as linear independent. In the space $\mathbb{R}^p$, the maximum number of independent vectors would be p.

From a statistical perspective, a set of linear independent vectors corresponds to variables not linearly related. Given a set of p linear independent vectors $B = \left(\vec{x}_1,\ldots,\vec{x}_p\right)$ in $\mathbb{R}^n$ ($p \leq$ n), we shall define the generated space by this family of vectors as the set of all of the $\vec{z}$ in $\mathbb{R}^n$ that can be expressed as a linear combination of B, known as generator base family of the space or simply base of the space. Therefore, if $\vec{z}$ pertains to the space, we have

$$\vec{z} = c_1\vec{x}_1 + \ldots + c_p\vec{x}_p \tag{15.3}$$

We define the dimension of the space E_p as the number of linear independent vectors able to generate it.

15.2 Matrices

We call matrix of m-rows and n-columns in R to every set of $m \times n$ elements distributed like this

$$\begin{pmatrix} a_{11} & \cdots & a_{1n} \\ \vdots & \ddots & \vdots \\ a_{m1} & \cdots & a_{mn} \end{pmatrix} \tag{15.4}$$

This matrix can be also represented[1] as $\left(a_{ij}\right)_{\substack{i=1\ldots m \\ j=1\ldots n}}$. The element a_{ij} of a matrix is located at row i and column j.

We denote as $M_{m\times n}(\mathbb{R})$ to the set of m-rows and n-columns matrices on $\mathbb{R}$. We call a row matrix any matrix belonging to $M_{1\times n}(\mathbb{R})$, whatever $n \in \mathbb{N}$ is. Similarly, we will say that a matrix is a column matrix when it belongs to the set $M_{m\times 1}(\mathbb{R})$ for any $m \in \mathbb{N}$.

15.2.1 Basic Definitions

We call transpose of a matrix and denote a A' or A^t to that matrix obtained by changing the rows into columns and columns into rows for a given matrix, $(a_{ij}) = (a_{ji})$, this means

[1] A comment on terminology: matrices are represented with [...] or (...) in different works of Mathematics, Physics or other fields, although we will preferently use (...).

J. Pastor, *Quantitative EEG Practical Guide*, https://doi.org/10.1007/978-3-032-05457-9_15

$$A = \begin{pmatrix} a_{11} & \cdots & a_{1n} \\ \vdots & \ddots & \vdots \\ a_{m1} & \cdots & a_{mn} \end{pmatrix} \to A' = \begin{pmatrix} a_{11} & \cdots & a_{m1} \\ \vdots & \ddots & \vdots \\ a_{1n} & \cdots & a_{mn} \end{pmatrix} \quad (15.5)$$

A basic property of a matrix is the rank (*rk*), defined as the maximum number of row or column vectors linearly independent. The rank of a matrix shows the necessary real dimension to represent the set of data or, in other words, the true number of variables we have.

The rank follows the next properties:

- $rk(A_{n \times p}) \leq \min(n,p)$, where min is the minimum value of the group included between brackets. When $rk(A_{n \times p}) = \min(n,p)$ the matrix is called of full rank.
- $rk(A + B) \leq rk(A) + rk(B)$
- $rk(AB) \leq \min(rk(A), rk(B))$
- $rk(AA') = rk(A'A) = rk(A)$

We need to define the basic operations with matrix. Let's have two matrices, $A, B \in M_{m \times n}(\mathbb{R})$,

$$A = \begin{pmatrix} a_{11} & \cdots & a_{1n} \\ \vdots & \ddots & \vdots \\ a_{m1} & \cdots & a_{mn} \end{pmatrix} \text{ and } B = \begin{pmatrix} b_{11} & \cdots & b_{1n} \\ \vdots & \ddots & \vdots \\ b_{m1} & \cdots & b_{mn} \end{pmatrix}$$

We define the sum or addition, C = A + B, where $C \in M_{m \times n}(\mathbb{R})$, as

$$C = \begin{pmatrix} c_{11} & \cdots & c_{1n} \\ \vdots & \ddots & \vdots \\ c_{m1} & \cdots & c_{mn} \end{pmatrix} = \begin{pmatrix} a_{11} & \cdots & a_{1n} \\ \vdots & \ddots & \vdots \\ a_{m1} & \cdots & a_{mn} \end{pmatrix} + \begin{pmatrix} b_{11} & \cdots & b_{1n} \\ \vdots & \ddots & \vdots \\ b_{m1} & \cdots & b_{mn} \end{pmatrix} \quad (15.6)$$

That is to say, $C = (c_{ij})_{\substack{i=1,\ldots m \\ j=1,\ldots n}}$ such as $c_{ij} = a_{ij} + b_{ij}$.

We define the subtraction as $c_{ij} = a_{ij} - b_{ij}$.

Both operations share the same properties than sum and subtraction for real numbers, i.e., commutative, associative, neutral element, and unitary element (*I*). We don't explain further because they are widely known.

The sum and subtraction are quite intuitive because we are adding or subtracting elements from the same row and column of both matrices. However, multiplication is more complex and not so straightforward. Suppose we have the next matrices, $A = (a_{ij}) \in M_{m \times p}(\mathbb{R})$ and $B = (b_{ij}) \in M_{p \times n}(\mathbb{R})$. Please pay attention that both matrices pertain to different sets and that the columns of A are equal to the rows of B. We define the product of both matrices C = AB, where $C = (c_{ij}) \in M_{m \times n}(\mathbb{R})$ and coefficients are defined as

$$c_{ij} = \sum_{k=1}^{p} a_{ik} b_{kj}; i = 1, \ldots m; j = 1, \ldots n \quad (15.7)$$

In other words, we are multiplying all the elements from the *i* row by the elements of the j column.

Let's see now some properties of the product of matrices.

The product is associative. Let's suppose the matrices $\in M_{m \times n}(\mathbb{R})$, $B \in M_{n \times p}(\mathbb{R})$ and $C \in M_{p \times q}(\mathbb{R})$, then it's verified

$$A(BC) = (AB)C \quad (15.8)$$

The product is distributive. Let's suppose the matrices $A \in M_{m \times n}(\mathbb{R})$ and $B, C \in M_{n \times p}(\mathbb{R})$, then it's verified

$$A(B+C) = (AB) + (AC) \quad (15.9)$$

However, the product is in general not commutative. That is, for matrices $A \in M_{m \times n}(R)$ and $B \in M_{n \times p}(\mathbb{R})$, if there exist $A \bullet B$, in general there doesn't exist $B \bullet A$, except for $m = n = p$, namely, that both matrices be squared (see below). But, even in this case when the product exist, not always will be true that $AB = BA$.

15.2.2 Square Matrices and Determinants

A matrix is called square when it has the same number of rows as columns, that is, when it belongs to the set $M_{m \times m}(\mathbb{R})$ for some $m \in \mathbb{N}$ and the number m is called order of the matrix. When a square matrix is equal to its transpose, i.e., A = A', it is called symmetric.

We call the main diagonal of the matrix the n-tuple $(a_{ii})_{i=1,\ldots n} = (a_{11}, \ldots a_{nn})$. A square matrix is said to be diagonal when the only nonzero elements are those on its main diagonal. A very special case is the unit or identity matrix

$$I = \begin{pmatrix} 1 & \dots & 0 \\ \vdots & 1 & \vdots \\ 0 & \dots & 1 \end{pmatrix} \quad (15.10)$$

The trace of a square matrix (tr):

$$\mathrm{tr}(A) = \sum_{i=1}^{n} a_{ii} \quad (15.11)$$

The trace is an overall measure of the size of the matrix, and it is a measure of variability. This linear operator has the following properties:

- $tr(A + B) = tr(A) + tr(B)$.
- $tr(\lambda A) = \lambda tr(A)$, where λ is a scalar.
- For a symmetric matrix, $tr(A^2) = tr(AA) = \sum_{i=1}^{n}\sum_{j=1}^{n} a_{ij}^{2}$.

The maximum rank of a square matrix A of order n is n. When the $rk(A) < n$ then the matrix is called singular.

Sometimes the properties of a system can be obtained from the features of matrices. For example, we can use the properties of positive real numbers to a specific kind of matrix called semidefine positive. First, we must introduce the concept of quadratic form. Let's transform vector $\vec{x}$ by means $\vec{y} = B\vec{x}$. In this case, the squared norm of the new vector will be

$$\vec{y}'\vec{y} = \vec{x}'B'B\vec{x} = \vec{x}'A\vec{x} \quad (15.12)$$

where $B'B = A$ is a squared and symmetric matrix. We shall call quadratic form to any scalar expression like this

$$\vec{x}'A\vec{x} = (x_1, \dots x_n)\begin{pmatrix} a_{11} & \dots & a_{1n} \\ a_{n1} & \dots & a_{nn} \end{pmatrix}\begin{pmatrix} x_1 \\ \vdots \\ x_n \end{pmatrix}$$

$$= \sum_{i=1}^{n} a_{ii}x_i^2 + 2\sum_{i=1}^{n}\sum_{j=i+1}^{n} a_{ij}x_i x_j \quad (15.13)$$

A matrix A is semidefinite positive if for any quadratic form with any vectors $\vec{x}$ and $\vec{x}'$ the result is a non-negative real number, e.g.

$$\vec{x}'A\vec{x} \geq 0 \quad (15.14)$$

However, when

$$\vec{x}'A\vec{x} > 0 \quad (15.15)$$

The matrix A is defined as definite positive.

A very important concept related to square matrices is the determinant. Let's have a 2 × 2 matrix $A = \begin{pmatrix} a_{11} & a_{12} \\ a_{21} & a_{22} \end{pmatrix}$. We define the determinant (det(A) or $|A|$) as the real number,

$$|A| = \begin{vmatrix} a_{11} & a_{12} \\ a_{21} & a_{22} \end{vmatrix} = a_{11}a_{22} - a_{12}a_{21} \quad (15.16)$$

In the case of rank 3 × 3 matrices, the determinant is more complex, e.g.,

$$\begin{vmatrix} a_{11} & a_{12} & a_{13} \\ a_{21} & a_{22} & a_{23} \\ a_{13} & a_{32} & a_{33} \end{vmatrix} = a_{11}\begin{vmatrix} a_{22} & a_{23} \\ a_{32} & a_{33} \end{vmatrix} - a_{21}\begin{vmatrix} a_{12} & a_{13} \\ a_{32} & a_{33} \end{vmatrix} + a_{31}\begin{vmatrix} a_{12} & a_{13} \\ a_{22} & a_{23} \end{vmatrix} \quad (15.17)$$

So, each determinant 2 × 2 is calculated as indicated above. Note that for the calculation of the determinant, the first element of a column is taken (it would be the same to use rows instead of columns) and it is multiplied by the submatrix resulting from removing from the matrix the elements located in the row and column to which that item belongs. In the first case, as we have taken the element a_{11}, located in row 1 and column 1, we will multiply this element by the determinant formed by the original matrix, removing row 1 and column 1 (leaving a determinant of 2 x 2). Second, we do the same thing, but change the sign of the element.

It is important to note that there are only determinants of those squared matrices.

Let's see some properties of determinants

- Interchanging two rows or two columns changes the sign of the determinant.
- A factor common to any row or column of a determinant can be factored out.
- If a column (or row) is changed by adding another column (or row), the value of the determinant does not change.
- $|AB| = |A||B|$. That is, the determinant of a matrix product is equal to the product of its determinants.

So far, we have seen how to calculate the determinants of square matrices 2 × 2 and 3 × 3, which, by the way, are the most interesting for us. However, to calculate the inverse of a matrix and for certain time series analyses, it is necessary to calculate determinants with a dimension greater than 3. Let us see, therefore, what is the general calculation method that, of course, includes the cases already commented. Although packs for numerical computing or statistical programs can easily calculate without so painful method.

First, we need to know some preliminary definitions.

Let $A = (a_{ij})$ be a square matrix of order K. We call the minor complement of an element $a_{i_0 j_0}$ to the determinant obtained by suppressing row i_0 and column j_0, without altering the order of the remaining. We will denote the complementary minor of the element $a_{i_0 j_0}$ and denote as $\alpha_{i_0 j_0}$. Let's see an example. If we have the matrix $A = \begin{pmatrix} a_{11} & a_{12} & a_{13} \\ a_{21} & a_{22} & a_{23} \\ a_{31} & a_{32} & a_{33} \end{pmatrix}$, the minor complementary of the element a_{11} will be the determinant remaining after removing the first row and column, namely, we will remove the colored numbers

$$\alpha_{11} = \begin{pmatrix} a_{11} & a_{12} & a_{13} \\ a_{21} & a_{22} & a_{23} \\ a_{31} & a_{32} & a_{33} \end{pmatrix} = \begin{vmatrix} a_{22} & a_{23} \\ a_{32} & a_{33} \end{vmatrix} = a_{22}a_{33} - a_{32}a_{23}$$

In the case of the element a_{23}, the complementary minor will be

$$\alpha_{23} = \begin{pmatrix} a_{11} & a_{12} & a_{13} \\ a_{21} & a_{22} & a_{23} \\ a_{31} & a_{32} & a_{33} \end{pmatrix} = \begin{vmatrix} a_{11} & a_{12} \\ a_{31} & a_{32} \end{vmatrix} = a_{11}a_{32} - a_{31}a_{12}$$

But in addition to the complementary minor, we need another element to be able to properly calculate a determinant. We call adjunct of the element $a_{i_0 j_0}$ of matrix A and we denote as $A_{i_0 j_0}$ to the product

$$A_{i_0 j_0} = (-1)^{i_0 + j_0} \alpha_{i_0 j_0} \qquad (15.18)$$

In the first of the above examples, the adjunct (*adj*) to the element a_{11} will be

$$A_{11} = (-1)^{1+1} \alpha_{11} = (-1)^2 \begin{vmatrix} a_{22} & a_{23} \\ a_{32} & a_{33} \end{vmatrix} = \begin{vmatrix} a_{22} & a_{23} \\ a_{32} & a_{33} \end{vmatrix}$$

And for the second case,

$$A_{23} = (-1)^{2+3} \alpha_{23} = (-1)^5 \begin{vmatrix} a_{11} & a_{12} \\ a_{31} & a_{32} \end{vmatrix} = -\begin{vmatrix} a_{11} & a_{12} \\ a_{31} & a_{32} \end{vmatrix}$$

With these both concepts we can already find a general expression for the determinant of any square matrix, which will be

$$|A| = \sum_{j=1}^{n} (-1)^{i+j} a_{ij}\alpha_{ij} = \sum_{j=1}^{n} a_{ij}A_{ij} \qquad (15.19)$$

In other words, it will be the sum over all j of each element of the i-th column, multiplied by its adjunct. Therefore, we can calculate a determinant using any column (and any row) and not just the first one.

Given a matrix A, the existence of inverse, denoted as A^{-1}, is a very important property of square matrices and can be defined as

$$AA^{-1} = A^{-1}A = I \qquad (15.20)$$

where I is the identity matrix.

The inverse can be calculated following the next steps:

1. Substitute every element of A by its adjunct.
2. Transpose the resultant matrix, obtaining the adjunct matrix A, denoted as Adj(A).
3. We divide every term of A by $|A|$.

For every non-singular matrix A, its inverse has the following properties:

$$(AB)^{-1} = B^{-1}A^{-1} \qquad (15.21)$$

$$(A')^{-1} = (A^{-1})' \qquad (15.22)$$

$$|A^{-1}| = |A|^{-1} \qquad (15.23)$$

If A is symmetric, A^{-1} is symmetric too. (15.24)

The inverse matrix is very important in statistics because it gathers the information about the co-joint dependence of several variables in a more complete way than the variance and covariance matrices.

We shall finish the chapter on square matrices with the concept of orthogonality. An orthogonal matrix is one representing a rotation in space, or the symmetry with respect to a plane. Let's suppose we rotate a vector $\vec{x}$ in the space by means of a linear transform with an orthogonal matrix C. We will obtain a new vector $\vec{y} = C\vec{x}$ which norm must be the same of $\vec{x}$, so

$$\vec{y}'\vec{y} = \vec{x}'C'C\vec{x} = \vec{x}'\vec{x} \tag{15.25}$$

Then, it must be true that $C'C = I$. From the definition of inverse, we know that $C^{-1}C = I$, then $C'C = C^{-1}C$, what means

$$C' = C^{-1} \tag{15.26}$$

What is the condition of orthogonality. Therefore, an orthogonal matrix must have rows (columns) formed by unitary orthogonal vectors, what means that these vectors form an orthonormal base of $\mathbb{R}^n$.

15.2.3 Inverses Generalized

We have obtained the inverse of a square matrix. However, it is interesting to obtain an inverse for rectangular matrices like this $A_{n\times p}$. We can obtain a Moore-Penrose (MP) inverse generalized matrix, denoted as A^- if we can have a matrix that obey this property,

$$AA^-A = A \tag{15.27}$$

Let's have a matrix like $A_{n\times p}$, such that $n > p$ and $rk(A) = p$. In this case the MP inverse generalized would be

$$A^- = \left(A'A\right)^{-1}A' \tag{15.28}$$

If we have $p > n$ and $rk(A) = n$, then the MP inverse matrix is

$$A^- = A'\left(AA'\right)^{-1} \tag{15.29}$$

However, in those cases where the rank of the matrix would be incomplete ($rk(A) < \min\{n,p\}$) the MP matrix will be

$$A^- = VD^{-1/2}U' \tag{15.30}$$

where $U_{n\times r}$ and $V_{p\times r}$ are orthogonal matrices and $D_{r\times r}$ is a diagonal matrix containing the eigenvalues (see below) different from zero. U and V are formed from the eigenvectors from eigenvalues different from zero and put in columns and can be obtained from these expressions

$$AA'U = UD \tag{15.31}$$

And

$$A'AV = VD \tag{15.32}$$

We will encounter this MP inverse generalized during the construction of the orthogonal projection.

We define an idempotent matrix A as a symmetric and squared matrix, verifying this property,

$$AA = A = A'A \tag{15.33}$$

Let's have a vector $\vec{y} \in \mathbb{R}^n$ and a matrix $X(n \times p)$ whose columns form a base of the subspace E_p. Then, the projection of the vector $\vec{y}$ over the space E_p is $A\vec{y}$, where A is symmetric and squared, idempotent, and $rk = p$, such $A = X(X'X)^{-1}X'$. Therefore, the orthogonal projection of $\vec{y}$ onto E_p gives rise to a vector $\vec{v} \in E_p$ by this transform.

$$\vec{v} = X\left(X'X\right)^{-1}X'\vec{y} = A\vec{y} \tag{15.34}$$

the only conditions to have $\vec{v} = A\vec{y}$ are (i) A be squared, (ii) idempotent, and (iii) symmetric.

15.3 Eigenvalues and Eigenvectors

These concepts are important in time series analysis, so we are going to look at them, albeit very briefly.

Let's have a square matrix $A = (a_{ij}) \in M_n(\mathbb{R})$ and $\lambda \in \mathbb{R}$. We say that λ is an eigenvalue (also known as characteristic value) of A when there exists a vector $\vec{x}$, denoted as eigenvector (or characteristic vector) that verifies the next relation,

$$A\vec{x} = \lambda\vec{x} \tag{15.35}$$

The dimensions of $\vec{x}$ must be obtained from the properties of the matrix product.

Matrices can be considered from a formal point of view (and not treated in this text), as functions of multidimensional vector spaces. To better understand what an eigenvector and an eigenvalue are, let's think about a one-dimensional function *f*. An eigenvector (λ) will have an eigenvector ($\vec{v}$) associated with it when the application of *f* over $\vec{v}$ gives rise to the eigenvector $\vec{v}$ multiplied by a constant that is λ. Namely

$$f(x)\vec{v} = \lambda\vec{v} \tag{15.36}$$

The matrix definition of eigenvectors/eigenvalues can be written like this, after passing the right part of the equality to the left and taking out the common factor

$$(A-\lambda I)\vec{x} = 0 \tag{15.37}$$

This is important for the calculation of the eigenvalues, because for this condition to be fulfilled in a non-trivial way (that is, for $\vec{x}$ not to be 0), it is necessary that the matrix $(A - \lambda I) = 0$, namely, it would be a singular matrix and, therefore, with determinant equals to zero, so

$$|A-\lambda I| = 0 \tag{15.38}$$

This expression is called the characteristic equation.

Evidently, for every eigenvector $\vec{x}$, also $a\vec{x}$, where $a \in \mathbb{R}$, will be another eigenvector with the same eigenvalue. To reduce this indeterminacy, we impose that eigenvectors have unitary norm.

We will better strengthen these concepts with an example. Suppose we have the matrix $A = \begin{pmatrix} 1 & 1 \\ 0 & 4 \end{pmatrix}$ for which we want to find the eigenvalues and eigenvectors. The unitary matrix for the dimension 2×2 is $I = \begin{pmatrix} 1 & 0 \\ 0 & 1 \end{pmatrix}$. For the matrix relationship $A - \lambda I$ we have

$$\begin{pmatrix} 1 & 1 \\ 0 & 4 \end{pmatrix} - \lambda \begin{pmatrix} 1 & 0 \\ 0 & 1 \end{pmatrix} = \begin{pmatrix} 1 & 1 \\ 0 & 4 \end{pmatrix} - \begin{pmatrix} \lambda & 0 \\ 0 & \lambda \end{pmatrix}$$

And subtracting both matrices, we have

$$\begin{pmatrix} 1-\lambda & 1 \\ 0 & 4-\lambda \end{pmatrix}$$

But above we have seen that for the existence of eigenvalues to be fulfilled, this matrix must be singular and, therefore, its determinant equals to zero, that is

$$\begin{vmatrix} 1-\lambda & 1 \\ 0 & 4-\lambda \end{vmatrix} = 0$$

And expanding the determinant we will obtain the following polynomial:

$$(1-\lambda)(4-\lambda) - 0 = 0$$

Operating

$$4 - 5\lambda + \lambda^2 = 0$$

This polynomial is called the characteristic polynomial of the matrix. Its resolution, using in this case the expression of the solution for second-degree linear equations, which is

$$\lambda = \frac{-b \pm \sqrt{b^2 - 4ac}}{2a} = \frac{5 \pm \sqrt{5^2 - 16}}{2} \left\{ \begin{matrix} 1 \\ 4 \end{matrix} \right.$$

Therefore, these two values will be the eigenvalues of the matrix. Each of them will have an associated eigenvector. We go to find the eigenvector associated with the eigenvalue $\lambda = 1$. The other follows the same process. This vector can be $\vec{x} \in M_{1\times2}(\mathbb{R})$, if precedes A in the product, or $\vec{x} \in M_{2\times1}(\mathbb{R})$ or follows A. In the last case, we will have from this equation

$$(A-\lambda I)\vec{x} = \vec{0} \tag{15.39}$$

In this case, $\vec{0} = \begin{pmatrix} 0 \\ 0 \end{pmatrix}$. Substituting for $\lambda = 1$

$$\left(\begin{pmatrix} 1 & 1 \\ 0 & 4 \end{pmatrix} - 1 \begin{pmatrix} 1 & 0 \\ 0 & 1 \end{pmatrix} \right) \begin{pmatrix} x_1 \\ x_2 \end{pmatrix} = \begin{pmatrix} 0 \\ 0 \end{pmatrix}$$

Operating

$$\begin{pmatrix} 1-1 & 1 \\ 0 & 4-1 \end{pmatrix} \begin{pmatrix} x_1 \\ x_2 \end{pmatrix} = 0 \rightarrow \begin{pmatrix} 0 & 1 \\ 0 & 3 \end{pmatrix} \begin{pmatrix} x_1 \\ x_2 \end{pmatrix} = \begin{pmatrix} 0 \\ 0 \end{pmatrix}$$

Then, we multiply the 2×2 matrix by the 2×1 vector, given a vector

$$\begin{pmatrix} 0x_1 + x_2 \\ 0x_1 + 3x_2 \end{pmatrix} = \begin{pmatrix} 0 \\ 0 \end{pmatrix}$$

And therefore $x_2 = 0$ meanwhile x_1 can be whatever real number. Obviously, the simplest vector will be $\vec{x} = \begin{pmatrix} 1 \\ 0 \end{pmatrix}$.

The eigenvectors of a square matrix are those vectors whose direction is not modified after a transform by the matrix. For every eigenvalue we shall have one eigenvector. However, it is not unusual to have one (or more) eigenvalues repeated, which is called multiplicity.

Among the most relevant properties of eigenvalues of a square matrix, we have

- If we have $A\vec{x} = \lambda\vec{x}$, then $A^m\vec{x} = \lambda^m\vec{x}$. And, specifically, $A^{-1}\vec{x} = \lambda^{-1}\vec{x}$.
- The eigenvalues of a matrix and its transpose are the same.
- $tr(A) = \sum_{i=1}^{n} \lambda_i$.
- $|A| = \prod_{i=1}^{n} \lambda_i$. See note.[2]

A very interesting kind of square matrices are those which are symmetric, so $A = A'$. These have the important properties that (1) eigenvalues are always real and (2) the eigenvectors are orthogonal. We can say that the eigenvalues of a symmetric matrix depict the magnitudes of the main axis of the ellipsoid centered at origin and defined by the extreme of the eigenvectors, meanwhile the eigenvectors show the directions of these main axes.

Suggested Reading

A rigorous and detailed description of these theories and concepts can be found in the next references.

Arfken GB, Weber HJ. Determinants and matrices. In: Mathematical methods for physicists. 6th ed. Amsterdam: Elsevier; 2005. ISBN-13978-0-12-059876-2.

Hassani S. Planar and spatial vectors. In: Mathematical methods. For students of physics and related fields. 2nd ed. New York: Springer; 2009. ISBN: 978-0-387-09503-5.s.

Peña D. Algebra matricial. In: Análisis de datos multivariantes. Madrid: McGraw-Hill; 2002. ISBN 84-481-3610-1.

[2]The use of the symbol $\prod_{n}^{i=1} x_i = x_1 x_2 \ldots x_n$, is similar to the well-known $\sum_{n}^{i=1} x_i = x_1 + x_2 + \ldots + x_n$, but multiplying the elements instead of summing.

16 Brief Reminder of Calculus

16.1 Introduction

Calculus is the branch of Mathematics dedicated to the study of differentiation and integration. Any rigorous treatment of Electromagnetism (EM), Biophysics and Electroencephalography requires both tools.

We will initially consider functions in $\mathbb{R}$, that is, functions that have only one variable. These functions are usually written as *f(x)*. However, many of the relevant functions in EM include functions defined in several dimensions.

16.2 Functions

A function is an application of a subset U of $\mathbb{R}$ on R and is written $f: U \subset \mathbb{R} \to \mathbb{R}$. The subset U where the function exists is known as the domain of definition. This means that for every $x \in U$ exists some $y \in \mathbb{R}$ such that $f(x) = y$. Every function, therefore, will have a range, which is the subset $U \subset \mathbb{R}$ and an image that is the set formed by all $y \in \mathbb{R}$.

The graph of a function is the set of all pairs belonging to $\mathbb{R}^2$ in the form $(x, f(x))$.

Suppose a function $f: U \subset \mathbb{R} \to \mathbb{R}$. We say that f tends to l or has a limit, when the function approaches l closer when x tends to a

$$\lim_{x \to a} f(x) = l$$

This definition is very important because it allows us to define the continuity of functions. In fact, a function $f(x)$ is continuous in a when

$$\lim_{x \to a} f(x) = f(a)$$

Suppose that the following functions $f(x)$ and $g(x)$ are well defined and continuous. In this case, we say that the following operations exist for the whole range of definition, whose results are, in turn, new functions, which we call $h(x)$:

$$\begin{gathered} h(x) = f(x) + g(x) \\ h(x) = f(x) - g(x) \\ h(x) = f(x) g(x) \\ h(x) = \frac{f(x)}{g(x)}, \text{where } g(x) \neq 0 \forall x \end{gathered} \tag{16.1}$$

So far, we have dealt with one-variable functions. However, there are many important functions that depend on more than one variable. Suppose a function that depends on two variables, $f: \mathrm{U} \subset \mathbb{R}^2 \to \mathbb{R}$, which we call x and y, whose image is an element $z \in \mathbb{R}$. This function is written as $f(x, y) = z$. We can encounter functions of three or even more variables.

16.3 Derivatives and Differentials

The concept of continuity of a function is very important because it is related to the concept of derivative. The function $f: U \subset \mathbb{R} \to \mathbb{R}$ can be

J. Pastor, *Quantitative EEG Practical Guide*, https://doi.org/10.1007/978-3-032-05457-9_16

derived at point a (and is indicated as $f(a)$ or $\frac{df(a)}{dx}$) when this limit exists and is finite

$$f'(a) = \lim_{x \to a} \frac{f(x) - f(a)}{x - a}$$

Changing the variable $x = a + h$, we can rewrite the definition as

$$f'(a) = \lim_{h \to 0} \frac{f(a+h) - f(a)}{h} \quad (16.2)$$

This means that when $x \to a$ then, obviously $h \to 0$.

For every point $a \in U$ we can define the limit at both sides, e.g., left to a or a^- (when $x < a$) and right or a^+ (when $x > a$). Both limits, therefore, can be defined as

$$f^-(a) = \lim_{\substack{x \to a^- \\ x<a}} \frac{f(x) - f(a)}{x - a}; f^+(a) = \lim_{\substack{x \to a^+ \\ x>a}} \frac{f(x) - f(a)}{x - a}$$

A condition to the existence of limit is $f^-(a) = f^+(a)$.

In general, the functions found in electroencephalography can be considered as continuous and derivable.

But, what is the meaning and utility of derivatives? Let $f: U \subset \mathbb{R} \to \mathbb{R}$ be a differentiable function at point a. A straight line (called secant, s) is characterized by passing through two points P and Q. Suppose that in a plane, those points are $(a, f(a))$ and $(a + h, f(a + h))$. The equation of a straight line is: $y = mx + b$, where m is the slope and b is the ordinate at the origin. To find m and b, we can substitute the values of the two points, so that we will obtain a system of two equations with two unknowns $f(a) = ma + b$ and $f(a + h) = m(a + h) + b$. Solving for the slope, which is what interests us now, we will obtain the expression $m = \frac{f(a+h) - f(a)}{h}$. Therefore, the slope of the secant is indeed the initial increment ratio (Fig. 16.1a).

If we reduce h, making that both points a and $a + h$ approach, the length of the secant will be shorter. At the limit we obtain a straight line that only contacts the function $f(x)$ at one point ($a = x$), being in fact no more a secant but a tangent (t). And the slope of this tangent is precisely the derivative of $f(a)$. If we repeat this process for all the points $x \in U$ we will have a function derived from $f(x)$ (whence the name of derivative) that indicates the slopes of tangents at every point and, therefore the instant change of the function.

Let's define the differential of a function in point a as the linear function that corresponds the increment (or differential) of the variable x to the increment of the tangent measured at a (Fig. 16.1b). From this figure, we note that dy is the increment of tangent of $y = f(x)$ at $x = a$ for an increment dx.

We know that the derivative is the slope of the tangent, so we have

$$\frac{dy}{dx} = \tan\alpha = f'(a) \Rightarrow dy = f'(a)dx \equiv f'(a)h$$

When h decreases, we can approximate the value of Δy by means of dy, because,

$$\lim_{h \to 0} \frac{\Delta y}{dy} = \lim_{h \to 0} \frac{f(a+h) - f(a)}{f'(a)h} = \frac{1}{f'(a)} \lim_{h \to 0} \frac{f(a+h) - f(a)}{h} = \frac{1}{f'(a)} f'(a) = 1$$

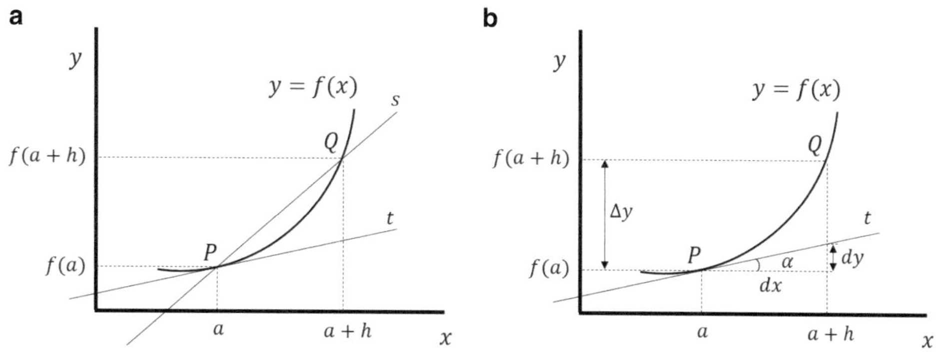

Fig. 16.1 Derivative and differential from continuous function. (**a**) A secant line (s) going through points P and Q is gradually converted into a tangent (t) when h tends to zero. At this point the line is the instant derivative of the function $y = f(a)$. (**b**) Differential of the function $y = f(x)$ at point a

So, $\Delta y = f(a+h) - f(a) \Longrightarrow f(a+h) = f(a) + \Delta y \equiv f(a) + dy \equiv f(a) + f(a)h$

And taking now $h = x - a$ we can approximate $f(x)$ by the tangent to the function at a by means of

$$f(x) = f(a) + f'(a)(x-a) \qquad (16.3)$$

This expression is directly related to the concept of differential introduced in Chap. 14. Linear algebra: vector analysis, Eq. 14.12.

Derivatives are extremely useful for analysis of functions because they allow us to identify points of maximum rate of change and points where the function changes its behavior, for example, from increase to decrease or changing from concave to convex.

Considering that derivatives are functions, we can repeat the same operations described above. In fact, suppose we have two derivable functions $f(x)$ and $g(x)$ in a range U. Then, addition, subtraction, product, and division (granted $g'(x) \neq 0$) are derivable and it's verified.

$$\begin{aligned}
(f+g)'(x) &= f'(x) + g'(x) \\
(f-g)'(x) &= f'(x) - g'(x) \\
(fg)'(x) &= f'(x)g(x) + f(x)g'(x) \qquad (16.4) \\
\left(\frac{f}{g}\right)'(x) &= \frac{f'(x)g(x) - f(x)g'(x)}{(g(x))^2}
\end{aligned}$$

Another very important property of differentiation is the so-called chain rule. To explain it, we need to know first what a composite function is. Suppose two functions, one of them $f: A \to B$ and another $g: B \to C \subset \mathbb{R}$. From the first of these definitions, we know that $\forall x \in A, \exists\, y \in B$ *such* $f(x) = y$. From the second, we know that $\forall y \in B, \exists\, z \in C$ *such* $g(y) = z$.

We will call the composite function of f over g the function $g \circ f : A \to C$ that verifies $\forall x \in A, \exists\, z \in C$ *such* $(g \circ f)(x) = z$. Considering the previous definitions, the composition of f and g implies that $(g \circ f)(x) = z \Rightarrow f(g(y)) = z$.

So, the derivative of a composite function is

$$(f \circ g)'(x) = g'(f(x)) \cdot f'(x) \qquad (16.5)$$

Next, we enlist some of the most useful and simple derivatives (Table 16.1).

16.3.1 Successive Derivatives

As we have seen, every differentiable function is continuous. Therefore, we can calculate the derivative of this new function.

If we substitute f by f' in the definition of derivative (Eq. 16.2) we will obtain

$$f''(x) = \lim_{h \to 0} \frac{f'(a+h) - f'(a)}{h} \qquad (16.6)$$

where $f''(x)$ is the second derivative of $f(x)$. We can also find the second derivative written as $\frac{d^2 f(x)}{dx^2}$.

Table 16.1 Derivatives

$\frac{d}{dx}c = 0$
$\frac{d}{dx}x = 1$
$\frac{d}{dx}\left[cf(x)\right] = cf'(x)$
$\frac{d}{dx}x^n = nx^{n-1}, n \in \mathbb{N}$
$\frac{d}{dx}\sin x = \cos x$
$\frac{d}{dx}\cos x = -\sin x$
$\frac{d}{dx}\tan x = 1 + (\tan x)^2$
$\frac{d}{dx}\ln x = \frac{1}{x}$
$\frac{d}{dx}\ln f(x) = \frac{1}{f(x)}f'(x)$
$\frac{d}{dx}e^x = e^x$
$\frac{d}{dx}e^{f(x)} = f'(x)e^x$
$\frac{d}{dx}a^x = a^x \ln a$
$\frac{d}{dx}a^{f(x)} = f'(x)a^x \ln a$

This process can be repeated for some functions to obtain derivatives of high order. Evidently, not all the functions can have an arbitrary number of derivatives, but there are types of functions (e.g., exponentials or trigonometric) that can be derived even infinitely.

16.3.2 Partial Derivatives

If we have a function like this $f : \mathbb{R}^2 \rightarrow \mathbb{R}$ we can consider the change of the function along one of the axes, maintaining the other variable constant. For a function $f(x, y)$, we can obtain the partial derivatives for x and y according to these definitions

$$\frac{\partial f(x,y)}{\partial x} = \lim_{h \to 0} \frac{f(x+h,y) - f(x,y)}{h} \quad (16.7)$$

$$\frac{\partial f(x,y)}{\partial y} = \lim_{h \to 0} \frac{f(x,y+h) - f(x,y)}{h} \quad (16.8)$$

Using a similar approach to the above-described, we can also obtain the second (and high-order) partial derivatives.

16.4 Integration

Let's have a continuous function defined in an interval $f : I \rightarrow \mathbb{R}$. We can say that F is the primitive function of f on I when F is derivable and $F' = f$ in I. The primitive is also known as antiderivative.

By definition, if f is a continuous function in I, any primitive function will verify

$$F(x) = \int f(x)dx \quad (16.9)$$

This theorem therefore relates derivatives and integrals so that the derivative of a primitive is the function f, while the integral of f is the primitive F. Therefore, derivation and integration are reciprocal operations.

If F is a primitive of f, then G, defined as $G = F + c$, where c is a constant, will also be a primitive of f. This is so because, as we saw above, the derivative of a constant is zero. We have seen in the definition that $F' = f$, so that $G' = F' + c' = F' + 0 = F'$.

The set of primitives of a continuous function is denoted by $\int f$ *or* $\int f(x)dx$ and will be equal to $\int f(x)dx = \{F + c : c \in \mathbb{R}\}$ and, with an obvious abuse of language, it is usually written as $\int f(x)dx = F(x) + c$.

We must not confuse these two expressions $\int f(x)dx$ and $\int_a^b f(x)dx$ since the former denotes a set of functions, the set of primitives of f, while the second is a real number that corresponds to the integral of f on the interval [a, b]. To distinguish them, the first is called an indefinite integral, while the second is called a definite integral.

16.4.1 Riemann's Integral

A rigorous demonstration will not be made, but we will justify why the definite integral between two points measures the area under the function.

Suppose a function f, defined on the interval [a,b], such that f: [a, b] → $\mathbb{R}$. We divide the entire interval into smaller, m intervals (partition), which are, in fact, rectangles. We set the height of the rectangles to the highest value of $f(x)$ in the interval considered ($f_U(x_i)$). All of these rectangles will lie above the function. We can do the same for the same partition, but defining the height of the rectangles for the minimum value of the function in the interval considered ($f_L(x_i)$) (Fig. 16.2).

We can define the upper sum of the rectangles that are above the function, which we define in this way

$$U = \sum_{i=1}^{m} f_U(x_i)(x_{i+1} - x_i) \qquad (16.10)$$

And for the partition rectangles below the function, the lower sum is defined:

$$L = \sum_{i=1}^{m} f_L(x_i)(x_{i+1} - x_i) \qquad (16.11)$$

As the width of the rectangles ($\Delta x = x_{i+1} - x_i$) becomes smaller and smaller, both sums will get closer, so that

$$\lim_{\Delta x \to 0} U = \lim_{\Delta x \to 0} L = \int_a^b f(x)dx \qquad (16.12)$$

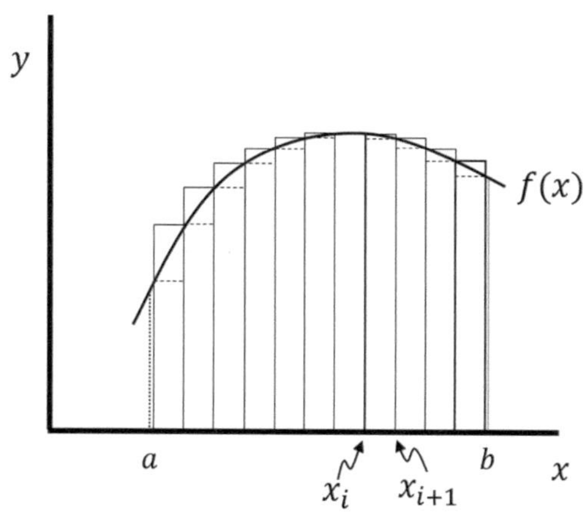

Fig. 16.2 Riemann's integral concept. The function is enclosed by a partition whose amplitude is higher, and the same partition with amplitude lower than the function

16.4.2 Properties of Integral

Let I be an interval, f and g two continuous functions on I, and a and b two real numbers not simultaneously zero. Then, an important property is

$$a\int f(x)dx + b\int g(x)dx = \int (af(x) + bg(x))dx \qquad (16.13)$$

Formulas for integrals are immediately checked by differentiation. Solving integrals in this way is called direct integration. Next, we show some of the most useful integrals (Table 16.2).

However, it is usual that we cannot solve a problem by direct integration. In this case, we must recourse to other strategies.

Integration by Part This is a powerful integration method when the primitive cannot be found directly. Let I be an interval, f and g two continuous functions with continuous derivatives on I. Then

$$\int f(x)g'(x)dx = f(x)g(x) - \int f'(x)g(x)dx \qquad (16.14)$$

Table 16.2 Integrals

$\int x^n dx = \frac{1}{n+1}x^{n+1} + C$
$\int \frac{dx}{x} = \ln\lvert x\rvert + C$
$\int e^x dx = e^x + C$
$\int a^x dx = \frac{a^x}{\ln a} + C$
$\int \sin x dx = -\cos x + C$
$\int \cos x dx = \sin x + C$
$\int \tan x dx = \ln\left\lvert\frac{1}{\cos x}\right\rvert + C$

Therefore, if we use these symbols (which do not change the mathematics, only simplify the formulae) $u = f(x)$; $dv = g'(x)dx$, by derivation and integration respectively, we will have $du = f(x)dx$; $v = g(x)$. With these symbols, the form of integration by parts is written

$$\int udv = uv - \int vdu$$

Integration by Change of Variable The basic idea of this method is to replace a function whose integration is complex, for another that is easier to integrate. The main problem with the method is that there are no fixed rules, not even guidelines for how to make the change, so, unlike the method of integration by parts, it depends a lot on the mathematical intuition of each one.

There are some types of integrals of great importance in various fields of Physics, so we are going to see them in a little more detail.

Line integral. Suppose a vector field $\vec{A}(t)$ and a trajectory $d\vec{s}$, from the initial point p_i, to the final point p_f. At each point of the trajectory, we compute the dot product $\vec{A}\cdot d\vec{s}$. When we add all these contributions, obtaining the so-called line integral (Γ)

$$\Gamma = \int_{p_i}^{p_f} \vec{A}\cdot d\vec{s} = \int_C \vec{A}\cdot d\vec{s} \qquad (16.15)$$

where C at the second integral represents the trajectory. When the line integral lies on a closed curve, it is called a circulation and can be written as

$$\oint_C \vec{A}\cdot d\vec{s} \qquad (16.16)$$

We can extend this same idea to a two-dimensional space $\mathbb{R}^2$, that is, to a surface. In this case, we will have to use the so-called double integral. Suppose a surface, S, which we can define as $S = \{(x, y) \in \mathbb{R}^2; a \le x \le b; c \le y \le d\}$. In this case, instead of using a lineal differential element $d\vec{s}$, we use a surface differential element, which is defined as $d\vec{a} = dx\vec{i} + dy\vec{j}$. We calculate the value of the dot product of the vector field $\vec{A}$ with the surface differential element in this way $\vec{A}\cdot d\vec{a}$ and then integrate over the segments $[a, b]$ for the x axis and $[c, d]$ for the y axis. This surface integral ($\varnothing$) will be

$$\varnothing = \int_a^b \int_c^d \vec{A}\cdot d\vec{a} = \int\int_S \vec{A}\cdot d\vec{a} \qquad (16.17)$$

Finally, we can generalize more, using the three dimensions of space, which will allow us to find the volume integral (τ). Suppose we have in this case a scalar function $u(\vec{r})$ defined on a volume such that $V = \{(x, y, z) \in \mathbb{R}^3; a \le x \le b; c \le y \le d; e \le z \le f\}$ and we calculate the product of the field with each of the points that define the volume differential element, which is defined as follows $dV = dxdydz$. In this case, the integral of volume, which is a triple integral, will be defined in this way

$$\tau = \int_a^b \int_c^d \int_e^f udV = \int\int\int_V udV$$

There are two very important theorems in vector analysis, which make use of multiple integrals like the ones we have seen.

Stokes theorem. Relates the circulation of a scalar function to the surface integral of its curl. Namely

$$\oiint_C \vec{A}\cdot d\vec{s} = \iint_S \nabla\times\vec{A}\cdot d\vec{a} \qquad (16.18)$$

Gauss-Ostrogradsky theorem. Relates the surface integral of a vector field to the volume integral of its divergence, thus

$$\oiint_S \vec{A}\cdot d\vec{a} = \iiint_S (\nabla\cdot A)dV \qquad (16.19)$$

There are some functions of special relevance because the frequency of appearing in several fields or by its relevance. Let's see some of them.

16.5 Trigonometric Functions

See Appendix 3_Trigonometry, for a reminder of these functions.

These functions are easily derivable, being their derivatives

$$(\sin x)' = \cos x, \quad (\cos x)' = -\sin x, \quad (\tan x)' = \cos^{-2} x, \qquad (16.20)$$

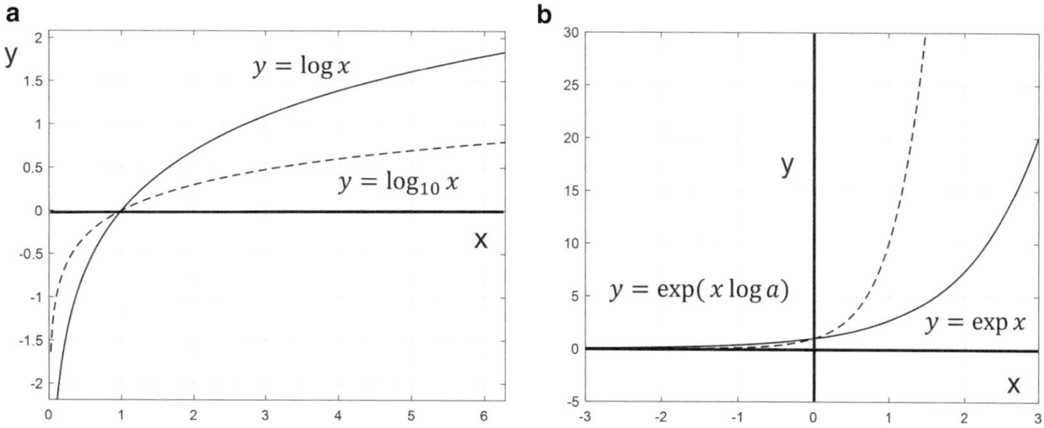

Fig. 16.3 Logarithmic and exponential functions

Some properties of harmonic functions are

- $\sin^2 x + \cos^2 x = 1$
- $\sin(x + y) = \sin x \cos y + \cos x \sin y$
- $\cos(x + y) = \cos x \cos y - \sin x \sin y$

We can obtain the inverse of these functions (arcsin, arccos, and arctan), all of which have derivatives

$$(\arcsin x)' = \frac{1}{\sqrt{1-x^2}}, \quad (\arccos x)' = -\frac{1}{\sqrt{1-x^2}}, \quad (\arctan x)' = \frac{1}{1-x^2},$$

16.6 Exponential and Logarithmic Functions

16.6.1 Natural or Neperian Logarithm and Exponential

Neperian logarithm function (log, Fig. 16.3a) is defined as $log : (0, +\infty) \to \mathbb{R}$,

$$\log x = \int_1^x \frac{1}{t} dt \tag{16.21}$$

Therefore, for every $x \in (0, +\infty)$ and for every $x > 0$, we have

$$\log' x = \frac{1}{x} \tag{16.22}$$

This function has very important properties. For any pair of positive real numbers, x and y, we have

- $\log(xy) = \log x + \log y$
- $\log \frac{x}{y} = \log x - \log y$
- $\log x^n = n \log x, \ for\ n \in N$

The inverse function of logarithm is the natural exponential (Fig. 16.3b), defined as $exp : \mathbb{R} \to (0, +\infty)$. Therefore, the expressions

$$\exp x = y \text{ and } \log y = x \tag{16.23}$$

are equivalent.

The derivative of the exponential is

$$\exp' x = \exp x$$

For every $x \in \mathbb{R}$

From this definition, we have the following properties:

- $\exp(x + y) = \exp x \exp y$
- $\exp 1 = e$. This equation is the definition of the real irrational number $e = 2.718218\ldots$
- $\exp x = e^x$.

16.6.2 Other Exponential and Logarithm Functions

Let's have $a > 0$. For every real number, it's defined the exponential function of base a to the function $f:\mathbb{R}\to(0,+\infty)$, as

$$f(x)=a^x=e^{x\log a} \tag{16.24}$$

This function has the following properties:

- $(a^x)' = a^x \log a$
- $a^0 = e^0 = 1$
- $a^1 = e^{\log a} = a$
- $a^{x+y} = a^x a^y$
- $(a^x)^y = a^{xy}$

If $a > 0$ and $a \neq 1$, its inverse function is the logarithm function of base a, such that f : $(0, +\infty) \to R$ defined as

$$\log_a x = \frac{\log x}{\log a} \tag{16.25}$$

The derivative of this function is $\left(\log_a x\right)' = \dfrac{1}{x\log a}$

16.6.3 Power Function

The function f: $(0, +\infty) \to \mathbb{R}$ defined as

$$f(x)=x^a=e^{a\log x} \tag{16.26}$$

is called power function of exponent a.

The derivative of this function is

$$\left(x^a\right)' = ax^{a-1} \tag{16.27}$$

16.6.4 Hyperbolic Functions

The functions hyperbolic sine (sinh), hyperbolic cosine (cosh), and hyperbolic tangent (tanh) are defined respectively as (Fig. 16.4)

$$\sinh x = \frac{e^x - e^{-x}}{2}, \quad \cosh x = \frac{e^x + e^{-x}}{2}, \quad \tanh x = \frac{e^x - e^{-x}}{e^x + e^{-x}} \tag{16.28}$$

The derivatives of these functions are

$$\left(\sinh x\right)' = \cosh x, \quad \left(\cosh x\right)' = \sinh x, \quad \left(\tanh x\right)' = \frac{1}{\cosh^2 x}, \tag{16.29}$$

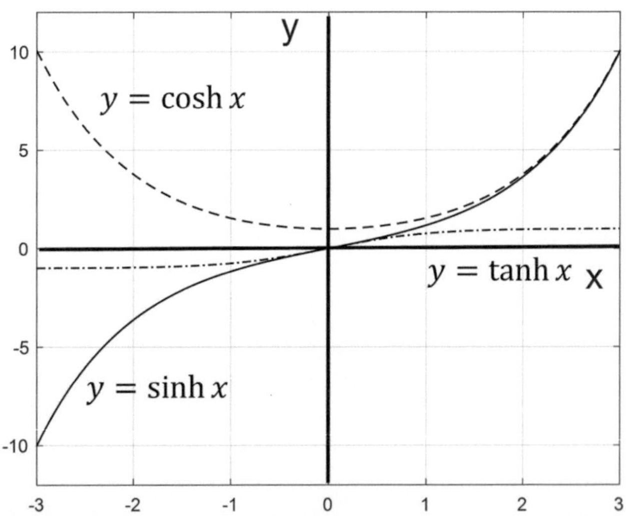

Fig. 16.4 Hyperbolic functions

Suggested Reading

A rigorous and detailed description of these theories and concepts are found in the next references.

Edwards CH, Penney DE. Vector analysis. In: Elementary differential equations. 6th ed. London: Prentice Hall. ISBN 0-13-239730-7.

Hassani S. Differentiation. In: Mathematical methods. For students of physics and related fields. 2nd ed. New York: Springer; 2009. ISBN: 978-0-387-09503-5.

Kuttler K. Functions. In: Calculus. Theory and applications, vol. I. London: World Scientific; 2011. ISBN-10 981-4324-26-4.

Kuttler K. Some important special functions. In: Calculus. Theory and applications, vol. I. London: World Scientific; 2011. ISBN-10 981-4324-26-4.

Kuttler K. The integral. In: Calculus. Theory and applications, vol. I. London: World Scientific; 2011. ISBN-10 981-4324-26-4.

Marsden JE, Tromba A. Double and triple integrals. In: Vector calculus. 6th ed. WH Freeman & Co Ltd.; 2011. ISBN 978-1429215084.

Differential Equations 17

17.1 Introduction

The general resolution of differential equations clearly exceeds the purpose of these notes, but first- and second-order equations will be used at various times, so it is important to know the most common resolution methods.

Since the derivative $dx/dt = f(t)$ of a function is the rate at which a magnitude $x(t)$ changes with respect to time t, it is natural that equations including derivatives are frequent to describe the changes. An equation that includes an unknown function and one or more of its derivatives is called a differential equation, for example.

$$f''(t) + f'(t) + f(t) = F(t) \tag{17.1}$$

This is a second-order differential equation. The order of a differential equation is defined by the highest derivative.

17.2 Mathematical Models and Differential Equations

Differential equations are the most common language (although not the only one) in which we can build mathematical models of physical or biological systems. These models are essential to propose experiments that can later be confirmed or discarded. Let's see an example that, from a biological system, we can find a differential equation that explains the behavior of the system.

Suppose a population susceptible to infection by SARS-CoV2. The number of infected patients is a function of time, that is $N = N(t)$. Contact between patients and non-infected people will increase the number of patients at a rate that we call α, i.e., the probability of infected patients/unit of time for a given person. At a delayed time, let's say Δt, the number of infected will be $N(t + \Delta t) = N + \alpha N \Delta t$, that is, the patients that were present initially, plus those that have appeared by contact. The difference between the two populations is

$$N(t + \Delta t) - N(t) = \alpha N \Delta t \tag{17.2}$$

But the left part of the equality is an increment, that is, it is $\Delta N = N(t + \Delta t) - N(t)$, so we can put

$$\Delta N = \alpha N \Delta t \tag{17.3}$$

So, solving for αN, we will have

$$\frac{\Delta N}{\Delta t} = \alpha N \tag{17.4}$$

This is an incremental ratio, but it's not yet a derivative. To do this, we must obtain the limit for increasingly smaller times. Exactly as we did when defining the derivatives, that is

$$\frac{dN}{dt} = \lim_{\Delta t \to 0} \frac{\Delta N}{\Delta t} \tag{17.5}$$

So the differential equation that governs the behavior of a group of patients will be

J. Pastor, *Quantitative EEG Practical Guide*, https://doi.org/10.1007/978-3-032-05457-9_17

$$\frac{dN}{dt} = \alpha N \quad (17.6)$$

In this way, knowing how the number of infected people is going to change over time, we can always know how many there will be at any desired time. In fact, if we know this equation, we can obtain the explicit function of $N(t)$. The process by which we obtain this function is known as solving differential equations. This is what we will see in the next section.

17.3 Solving First-Order Differential Equations

The equations involving the function $y = f(x)$ and its first derivative are equations of the first degree and are written as

$$\frac{dy}{dx} = f(x) \quad (17.7)$$

This expression is rearranged as

$$dy = f(x)dx \quad (17.8)$$

We can integrate both sides of the equality, that is

$$\int dy = \int f(x)dx \quad (17.9)$$

So (see Chap. 16 Brief Reminder of Calculus)

$$y = \int f(x)dx + c \quad (17.10)$$

It will be the general solution of the equation. To do this, we obtain the integral, according to the rules seen above. The constant c is arbitrary, and its function is to obtain a particular result when the initial conditions of the problem are known. That is, an equation describes the dynamics of the system, but the specific value obtained will depend on the moment in which the system is considered.

Therefore, we can say that $y = G(x) + c$, where $G(x)$ is the primitive function of $f(x)$.

To obtain a particular solution, we must know the initial conditions, usually given as $y(x_0) = y_0$.

This resolution method is called separation of variables and consists of placing the variables that depend on y on different sides of the equality and those that depend on x, on the other.

Let's see an example with the SARS-CoV2-infected model described above. If we have an initial number of patients N_0 and an infection rate of α infected/person. Using the equation obtained above, we proceed to the separation of variables. We must try to put on the same side the expressions that involve N, and on the other side those that involve t, so that we will have

$$\frac{dN}{dt} = \alpha N \rightarrow \frac{dN}{N} = \alpha dt \quad (17.11)$$

We can integrate both sides of this equation,

$$\int \frac{dN}{N} = \int \alpha dt + \ln c \quad (17.12)$$

In Chapter 16, we observed that an indefinite integral has an associated constant. In this case, we have chosen $\ln c$ as constant, where $c \in R$ and $c > 0$. The reason for this choice will be evident right below.

Let's see each of these integrals separately. From tables, it is easy to obtain

$$\ln N = \int \frac{dN}{N}$$

The integral on the right-hand side is even simpler, since the constant α can be taken out of the integral, leaving

$$\int \alpha dt = \alpha \int dt$$

where $\int dt = t$. Therefore, we obtain the equation

$$\ln N = \alpha t + \ln c \quad (17.13)$$

To eliminate natural logarithms at both sides of equal, we take exponents

$$\exp(\ln N) = \exp(\alpha t + \ln c) = \exp(\alpha t)\exp(\ln c) \quad (17.14)$$

And remembering that exponential and logarithm are inverse functions, we obtain $\exp(\ln N) = N$ and $\exp(\ln c) = c$, where c is a constant and this is the reason why we initially choose a logarithm in the computation of the primitive. Finally, we have

$$N(t) = ce^{\alpha t} \quad (17.15)$$

An especially important equation is the first-order linear differential equation of this type

$$\frac{dy}{dx}+P(x)y(x)=Q(x) \qquad (17.16)$$

Where $P(x)$ and $Q(x)$ are continuous functions. To solve this equation, let's make the change of variable regarding $P(x)$

$$\rho(x)=\exp\int P(x)dx \qquad (17.17)$$

Now we multiply both sides of the equation by this term

$$\rho(x)\frac{dy}{dx}+P(x)\rho(x)y(x)=\rho(x)Q(x)$$

But, from the differentiation property of a product of functions (see Eq 16.5 in Chap. 16), we can see that

$$\frac{d}{dx}\rho(x)y(x)=\rho(x)\frac{dy(x)}{dx}+y(x)\frac{d\rho(x)}{dx}$$

To make the process clearer, we will use the definition of $\rho(x)$

$$\frac{d}{dx}e^{\int P(x)dx}y(x)=e^{\int P(x)dx}\frac{dy(x)}{dx}+y(x)\frac{d}{dx}e^{\int P(x)dx}$$

Now we apply the property of differentiation of exponential functions

$$\frac{d}{dx}e^{\int P(x)dx}=P(x)e^{\int P(x)dx}=P(x)\rho(x)$$

Obtaining

$$\rho(x)\frac{dy(x)}{dx}+y(x)P(x)\rho(x)=\rho(x)Q(x)$$

So actually, the above left-hand side of the equation can be written as

$$\rho(x)\frac{dy(x)}{dx}+y(x)P(x)\rho(x)=\frac{d}{dx}\rho(x)y(x)$$

And

$$\frac{d}{dx}\rho(x)y(x)=\rho(x)Q(x)$$

And this is important because it allows us to integrate directly through the separation of variables

$$d(\rho(x)y(x))=\rho(x)Q(x)dx \qquad (17.18)$$

So we can integrate both terms, obtaining

$$\int d(\rho(x)y(x))=\int \rho(x)Q(x)dx+C$$

And

$$\rho(x)y(x)=\int \rho(x)Q(x)dx+C$$

What implies,

$$y(x)=\rho(x)^{-1}\int \rho(x)Q(x)dx+\rho(x)^{-1}C \qquad (17.19)$$

Suppose an equation of like this. $\frac{dV(t)}{dt}+\frac{1}{\tau}V(t)=i_{max}$. This equation governs the change in voltage when a square pulse of voltage is applied to a part of the neuronal membrane. This is a first-order linear differential equation where $P(t)=\frac{1}{\tau}$ and $Q(t) = i_{max}$. The symbol τ is the membrane constant (s^{-1}) and i_{max} is the maximum current going through the circuit (A). First, we make the change of variable.

$$\rho(t)=e^{\int\frac{dt}{\tau}}=e^{t/\tau}$$

Where we have used the property of integration by a constant. After the process followed above, we can substitute in Eq. 17.19, to find the function $V(t)$

$$V(t)=e^{-t/\tau}\int\frac{i_{max}}{\tau}e^{t/\tau}dt+Ce^{-t/\tau}$$

We can take the constant out of the integral and operate to properly integrate the exponent, that is

$$V(t)=e^{-t/\tau}i_{max}\int\frac{1}{\tau}e^{t/\tau}dt+Ce^{-t/\tau}$$

And using Ohm's law $i_{max}R = V_{max}$

$$V(t)=e^{-t/\tau}V_{max}e^{t/\tau}+Ce^{-t/\tau}=V_{max}+Ce^{-t/\tau}$$

To find the value of the constant, we use the initial condition by which we assume that the voltage before the application of the current is zero, so that at that time $e^{-t/\tau}=e^{0}=1$

$$V(0)=V_{max}+C=0\rightarrow C=-V_{max}$$

And substituting

$$V(t)=V_{max}-V_{max}e^{-t/\tau}$$

Therefore, by taking out the common factor, we will finally have the evolution of the voltage over time when applying a current pulse

$$V(t) = V_{\max}\left(1 - e^{-t/\tau}\right)$$

This equation is very important in basic neurophysiology.

17.4 Solving Second-Order Differential Equations

We only limit ourselves to the most common types of equations because these are extremely frequent in EM and other fields of physical and biological sciences. A differential equation is said to be linear when the variable y and its derivatives y' and y'' are linear (i.e., do not depend on powers). Thus, a linear second-order equation is written in the form,

$$A(x)y''(x) + B(x)y'(x) + C(x)y(x) = F(x) \tag{17.20}$$

When the function $F(x) = 0$ we say that the equation is homogeneous.

In general, any nonhomogeneous equation is associated with a homogeneous one.

The only equations in which we are interested in are those where functions $A(x)$, $B(x)$, and $C(x)$ are constant.

Therefore, we go to solve initially homogeneous second-order linear equations like this

$$y''(x) + py'(x) + qy(x) = 0 \tag{17.21}$$

To do that, we make use of the principle of superposition. Let y_1 and y_2 be two solutions of the homogeneous linear equation. If c_1 and c_2 are constants, then the linear combination

$$y = c_1 y_1 + c_2 y_2 \tag{17.22}$$

It is also a solution of the equation.

It can be probed that any equation like this, with two linear independent solutions (y_1, y_2) have a general solution. Firstly, we must keep in mind that,

$(e^{rx})' = re^{rx}$ and $(e^{rx})'' = r^2e^{rx}$

Hence, supposing $y = e^{rx}$ and substituting its derivatives above, we can obtain,

$$r^2 e^{rx} + pre^{rx} + qe^{rx} = 0 \tag{17.23}$$

From this equation, we extract the term e^{rx}, so $(r^2 + pr + q)e^{rx} = 0$. Obviously $e^{rx} \neq 0$, therefore the only possibility to obey that equation is $r^2 + pr + q = 0$, what is called characteristic equation. This is a linear second-grade equation, whose solutions are,

$$r = \frac{-p \pm \sqrt{p^2 - 4q}}{2}$$

Let's see an example,

$$y'' - 5y' + 6y = 0$$

Supposing that $y = e^{rx}$ will be a solution, we obtain the first two derivatives, $y' = re^{rx}$ and $y'' = r^2e^{rx}$ and then substitute in the equation,

$$r^2 e^{rx} - 5re^{rx} + 6e^{rx} = 0$$

Extracting the common term

$$\left(r^2 - 5r + 6\right)e^{rx} = 0$$

Hence, the only possibility to solve this equation is,

$$r^2 - 5r + 6 = 0$$

And solving as usual

$$r = \frac{5 \pm \sqrt{5^2 - 24}}{2} = \frac{5 \pm 1}{2} = \begin{cases} 2 \\ 3 \end{cases}$$

Therefore, the general solution of this equation is

$$y(x) = c_1 e^{2x} + c_2 e^{3x}$$

The exact values of the constants c_1 and c_2 depend on the initial values of the system at $x = 0$, e.g., $y(x = 0) = y_0$ and $y'(x = 0) = y'_0$. We obtain the derivative of the previous general solution as

$$y'(x) = 2c_1 e^{2x} + 3c_2 e^{3x}$$

And next we use the initial values at $x = 0$

$$y_0 = c_1 + c_2$$

$$y'_0 = 2c_1 + 3c_2$$

This is a system of two equations with two variables, solved as

$$c_1 = 3y_0 - y'_0 \text{ and } c_2 = -2y_0 + y'_0$$

And, the particular solution will be

$$y(x) = \left(3y_0 - y_0'\right)e^{2x} + \left(-2y_0 + y'_0\right)e^{3x}$$

17.5 Exact Differential Equations

Consider the first-order differential equation,

$$P(x,y) + Q(x,y)\,y' = 0, \qquad (17.24)$$

where P and Q are functions with partial derivatives in an open rectangle $D = (a,b) \times (c,d)$. The equation will be exact if there exists a function $f: D \to R$ such that

$$\nabla f(x,y) = \left(P(x,y), Q(x,y)\right), \text{for every}(x,y) \in D \qquad (17.25)$$

Or, equivalently

$$\frac{\partial f(x,y)}{\partial x} = P(x,y); \frac{\partial f(x,y)}{\partial y} = Q(x,y) \qquad (17.26)$$

If the equation is exact, the general solution will be

$$f(x,y) = C$$

See Laplacian at Chap. 14.

From the properties of the partial derivatives, if such a function exists, then it must be verified that,

$$\frac{\partial^2 f(x,y)}{\partial x \partial y} = \frac{\partial P(x,y)}{\partial y}; \frac{\partial^2 f(x,y)}{\partial x \partial y} = \frac{\partial Q(x,y)}{\partial x} \qquad (17.27)$$

So the equation will be exact if, and only if

$$\frac{\partial P(x,y)}{\partial y} - \frac{\partial Q(x,y)}{\partial x} = 0, \text{for every}(x,y) \in D \qquad (17.28)$$

Once we know that the equation is exact, we must find the function that makes it so. It must be verified that

$$\frac{\partial f(x,y)}{\partial x} = P(x,y) \qquad (17.29)$$

Fixing the variable y and integrating over x

$$f(x,y) = \int P(x,y)\,dx + h(y) \qquad (17.30)$$

where h is a function that depends only on y. Deriving now this expression with respect to y, together with the previous equality, we will have,

$$Q(x,y) = \frac{\partial f(x,y)}{\partial y} = \frac{\partial}{\partial y}\left(\int P(x,y)\,dx + h(y)\right)$$
$$= \frac{\partial}{\partial y}\int P(x,y)\,dx + h'(y) \qquad (17.31)$$

Then, obtaining $h'(y)$

$$\frac{dh(y)}{dy} = Q(x,y) - \frac{\partial}{\partial y}\int P(x,y)\,dx, \qquad (17.32)$$

And integrating, we can obtain $h(y)$

$$h(y) = \int Q(x,y)\,dy - \int\left(\frac{\partial}{\partial y}\int P(x,y)\,dx\right)dy \qquad (17.33)$$

17.6 Partial Differential Equations

A general solution of partial differential equations can be extremely complex and is not covered in this work. But, let us give a brief description of the method of separation of variables. The most interesting partial differential equation for our work is the wave equation. It encompasses finding a function that depends on both x and t and represents the displacement of a particle (e.g., on a string or on the surface of a pond) that has moved from its equilibrium position. The unidimensional wave equation is given by

$$\frac{\partial^2 f(x,t)}{\partial x^2} = \frac{1}{v^2}\frac{\partial^2 f(x,t)}{\partial t^2} \qquad (17.34)$$

We can try a solution by assuming that the position ($\Phi(x)$) and time ($T(t)$) functions are actually independent functions and that they are combined in this way

$$f(x,t)=T(t)\varnothing(x) \tag{17.35}$$

And replacing this equation in Ap 17.34, we will have

$$\frac{d^2\varnothing}{dx^2}T=\frac{1}{v^2}\frac{d^2T}{dt^2}\varnothing \tag{17.36}$$

Now we can divide both parts of the equation by $T(t)\ \varnothing\ (x)$

$$\frac{d^2\varnothing}{dx^2}\frac{T(t)}{T(t)\varnothing(x)}=\frac{1}{v^2}\frac{d^2T}{dt^2}\frac{\varnothing(x)}{T(t)\varnothing(x)},$$

$$\text{so}\frac{d^2\varnothing}{dx^2}\frac{1}{\varnothing(x)}v^2=\frac{d^2T}{dt^2}\frac{1}{T(t)} \tag{17.37}$$

So, we have obtained an equation with simple derivatives in which the left half depends only on x, while the right half depends on t. Therefore, for both halves to be satisfied, each of them must be equal to a constant, that is, not dependent on x or t. We can test this with the following expressions

$$\frac{d^2\varnothing}{dx^2}\frac{1}{\varnothing(x)}v^2=-\omega^2\text{and}\frac{d^2T}{dt^2}\frac{1}{T(t)}=-\omega^2 \tag{17.38}$$

We can solve these second-order differential equations as discussed above, obtaining

$$T(t)=A\sin\omega t+B\cos\omega t \tag{17.39}$$

$$\varnothing(x)=C\sin kx+D\cos kx \tag{17.40}$$

where $k=\frac{\omega}{v}$ and A, B, C, and D are arbitrary constants.

We can use these equations in Eq. 17.35 and

$$f(x,t)=(A\sin\omega t+B\cos\omega t)\times(C\sin kx+D\cos kx) \tag{17.41}$$

The solution depends largely on whether we are analyzing a finite or infinite string. This is important because in the first case the allowed frequencies will be discrete, while in the second, they will be continuous. It is possible to obtain a general solution for the wave equation by adding solutions, obtaining the following expression:

$$f(x,t)=\sum_n\left\{A_n\cos\left[\frac{\omega_n}{v}(x+vt)\right]+B_n\sin\left[\frac{\omega_n}{v}(x+vt)\right]+C_n\cos\left[\frac{\omega_n}{v}(x-vt)\right]+D_n\sin\left[\frac{\omega_n}{v}(x-vt)\right]\right\} \tag{17.42}$$

In the case of a finite string or, in the case of an infinite string, we shall have

$$f(x,t)=\int_{-\infty}^{\infty}F(\omega)e^{\left[i\frac{\omega}{v}(x+vt)\right]}d\omega+\int_{-\infty}^{\infty}G(\omega)e^{\left[i\frac{\omega}{v}(x-vt)\right]}d\omega \tag{17.43}$$

Therefore, to solve these types of equations, it is necessary to develop the function as a sum of sine and cosine functions or as an integral of exponential functions. In analogy with the general solution of an ordinary second-order differential equation, which requires two constants for its general solution, in the case of a second-order equation in partial derivatives, it depends on two arbitrary functions (Fig. 17.1). The best method is to represent the solution starting from two initial conditions, such as the initial shape and velocity of each point on the string. The general solution depends on a function that propagates to the right and another that moves to the left.

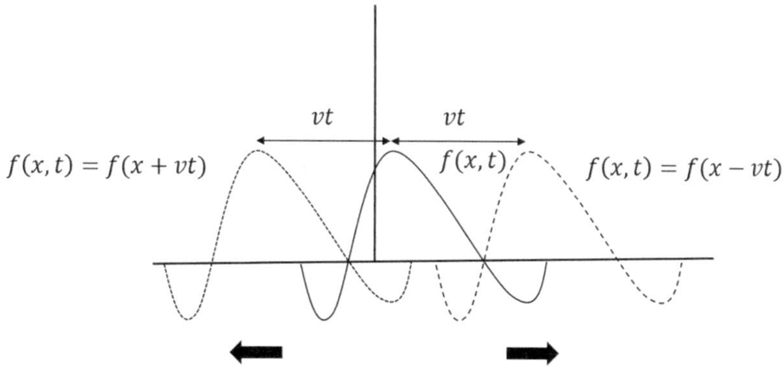

Fig. 17.1 Wave propagation referred to a Cartesian coordinate system

Suggested Readings

A detailed description of these theories and concepts can be found in the next references.

Arfken GB, Weber HJ. Differential equations. In: Mathematical methods for physicists. 6th ed. Amsterdam: Elsevier Academic Press; 2005. ISBN 978-0-12-059876-2.

Edwards CH, Penney DE. First-order differential equations. In: Elementary differential equations. 6th ed. London: Prentice Hall; 2008. ISBN 0-13-239730-7.

Edwards CH, Penney DE. Linear equations of higher order. In: Elementary differential equations. 6th ed. London: Prentice Hall; 2008. ISBN 0-13-239730-7.

King AC, Billingham J, Otto SR. Variable coefficient, second order, linear, ordinary differential equations. In: Differential equations. Linear, nonlinear, ordinary. New York: Partial Cambridge University Press; 2003. ISBN 978-0-521-01687-2.

18 Complex Numbers

18.1 Introduction to the Complex Numbers

For all $x \in \mathbb{R}$ such that $x \neq 0$ it is verified that $x^2 > 0$ and if $x = 0$, then $x^2 = 0$. That is, the square of any real number is always positive and, therefore, there are no square roots of negative numbers.

To have the possibility to obtain of square roots from negative numbers, it was developed the set of complex numbers, named as C, where it is defined the symbol $i = \sqrt{-1}$ or, what is the same $i^2 = -1$. Therefore, with this definition, we can find the root of a negative number. The set C is built from elements of $\mathbb{R}$. Suppose $z \in C$ and let be $x, y \in \mathbb{R}$. We define

$$z = x + iy, \tag{18.1}$$

The part not associated with i is called the real part of z, that is $Re(z) = x$, while the number associated with i is called the imaginary part, or $Im(z) = y$.

Complex numbers are represented on a two-dimension plane, called the Argand's diagram. In this representation, the value of the real part is the magnitude on the x-axis (real axis), while the imaginary magnitude is represented on the y-axis. Let's see two examples: $z_1 = 2 - i$ and $z_2 = -1 + 2i$ (Fig. 18.1a).

J. Pastor, *Quantitative EEG Practical Guide*, https://doi.org/10.1007/978-3-032-05457-9_18

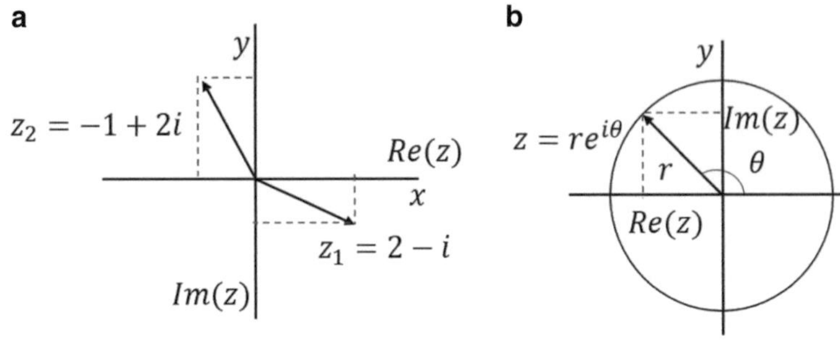

Fig. 18.1 Representation of complex numbers and Argand's diagram. (**a**) Representation of two numbers, z_1 and z_2. (**b**) Representation of a complex number as a polar form. Im = imaginary; Re = real.

18.2 The Complex Exponent. Polar form

The Euler's formula establish the theorem $e^{ix} = \cos x + i \sin x$. From this equation, it is easy to see that for every complex number $z \in C$, such that $z = x + iy$, it can be expressed in polar form as $z = re^{i\theta}$ (Fig. 18.1b). Using Euler's formula, we can put $z = r(\cos \theta + i \sin \theta)$ and since both expressions refer to the same complex number, the real part $x = r \cos \theta$ and the imaginary part $y = r \sin \theta$.

The term r is called the modulus of the complex number, while the term θ is called the argument and is the angle formed by the modulus z and the (real) $x - axis$

18.3 Complex Algebra

Let's have two complex numbers, like these $z_1 = x_1 + iy_1$ and $z_2 = x_2 + iy_2$ in Cartesian coordinates, which can also be expressed in exponential forms $z_1 = r_1 e^{\theta_1}$ and $z_2 = r_2 e^{\theta_2}$

We can define addition (or subtraction in a straightforward way) of complex numbers in terms of their Cartesian coordinates as

$$\begin{aligned} z_1 + z_2 &= (x_1 + iy_1) + (x_2 + iy_2) \\ &= (x_1 + x_2) + i(y_1 + y_2) \end{aligned} \quad (18.2)$$

Or in using their exponential forms

$$z_1 + z_2 = r_1 e^{i\theta_1} + r_2 e^{i\theta_2} = r_1(\cos\theta_1 + i\sin\theta_1) + r_2(\cos\theta_2 + i\sin\theta_2) = r_1\cos\theta_1 + r_2\cos\theta_2 + i(r_1\sin\theta_1 + r_2\sin\theta_2) \quad (18.3)$$

Multiplication will be

$$\begin{aligned} z_1 z_2 &= (x_1 + iy_1)(x_2 + iy_2) \\ &= (x_1x_2 + ix_1y_2 + iy_1x_2 - y_1y_2) \end{aligned} \quad (18.4)$$

The last term comes from the $i^2 = -1$. So, we can assembly in a more convenient form

$$z_1 z_2 = (x_1x_2 - y_1y_2) + i(x_1y_2 + y_1x_2) \quad (18.5)$$

Usually, multiplication is easiest in the polar form

$$z_1 z_2 = r_1 e^{i\theta_1} r_2 e^{i\theta_2} = r_1 r_2 e^{i\theta_1} e^{i\theta_2} = r_1 r_2 e^{i(\theta_1 + \theta_2)} \quad (18.6)$$

Although defined in Cartesian coordinates too, division is easiest in polar form

$$\frac{z_1}{z_2} = \frac{r_1 e^{i\theta_1}}{r_2 e^{i\theta_2}} = \frac{r_1}{r_2} e^{i(\theta_1 - \theta_2)} \quad (18.7)$$

For every complex number $z_1 = x_1 + iy_1$ we can obtain its complex conjugate, denoted as

$$z_1^* = x_1 - iy_1 \quad (18.8)$$

For a complex number in polar terms, $z_1 = r_1 e^{i\alpha_1}$ its conjugate is

$$z_1^* = r_1 e^{-i\alpha_1}, \quad (18.9)$$

That is, the complex conjugate is obtained by changing the sign of the imaginary part of the original complex.

To find the module r of a complex number z, multiply it by its conjugate z*, and find the square root of its absolute value, that is

$$\left|zz^*\right|^{1/2} = \left|re^{i\theta} re^{-i\theta}\right|^{1/2} = r \quad (18.10)$$

18.4 Complex Functions

A complex function is an application of a subset U of C on C and is written $f: U \subset C \rightarrow C$. This means that for every $z \in U$ exists some $w \in C$ such that $f(z) = w$. For example, $f(z) = e^{xy} + ix$ or $f(z) = \sin xy - i \cos xy$.

With some cautions and for the simple expressions, we can obtain the derivatives and integrals of complex functions from the same expressions obtained for real numbers (see Chap. 16_Brief reminder of calculus). By resume, we indicate some of these derivatives

$$\frac{d}{dz}c = 0$$
$$\frac{d}{dz}z = 1$$
$$\frac{d}{dz}\left[cf(z)\right] = cf'(z)$$
$$\frac{d}{dz}z^n = nz^{n-1}, n \in \mathbb{N}$$
$$\frac{d}{dz}\left[f(z) + g(z)\right] = f'(z) + g'(z)$$
$$\frac{d}{dz}\left[f(z)g(z)\right] = f(z)g'(z) + f'(z)g(z)$$
$$\frac{d}{dz}\left[f(z)/g(z)\right] = \frac{g(z)f'(z) - f(z)g'(z)}{g(z)^2}, when\ g(z) \neq 0 \quad (18.11)$$

Suggested Reading

A detailed description of these theory and concepts can be found in the next references.

Arfken GB, Weber HJ. Functions of a complex variable I. Analytic properties, Mapping. Mathematical methods for physicists. 6th ed. Amsterdam: Elsevier/Academic; 2005. ISBN 978-0-12-059876-2.

Arfken GB, Weber HJ. Functions of a complex variable II. Analytic properties, mapping. Mathematical methods for physicists. 6th ed. Amsterdam: Elsevier/Academic; 2005. ISBN 978-0-12-059876-2.

Spiegel M, Lipschutz S, Schiller J, Spellman D. Complex numbers. Schaum's outline of complex variables. 2nd ed. New York: Mc Graw Hill; 2009. ISBN 978-0-07-161570-9.

Spiegel M, Lipschutz S, Schiller J, Spellman D. Functions, limits and continuity. Schaum's outline of complex variables. 2nd ed. New York: Mc Graw Hill; 2009. ISBN 978-0-07-161570-9.

19 Laplace Transform, LTI Systems, and Convolution

19.1 Laplace Transform

We have seen in Chap. 17 how to solve linear second-order differential equations. However, there are some situations in which it is preferable to use the method we are going to describe. In addition, the method is a transformation method and has some similarities with the Fourier transform. For all these reasons, it is of great interest to know this method.

The integral in the definition of the Laplace transform probably appeared first in the work of Euler. However, it was the French mathematician Pierre Simon de Laplace (1749–1827), who employed such integrals in his work on probability theory. The methods for solving differential equations based on Laplace transforms were discovered and popularized by practicing engineers—notably the English electrical engineer Oliver Heaviside (1850–1925), although their rigorous justification was not made until well into the twentieth century.

The Laplace transform can be defined for a function $f(t)$ defined for all $t \geq 0$; the Laplace transform of f is the function F defined as follows:

$$F(s) = \mathcal{L}\{f(t)\} = \int_0^\infty f(t)e^{-st}dt \quad (19.1)$$

We must remind that an improper integral over an infinite interval is defined as a limit of integrals over bounded intervals, that is

$$\int_a^\infty f(t)dt = \lim_{b\to\infty}\int_a^b f(t)dt$$

Therefore, when the integral in Eq. 19.1 converges, it converges not simply to a number (such as the Riemann integral, see Chap. 16, Brief remainder of calculus), but to a function F of s.

For example, let's find the transform of $f(t) = 1$, for $t \geq 0$

$$\mathcal{L}\{1\} = \int_0^\infty e^{-st}dt = \left[-\frac{1}{s}e^{-st}\right]_0^\infty = \lim_{b\to\infty}\left[-\frac{1}{s}e^{-bs} + \frac{1}{s}\right]$$

And considering that $\lim_{b\to\infty} e^{-bs}$ =0, then $\mathcal{L}\{1\} = \frac{1}{s.}$

Just as with the Fourier transformation, which has an inverse transformation, we can find the inverse Laplace transformation, although it is rarely used.

Some interesting and frequently used transform are shown next. Their expressions are easy to deduce and can be seen in detail in Suggested readings

$$\mathcal{L}\{\delta(t)\} = \int_0^\infty \delta(t)e^{-st}dt = 1 \quad (19.2)$$

where $\delta(t)$ is the unit impulse (see Chap. 3).

$$\mathcal{L}\{U(t)\} = \int_0^\infty 1e^{-st}dt = \frac{1}{s} \quad (19.3)$$

where $U(t)$ is the step function, defined as follows:

$$U(t) = \begin{cases} 0, & t \leq 0 \\ 1, & t > 0 \end{cases}$$

J. Pastor, *Quantitative EEG Practical Guide*, https://doi.org/10.1007/978-3-032-05457-9_19

In fact, this close relationship between the two functions emerges from the following relationship:

$$\delta(t) = \frac{dU(t)}{dt} \tag{19.4}$$

Thus, the $\delta(t)$ function is the derivative of the step function.

The exponential, sine, and cosine functions are especially important.

$$\mathcal{L}\{e^{at}\} = \int_0^\infty e^{at} e^{-st} dt = \int_0^\infty e^{-(s-a)t} dt = \frac{1}{s-a} \tag{19.5}$$

$$\mathcal{L}\{\sin \omega t\} = \int_0^\infty \sin \omega t e^{-st} dt = \int_0^\infty \left(\frac{e^{i\omega t} - e^{-i\omega t}}{2i} \right) e^{-st} dt = \frac{\omega}{s^2 + \omega^2} \tag{19.6}$$

$$\mathcal{L}\{\cos \omega t\} = \int_0^\infty \cos \omega t e^{-st} dt = \int_0^\infty \left(\frac{e^{i\omega t} + e^{-i\omega t}}{2} \right) e^{-st} dt = \frac{s}{s^2 + \omega^2} \tag{19.7}$$

19.2 The Effect of Delay

A property of the Laplace transform that is of extraordinary importance in signal analysis theory is the delay effect of the transform. So, let's look at it in some detail. Suppose a function $f(t)$. The Laplace transform is given by the Eq. 19.1. By means of this expression, we can obtain the right-shifted version of the function displaced a period of τ, as

$$\mathcal{L}\{f(t-\tau)\} = \int_0^\infty f(t-\tau) e^{-st} dt \tag{19.8}$$

We can change the variable, according to $T = t - \tau$ (obviously $t = T + \tau$) and, consequently $dT = dt$. Substituting in the precedent equation

$$\mathcal{L}\{f(T)\} = \int_0^\infty f(T) e^{-s(T+\tau)} dT = e^{-s\tau} \int_0^\infty f(T) e^{-sT} dT = e^{-s\tau} \mathcal{L}\{f(T)\} \tag{19.9}$$

This means that a delay τ in the time domain is transformed as a multiplication factor $e^{-s\tau}$ in the s-domain. As we shall see later (Chap. 20), this result is fundamental for understanding the z-transform.

By resume, this mathematical tool, the Laplace transform, is extremely useful. It can be used to determine the parameters of probability density functions or solve differential equations or process signals.

19.3 Linear Time-Invariant Systems

A linear time-invariant (LTI) system is completely characterized by the knowledge of its response, called as r, to a simple and basic input, termed as stimulus or s. Therefore, the response to any arbitrary summation of stimulus, forming a complex stimulus, will be obtained. LTI systems are defined by these properties:

1. Linearity: a system is linear if its output y is linearly related to the input x, e.g., $y = mx$, where m is a constant.
2. Superposition: if we have these two outputs, $y_1 = mx_1$ and $y_2 = mx_2$, then $y_1 + y_2 = m(x_1 + x_2)$.
3. Time invariance: what means that the system's response does not depend on time and, therefore, the same input at different times gives the same output, that is, if $x(t) \rightarrow y(t)$, then $x(t - \tau) \rightarrow y(t - \tau)$.
4. Convolution. We shall describe it in more detail below.

The relationship between input $x(t)$ and output $y(t)$ is usually described by ordinary differential equations. A singular case is the response to the unit impulse δ (see Chap. 3), named impulse response or weighting function h.

19.4 Convolution

Before addressing the interaction between signal pairs, we must briefly introduce a mathematical operation called convolution. It is a method to combine two signals, one of which is shifted in time by a defined amount (τ). For signals $x(t)$ and $y(t)$, the convolution ($\otimes$) is defined as follows:

$$x(t)\otimes y(t)=\int_{0}^{t}x(\tau)y(t-\tau)d\tau \quad (19.10)$$

In essence, a signal ($x(\tau)$) is multiplied by the offset from the other version (i.e., $y(t - \tau)$), and this produces a unique value that is the area (i.e., the defined integral) under the resulting function of the product.

However, the convolution computation is often complex, and it is consequently transformed into frequency domain to facilitate calculation. Using equation Eq. 19.1 and the change of variable $T = t - \tau$, the transform of the convolution of two time series can be shown to be the product of their transforms ($X(s)$ and $Y(s)$)

$$\mathcal{F}\{x(t)\otimes y(t)\}=X(s)Y(s) \quad (19.11)$$

The mathematical operation of convolution is the basis of techniques such as cross and auto-correlation and time-frequency analysis using wavelets, so we will briefly look at how it is applied (see Fig. 19.1). Suppose we want to find the convolution at different periods of a function $x(t)$ with another test function, which we call $y(t)$ (Fig. 19.1a). This last function will shift different time intervals (which is the meaning of the parameter τ in equation 19.10). For every τ, we compute the product of both functions. Obviously, there will be regions where one (or both) of the functions is zero, so the product will also be zero. However, there will be other regions where $x(t) \neq 0$ and $y(t) \neq 0$, so the product $x(t)y(t) \neq 0$. However, the operation performed renders another function that, obviously, will not have a unique value, but will be defined in the interval $[0, t]$. Therefore, to obtain a unique numerical value of the product of functions, we find the definite integral between the same limits. This integral (see Chap. 16_Brief reminder of calculus) gives precisely the value of the area.

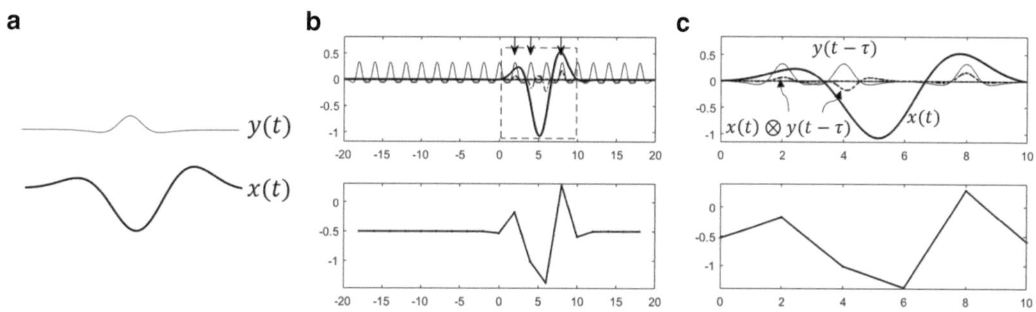

Fig. 19.1 Convolution of two functions defined for $t \in [-20, 20]$. (**a**) Function test ($y(t)$, with the shape of a typical wavelet) at top and the function constant with time ($x(t)$) at bottom. (**b**) The test function changes through the time and the product of both functions at any time is accomplished. (**c**) Detail of the dashed square with only three $y(t - \tau)$ (ligth grey lines) indicated by arrows, $x(t)$ (black thick line) and function products (dashed black lines). Also in **b** and **c** the bottom graph indicates the discrete values of convolution at every time

Suggested Readings

A rigorous and detailed description of these theories and concepts is showed in the next references.

Edwards CH, Penney DE. Laplace transform methods. In: Elementary differential equations. 6th ed. London: Prentice Hall; 2008. ISBN 0-13-239730-7.

Van Drongelen W. Laplace and z-transform. In: Signal processing for neuroscientists. Amsterdam: Elsevier; 2007. ISBN 13:978-0-12-370867-0.

Van Drongelen W. LTI systems, convolution, correlation and coherence. In: Signal processing for neuroscientists. Amsterdam: Elsevier; 2007. ISBN 13:978-0-12-370867-0.

The z-Transform

20

20.1 Introduction

The z-transform translates a discrete-time signal into a complex-valued frequency-domain representation in z-plane. See Chap. 18_Complex numbers.

In this chapter, we will briefly analyze the z-transform, the equivalent of the Laplace's transform for discrete variable. The effect of delay on the Laplace transform was shown in Chapter 19_Laplace transform, convolution, LTI systems, showing that $\mathcal{L}\{f(t-\tau)\} = e^{-s\tau}\mathcal{L}\{f(T)\}$. This equality implies that a delay τ in the time domain transforms in the s-domain by multiplying the function's transform by an exponential factor. This result is crucial for understanding the z-transform.

We have seen how the dynamics in continuous systems is described by differential equations that can be solved using the Laplace (or Fourier) transform. In discrete systems, the equations describing the dynamics of a system are given by difference equations, as

$$\sum_{p=0}^{n} a_p y(k-p) = \sum_{p=0}^{m} b_p x(k-p) \tag{20.1}$$

where x are the input and y the output of the system. Usually $n < m$.

Let's consider a discrete time series like this (see Chap. 3)

$$x(n) = \sum_{k=0}^{n} x(k)\delta(t-k\tau) \tag{20.2}$$

And the Laplace transform

$$\mathcal{L}\{x(n)\} = \mathcal{L}\left\{\sum_{k=0}^{n} x(k)\delta(t-k\tau)\right\}$$

And, applying the properties of Laplace transform

$$\mathcal{L}\{x(n)\} = \sum_{k=0}^{n} x(k)e^{-ks\tau} \tag{20.3}$$

And renaming according to this

$$e^{s\tau} \equiv z \text{ or } e^{-s\tau} \equiv z^{-1} \tag{20.4}$$

Rewriting Eq. 20.3 as

$$X(z) = \sum_{k=0}^{n} x(k)z^{-k} \tag{20.5}$$

The $(n-1)$ notation is shorthand for indicating a one-unit change in the entire time series, shifting one position to the right. We can now use a new notation to indicate this change in the z-domain, using the expressions $X(z)$ and z:

J. Pastor, *Quantitative EEG Practical Guide*, https://doi.org/10.1007/978-3-032-05457-9_20

$$\begin{array}{llll} \text{Original time series} & x(n) & \Leftrightarrow & X(z) \\ \text{Time series shifted by 1 sample} & x(n-1) & \Leftrightarrow & z^{-1}X(z) \\ \vdots & \vdots & \vdots & \vdots \\ \text{Time series shifted by } m \text{ samples} & x(n-m) & \Leftrightarrow & z^{-m}X(z) \end{array} \tag{20.6}$$

It is common to find operations like this in difference equations

$$\ldots x(n)-x(n-1)\ldots$$

Using the new variables introduced above, this difference is written as follows:

$$\ldots X(z)-z^{-1}X(z)\ldots$$

This is a highly relevant method because it allows any discrete LTI system to be described as a transformed system in the z-domain.

Using this method, equation Eq. 20.1 will be written as

$$\sum_{p=0}^{n} a_p z^{-p} Y(z) = \sum_{p=0}^{m} b_p z^{-p} X(z) \tag{20.7}$$

20.2 The z-Transform

We have introduced the z-transform in equation 20.6, but the analytical derivation of $X(z)$ has not been explicitly defined. To do this, we start from the analytical definition of the Laplace transform, in which the integral is replaced by a summation

$$F(s) = \int_0^{\infty} f(t) e^{-st} dt \Rightarrow \sum_{n=0}^{\infty} x(n) e^{-st}$$

And changing $F(s) \Leftrightarrow X(z)$ and using Eq. 20.4, we get

$$X(z) = \sum_{n=0}^{\infty} x(n) z^{-n} \tag{20.8}$$

Suggested Readings

A detailed description of these theory and concepts can be found in the next reference.

Al-Nashsh H, Sabesan S, Krihann B, George JS, Tsakalis K, Iasemidis L, Tong S. Single-channel EEG analysis. In: Tong S, Nithsh V, editors. Quantitative EEG analysis methods and clinical applications. Boston: Thakor Artech House; 2009. ISBN-13: 978-1-59693-204-3.

Lynn PA. The laplace transform and the z-transform. In: Electronic signals and systems. London: Macmillan Education UK; 1986. p. 225–72. https://doi.org/10.1007/978-1-349-18461-3_6. ISBN 978-0-333-39164-8.

Van Drongelen W. Laplace and z-transform. In: Signal processing for neuroscientists. Amsterdam: Elsevier; 2007. ISBN 13:978-0-12-370867-0.

21 Trigonometry

21.1 Introduction

Trigonometry is the division of mathematics concerned with relationships between angles and triangles. More specifically, the trigonometric functions relate the angles of a right triangle to ratios of its side lengths. Trigonometric functions are the fundamental tool of Fourier analysis and are extensively used to describe EEG patterns, so a brief description of them is essential.

Let's assume a circle of unitary radius. Let us also suppose a point of coordinates (x, y) located on this circumference. The radius vector is moving in a counterclockwise direction and is represented in Fig. 21.1a.

The angle α is formed by the positive part of the x-axis and the radius vector. Note that a right triangle has been formed with two catheti: A, which is the cathetus opposite to angle α and B, which is the adjacent cathetus. Logically, C is the hypotenuse of the triangle. The following definitions involve only the relationships between the sides of a right triangle as a function of the angle α:

1. Sine: is defined as the opposite side divided by hypotenuse.

$$\sin\alpha = \frac{y}{C} \tag{21.1}$$

2. Cosine: is defined as the adjacent side divided by hypotenuse.

$$\cos\alpha = \frac{x}{C} \tag{21.2}$$

3. Tangent: is defined as the quotient between sine and cosine, i.e., opposite side divided by the adjacent one.

$$\tan\alpha = \frac{\sin\alpha}{\cos\alpha} = \frac{y}{x} \tag{21.3}$$

4. Cotangent: is defined as the quotient between cosine and sine, i.e., adjacent side divided by the opposite one

$$\text{cotan}\,\alpha = \frac{\cos\alpha}{\sin\alpha} = \frac{x}{y} \tag{21.4}$$

These definitions are given for a fixed angle. Now suppose that the angle changes in time in a counterclockwise. If we represent the amplitude, i.e., the ordinate y as a function of the angle, we will have the following graph (Fig. 21.1b).

A radian is defined as the angle such that the length of the arc it sweeps is equal to the length of the radius that forms it. What is important in our case is that radians, like degrees, are ways of measuring angles. Table 21.1 shows the value (in rads) of sine and cosine functions for different angles in degrees and their equivalence in rads.

J. Pastor, *Quantitative EEG Practical Guide*, https://doi.org/10.1007/978-3-032-05457-9_21

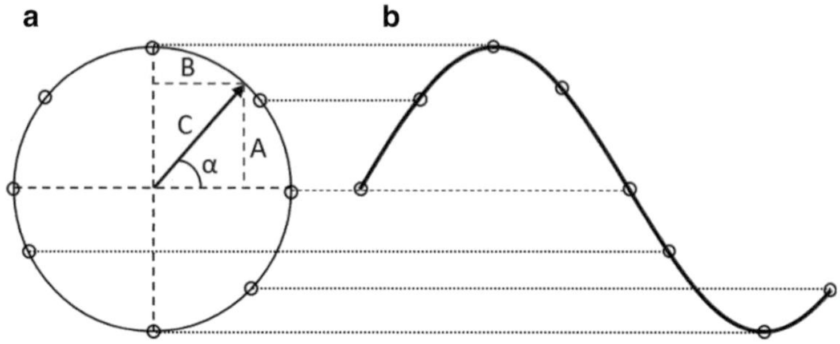

Fig. 21.1 Angles in a unit circumference. (**a**) The radius vector (C) indicating the coordinate point (B,A) as a function of the angle α. (**b**) Function sin α

Table 21.1 Value of the cosine and sine functions for especially relevant angles of the unit circle, measured in degrees and radians

α (degrees)	α (radians)	cosα	sinα
0	0	1	0
30	$\pi/6$	$\sqrt{3}/2$	1/2
45	$\pi/4$	$\sqrt{2}/2$	$\sqrt{2}/2$
60	$\pi/3$	1/2	$\sqrt{3}/2$
90	$\pi/2$	0	1
180	π	−1	0
270	$3\pi/2$	0	−1
360	2π	1	0

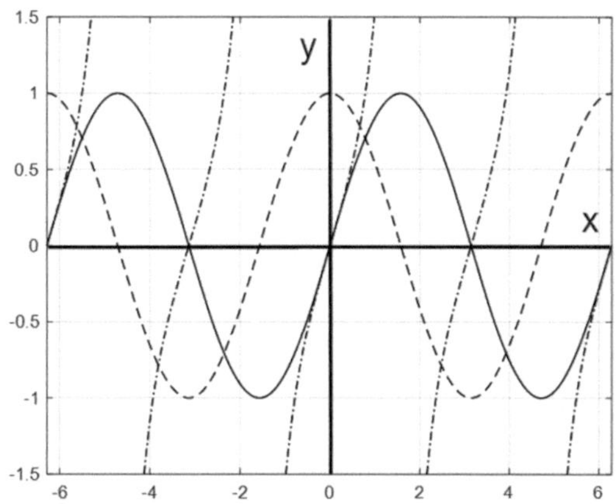

Fig. 21.2 Trigonometric functions sine (continuous line), cosine (dotted line), and tangent (dashed lines)

Figure 21.2 shows the sine, cosine, and tangent functions. The name of the independent variable is not important, and instead of alpha (as above), we have used x.

These functions are of unimaginable importance in all of Science, of course, in Neurophysiology and specially in EEG, since they are very useful in the definition of periodic functions, which are the basis of Fourier analysis, and they are used as models of cortical functioning and the definition of EEG bands.

The sine and cosine functions have a range of definition $(-\infty, \infty)$ and their domain being ± 1. Both functions have the same morphology (Fig. 21.3a), but they are displaced by an amount of $\pi/2$ radians. However, the tangent function (Fig. 21.3b) has a range of definition $(-\pi/2, \pi/2)$ and a domain $\pm\infty$.

These functions can be modified by parameters such as amplitude (A), angular frequency (ω), or phase (φ). Given the importance of these types of modified functions, we will briefly explain their meaning (Fig. 21.3c). A function like this is written as

$$y(t) = A\sin(\omega t + \varphi) \tag{21.5}$$

The parameter amplitude modifies the maximum distance between the extreme values, so that it will be greater than the sine function if $A > 1$ or less if $A < 1$. The angular frequency is the ordinary frequency (f), multiplied by 2π, which converts it from cycles/s or Hz to radians/s. The period (T) is the time required to complete one cycle, that is, the time that elapses between two consecutive equal values of the function. It is related to the frequency by this expression $T = 1/f$, and how $\omega = 2\pi f$, the relationship with the angular frequency will be $T = 2\pi/\omega$. Finally, the phase can be defined as the fraction of the period elapsed since the instant corresponding to the state taken as reference.

Each of the functions we have defined above has its inverse function, called arcsine, arccosine, and arctangent respectively for sine, cosine, and tangent. See Fig. 21.4. The definition ranges and domains of these functions are interchanged with those of the original functions.

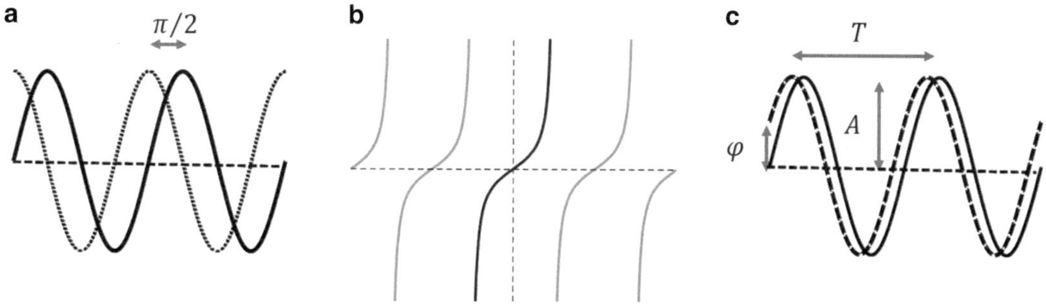

Fig. 21.3 Trigonometric functions. (**a**) Sine and cosine functions $\pi/2$ radians apart. (**b**) Several tangent functions, although one of them has been highlighted (black line). (**c**) Sine function modified by parameters of amplitude (for simplicity, only the positive half of the amplitude is shown), period (T), and phase

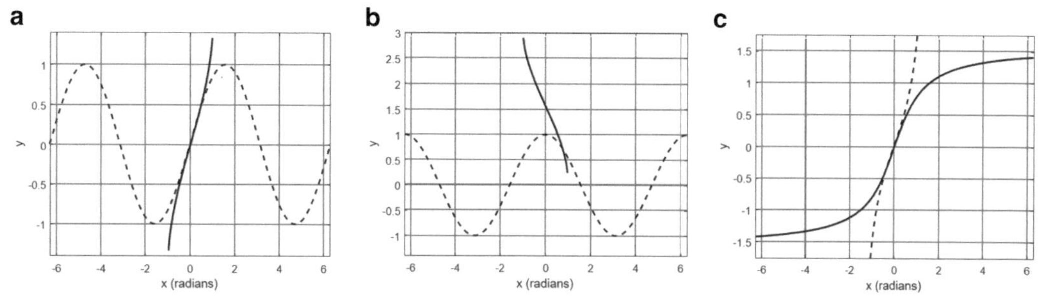

Fig. 21.4 Inverse functions (solid line). The direct function (dashed line) is also shown. (**a**) Sine and arcsine. (**b**) Cosine and arccosine. (**c**) Tangent and arctangent

Finally, some trigonometric relationships of special importance are indicated.

$$\begin{aligned}\sin^2(x)+\cos^2(x)&=1\\ \sin(x+y)&=\sin(x)\cos(y)+\cos(x)\sin(y)\\ \cos(x+y)&=\cos(x)\cos(y)-\sin(x)\sin(y)\end{aligned} \quad (21.6)$$

And, from Eq. 21.6, we obtain for the case of $x = y$

$$\cos(2x)=\cos^2(x)-\sin^2(x) \quad (21.7)$$

This equation is very useful to solve integrals with squared trigonometric functions, operating and using Eq. 21.6

$$\begin{aligned}\cos^2(x)&=\sin^2(x)+\cos(2x)\\ &=1-\cos^2(x)+\cos(2x)\rightarrow\end{aligned}$$

$$2\cos^2(x)=1+\cos(2x)\rightarrow$$

$$\cos^2(x)=\frac{1}{2}(1+\cos(2x)) \quad (21.8)$$

And, using the same approach for the sine function

$$\sin^2(x)=\frac{1}{2}(1-\cos(2x)) \quad (21.9)$$

Suggested Readings

A detailed description of these theory and concepts can be found in the next reference.

Coburn JW, Herdlick. Trigonometry. 2nd ed. McGraw Hill; 2011. ISBN 9780073519487

Ferreira HG, Marshall MW. Trigonometric functions. In: The biophysical basis of excitability. London: Cambridge University Press; 2008. ISBN 978-0-521-06727-0.

22 Probability

22.1 Definitions of Probability

There are numerous problems with the definition of probability (denoted as p). One of the most intuitive (but not mathematically rigorous) is Laplace's definition (also known as estimated or empirical) that can be stated as the limit of the relative frequency of a success. It's easy to suppose that when the number of observations increases, probability can be obtained as

$$p = \lim_{N \to \infty} \left(\frac{h}{N} \right), \tag{22.1}$$

where h is the number of favorable cases and N is the number of total cases. The ratio h/N defines the relative frequency. Unfortunately, we have no way to assure that this limit exists. Therefore, a different definition of probability was introduced, known as the axiomatic definition of probability by Kolmogorov. Let be the sample space (Ω) of a random experiment, $\Omega = \{x_1, x_2, \ldots x_n\}$. The space includes all the possible results of the experiment. We can define an event (A) as any subset of Ω, i.e., $A \subset \Omega$ (e.g., odds in a dice or diamonds from a maze). Obviously, an event can also include one element from Ω, which is called elementary success. Besides these events, we have two more, called as empty event (ϕ), which is not included at Ω, and the certain event, in fact Ω.

The sample space may be finite or countable or infinite. It would be finite when it is formed for a limited number of elements. Countable is a space in which a unique relationship can be established with the natural numbers (N). Finally, a sample space would be infinite if it has as many members as the infinite number of elements of the interval $0 \leq x \leq 1$.

The axiomatic categorization defines an application (p) that assigns a real number to every event ($p(A)$) that fits the following features:

1. $$\text{for every event } A, 0 \leq p(A) \leq 1. \tag{22.2}$$

2. $$p(\Omega) = 1. \tag{22.3}$$

If $A_1, A_2, \ldots, A_n, \ldots$ are mutually exclusive events (i.e., $A_i \cap A_j = \phi$, $i, j = 1, 2, \ldots n$), then

$$p(A_1 \cup A_2 \ldots \cup A_n \ldots) = p\left(\bigcup_{i=1}^{\infty} A_i \right) = \sum_{i=1}^{\infty} p(A_i) \tag{22.4}$$

The properties of probability are analogous to those of relative frequency. However, while relative frequency is an empirical measure of the occurrence of an event, probability is a theoretical measure.

In statistics, the word experiment is used to designate any act that provides data. Two types of experiments are distinguished:

Deterministic. They occur in those situations in which the successive performance of an experiment under the same conditions produces the same result (for centuries, Science has dealt with phenomena that gave rise to deterministic situations, for example, the free fall of bodies is a deterministic phenomenon). In these experiments, the same causes produce the same effects.

J. Pastor, *Quantitative EEG Practical Guide*, https://doi.org/10.1007/978-3-032-05457-9_22

Random. They occur in those situations in which the successive performance of an experiment under the same conditions produces different results. Random phenomena are the yield of a seed, the duration of the life of a component, monthly energy consumption in a house, the result of a coin toss, etc. In random experiments, the same causes produce different effects.

22.2 Operations Between Events

Inclusion of events (Fig. 22.1a). Let A and B be such that A, $B \subset \Omega$, event A will be said to be contained or included in event B, if all sample points of A belong to B

$$A \subset B \text{ if } x \in A \rightarrow x \in B \tag{22.5}$$

Complementary of an event (Fig. 22.1b). Let A be an event, such that $A \subset \Omega$, we will call complementary event of A, and it is denoted by A^C or also by $\overline{A}$, that formed by the sample points that do not belong to A:

$$A^c = \{x \in \Omega / x \notin A\} \tag{22.6}$$

The symbol/means such that.

Union of two events (Fig. 22.1c). Let A and B be such that A, $B \subset \Omega$, we define the union of the events A and B and denote by $A \cup B$, the event formed by all the sample points that belong to at least one of the events:

$$A \cup B = \{x \in \Omega / x \in A \text{ or } x \in B\} \tag{22.7}$$

Intersection of events (Fig. 22.1d). Let A and B be such that A, $B \subset \Omega$, we define the intersection of events A and B and denote by $A \cap B$ or by AB, the event formed by all sample points belonging to both events:

$$A \cap B = \{x \in \Omega / x \in A \text{ and } x \in B\} \tag{22.8}$$

Incompatible, disjoint, or mutually exclusive events. Let A and B be such that A, $B \subset \Omega$, it will be said that the event A is incompatible with the event B, if they do not have sample points in common: $A \cap B = \varnothing$.

Morgan's Laws. Let A and B be such that A, $B \subset \Omega$, then:

$$(A \cup B)^c = A^c \cap B^c \tag{22.9}$$

$$(A \cap B)^c = A^c \cup B^c \tag{22.10}$$

22.3 Properties of Probability

The following properties are verified for probability.

$$p(\varnothing) = 0 \tag{22.11}$$

$$p(A^c) = 1 - p(A) \tag{22.12}$$

$$\text{If } A \subset B \text{ then } p(A) \le p(B) \tag{22.13}$$

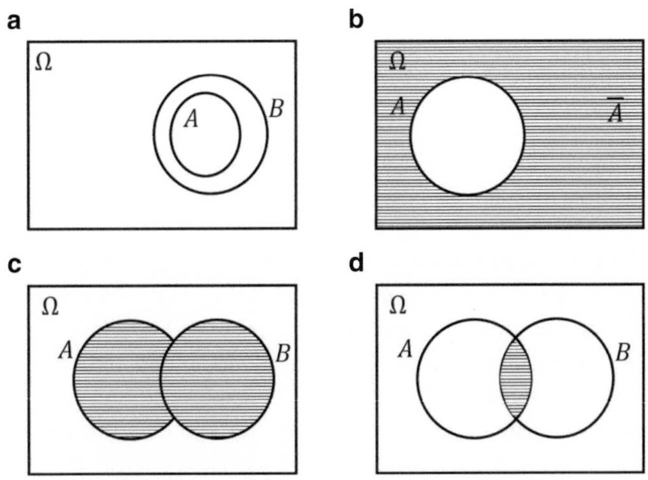

Fig. 22.1 Operations with events A and B. (**a**) Inclusion; (**b**) complementary of **a**; (**c**) union, and (**d**) intersection

1. For any pair of events, A and B, $p(A \cup B) = p(A) + p(B) - p(A \cap B)$. Obviously, in the case of mutually exclusive events, where $A \cap B = \emptyset$, $p(A \cup B) = p(A) + p(B)$

22.4 Conditional Probability. Mutually Exclusive Events

If A and B are two events, the probability of B occurring given that A has occurred is denoted as $p(B/A)$ and is called the conditional probability of B given A.

If we denote the event $A \cap B$ as the event of both A and B occurring, this probability will be

$$p(A \cap B) = p(A)p(B/A) \quad (22.14)$$

In particular, if they are independent events, it will be verified that $p(B) = p(B/A)$, so that the conditional probability will be the product of the two independent probabilities:

$$p(A \cap B) = p(A)p(B) \quad (22.15)$$

Together with the concept of conditional probability appears the concept of independence of events. Intuitively, two events A and B of the sample space Ω are said to be independent if the occurrence of one does not modify the probability of the other occurring. We will say that two events A and B of the sample space Ω are independent, if and only if $p(B/A) = p(B)$. Therefore, two events A and B are said to be dependent if they are not independent.

22.5 Probability Distributions

We have seen that a sample space can take a discrete set of sample points, that is $\Omega = \{x_i\}$, $i = 1, 2 \ldots N$. We have seen that each of these points (actually, each of the events, but that will not change the concept) has an associated probability, that is, there is an application between $x_i \rightarrow p_i$.

The function $p(\Omega) = \{p_i\}_{i=1,\ldots N}$ is called a discrete probability function. We can think of probability distributions as theoretical forms of relative frequency distributions when the number of observations is very large. We must not forget that, in order to be true probabilities, the area under the polygon must be equal to unity, that is, it must be verified that

$$\sum_{i=1}^{N} p_i = 1 \quad (22.16)$$

When the number of observations increases a lot, the width of the intervals gets smaller and smaller, so that the discrete distribution, at the limit, becomes a continuous one (Fig. 22.2). In this case the characteristic of a continuous distribution of probability is similar to the one indicated above, but using integral instead of summation.

$$\int_{-\infty}^{\infty} p(x)dx = 1 \quad (22.17)$$

So the area under the curve is equal to unity. The function $p(x)$ is also called the probability density function (pdf), since it actually has units of dx^{-1}, so the expression $p(x)dx$ is dimensionless.

Another associated distribution of great importance is the cumulative probability function (F), defined as.

$$F(x) = \int_{-\infty}^{x} p(y)dy \quad (22.18)$$

In Fig. 22.2, we can observe both kinds of functions. It is interesting to note that a specific function for $p(x)$ was used to create this figure. This function is called the Gaussian function and will be discussed further below.

Both types of functions are the true key concepts of statistical test for the contrast of hypothesis.

The mathematical characterization of distributions is achieved by different magnitudes. The mean of a random variable, x, is generally addressed by the expected value, and denoted as $E(x)$. If the variable is discrete with probability distribution $p(x)$, the expression is

$$E(x) = \sum_{i=1}^{N} p(x_i)x_i \quad (22.19)$$

If the variable is continuous the summation is replaced by an integral, and the expected value of a function of a random variable, x, is defined as

$$E(x) = \int_{-\infty}^{\infty} xp(x)dx \quad (22.20)$$

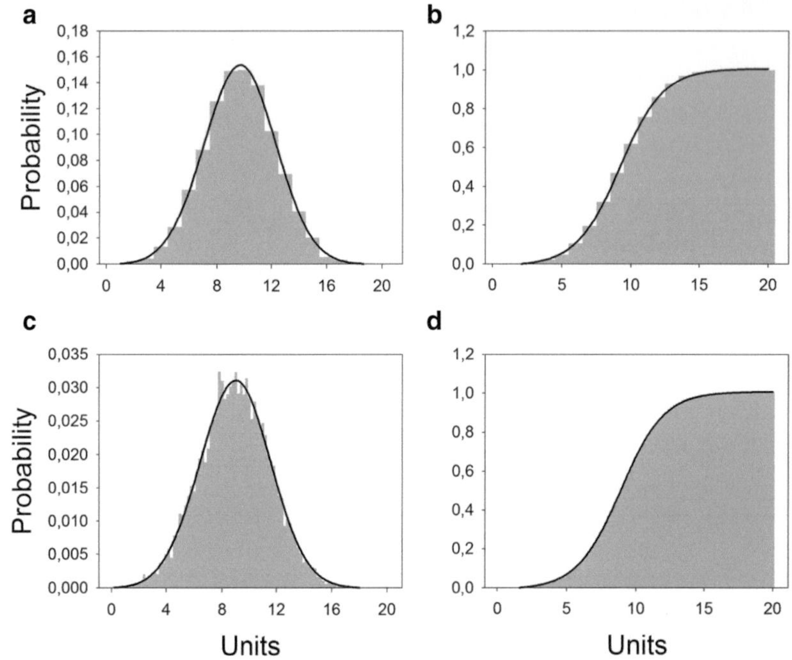

Fig. 22.2 Probability density (*A* and *C*) and cumulative functions (*B* and *C*) for the same random Gaussian noise at different bins. It can be observed that the shorter width of bins *C* and *D* approaches more and more to the continuous function (black lines)

Other values used to characterize the probability distributions of random variables are called *k*th-moment about the origin (μ'_k) for a variable x is defined as

$$\mu'_k = E\left(x^k\right) = \sum_{i=1}^{N} p\left(x_i\right) x_i^k \qquad (22.21)$$

In the case of a discrete variable. If the variable is continuous, the moment about the origin

$$\mu'_k = E\left(x^k\right) = \int_{-\infty}^{\infty} x^k p\left(x\right) dx \qquad (22.22)$$

So that μ'_1 is simply the mean and generally denoted by μ.

The *kth* moment about the mean, μ_k, is defined for discrete variables as

$$\mu_k = E\left(x-\mu\right)^k = \sum_{i=1}^{N} \left(x_i-\mu\right)^k p\left(x_i\right) \qquad (22.23)$$

And for continuous variable

$$\mu_k = E\left(x-\mu\right)^k = \int_{-\infty}^{\infty} \left(x-\mu\right)^k p\left(x\right) dx \qquad (22.24)$$

If $k = 2$, we will have $\mu_2 = s^2$, that is, the second moment about the mean is equal to the variance.

To avoid particular units when analyzing distributions, dimensionless moments about the mean (a_k) are introduced, which are defined as:

$$a_k = \frac{\mu_k}{s^k} = \frac{\mu_k}{\left[\left(\mu_2\right)^{1/2}\right]^k} \qquad (22.25)$$

We have very important dimensionless moments characterizing the distribution. These are:

Skewness. It is a measure of the lack of symmetry in a probability distribution. Usually quantified by the index, a_3 (sometimes denoted as s), given by

$$a_3 = \frac{\mu_3}{s^3} = \frac{\mu_3}{\mu_2^{3/2}} = \frac{\sum_{i=1}^{N}\left(x_i-\mu\right)^3 p\left(x_i\right)}{\left(\sum_{i=1}^{N}\left(x_i-\mu\right)^2 p\left(x_i\right)\right)^{3/2}} \qquad (22.26)$$

The index takes the value zero for a symmetrical distribution. A distribution is said to have positive skewness when it has a long thin tail to the right, and to have negative skewness when it has a long thin tail to the left

Kurtosis: The extent to which the peak of a unimodal probability distribution or frequency distribution departs from the shape of a normal distribution. Usually measured for a probability distribution as a_4, given by

$$a_4 = \frac{\mu_4}{s^4} = \frac{\mu_4}{\mu_2^2} = \frac{\sum_{i=1}^{N}(x_i - \mu)^4 p(x_i)}{\left(\sum_{i=1}^{N}(x_i - \mu)^2 p(x_i)\right)^2} \quad (22.27)$$

Distributions can be either more pointed (leptokurtic, $a_4 > 3$) or flatter (platykurtic, $a_4 < 3$), with zero kurtosis being called mesokurtic ($a_4 = 3$).

These definitions can be easily formulated for continuous variables substituting summation by integrals.

Probability distributions are the foundation of statistical significance tests. Although we won't cover any of them, we will briefly review their foundation, starting with the most important distribution: the Gaussian distribution, or normal distribution.

22.6 Gaussian Distribution

It is the most widely used distribution in any field of Statistics or Physics. Even when we make a probability estimate in our daily lives (e.g., about a person's height or the result of an exam), the intuitive concept of probability distribution (without the need for a rigorous formulation) that we use is Gaussian.

The density function for this distribution is

$$p(x) = \frac{1}{\sigma\sqrt{2\pi}} e^{-(x-\mu)^2/2\sigma^2} \quad (22.28)$$

where μ and σ are the mean and standard deviation of the population, respectively.

The cumulative probability function ($F(x)$) is defined as.

$$F(x) = \frac{1}{\sigma\sqrt{2\pi}} \int_{-\infty}^{x} e^{-(y-\mu)^2/2\sigma^2} dy \quad (22.29)$$

Both functions are represented in Fig. 22.2.

We can make a variable change that is very useful for these calculations. We will normalize the distribution according to the σ

$$z = \frac{x - \mu}{\sigma} \quad (22.30)$$

The mean and standard deviation values for this new variable are 0 and 1. And, substituting in Eq. 22.29, we get

$$F(z) = \frac{1}{\sqrt{2\pi}} \int_{-\infty}^{z} e^{-u^2/2} du \quad (22.31)$$

This is the so-called standardized normal density function, where z is the standard value. From this expression, it is possible to find the probability between two points by integrating the normalized probability density function. For example, if you want to find the probability that an event is less than the mean value (which for the normalized variable is zero), you would obtain it by integrating this expression

$$F(z) = \frac{1}{\sqrt{2\pi}} \int_{-\infty}^{0.5} e^{-u^2/2} du$$

which corresponds to the area of the curve below 0.5. This integration requires a special integral, called the Gauss gamma function, which we will not introduce here. However, the important thing is to see how, from these expressions, which allow the areas of the curve to be calculated for defined values, it is possible to determine the probability of any event.

From this integral, it is easy to see that the area between $\pm\sigma$ is 0.6827, between $\pm 2\sigma$ is 0.9545 and between $\pm 3\sigma$ is 0.9973.

In Table 22.1, we show the most important properties of the normal distribution.

Table 22.1 Properties of the normal distribution

Mean	μ
Variance	σ^2
Standard deviation	σ
Skewness	$a_3 = 0$
Kurtosis	$a_4 = 3$

Suggested Readings

A rigorous and detailed description of these theories and concepts can be found in the next references.

Edwards CH, Penney DE. Probability. In: Elementary differential equations. 6th ed. London: Prentice Hall; 2007. ISBN 0-13-239730-7.

Everitt BS, Skrondal A. The Cambridge dictionary of statistics. 4th ed. Cambridge: Cambridge University Press; 2010. ISBN 978-0-511-78827-7

Spiegel MR, Schiller JJ, Srinivasan RA. Basic probability. In: Theory and problems of probability and statistics. 2nd ed. New York: McGraw-Hill; 2000. ISBN 0-07-135004-7.

Spiegel MR, Schiller JJ, Srinivasan RA. Random variables and probability distributions. In: Theory and problems of probability and statistics. 2nd ed. New York: McGraw-Hill; 2000. ISBN 0-07-135004-7.

MIX
Papier aus verantwortungsvollen Quellen
Paper from responsible sources
FSC® C105338

If you have any concerns about our products,
you can contact us on
ProductSafety@springernature.com

In case Publisher is established outside the EU,
the EU authorized representative is:
Springer Nature Customer Service Center GmbH
Europaplatz 3, 69115 Heidelberg, Germany

Printed by Libri Plureos GmbH
in Hamburg, Germany